ANIMAL DOMESTICATION AND BEHAVIOR

Animal Domestication and Behavior

EDWARD O. PRICE

Professor
Department of Animal Science
University of California
Davis
CA 95616
USA

CABI *Publishing*

CABI *Publishing* is a division of CAB *International*

CABI Publishing
CAB International
Wallingford
Oxon OX10 8DE
UK

CABI Publishing
10E 40th Street
Suite 3203
New York, NY 10016
USA

Tel: +44 (0)1491 832111
Fax: +44 (0)1491 833508
E-mail: cabi@cabi.org
Website: www.cabi-publishing.org

Tel: +1 212 481 7018
Fax: +1 212 686 7993
E-mail: cabi-nao@cabi.org

A catalogue record for this book is available from the British Library, London, UK.

Library of Congress Cataloging-in-Publication Data

Price, Edward O.
    Animal domestication and behavior / by Edward O. Price.
            p.   cm.
Includes bibliographical references (p.   ).
    ISBN 0-85199-597-7 (alk. paper)
    1. Domestication.   2. Animal behavior.   3. Domestic animals.   I. Title.
    SF41 .P75 2002
    636.08′2--dc21   2002001333

ISBN 0 85199 597 7

Typeset by AMA DataSet Ltd, UK.
Printed and bound in the UK by Biddles Ltd, Guildford and King's Lynn.

Contents

Preface   vii
Acknowledgments   x

Part I: General Aspects
1   Why Study Domestication?   1
2   Domestication Defined   10
3   Approaches to the Study of Domestication   13
4   Pre-adaptations for Domestication   21

Part II: Genetic Mechanisms Influencing Domestication
5   Inbreeding   30
6   Genetic Drift   38
7   Artificial Selection   43
8   Natural Selection in Captivity   51
9   Relaxation of Natural Selection   63

Part III: Variation Under Domestication
10 Genetic Variability and Behavior   72
11 Morphological and Physiological Traits   83

Part IV: Adaptation to the Biological Environment
12 Feeding and Drinking   95
13 Predation, Infectious Diseases and Parasites   107
14 Interactions with Humans   113
15 Social Environment   130

Part V: Adaptation to the Physical Environment
16 Climate and Shelter   142
17 Use of Space   149

Part VI: Behavioral Development, Feralization and Animal Welfare

18 Behavioral Development in Captive Animals   161
19 Reintroductions and Feralization   182
20 Welfare and Ethics   204

References   230
Index   283

Preface

My appreciation for domestic animals started very early in life since I was born into a lineage of New Jersey dairy farmers. In addition to our 60-plus Guernsey cows, I had a dog, a pony, sheep, goats and rabbits to care for as I was growing up. Each cow had a formal name (they were registered) and some had nicknames. One of my many chores on the farm was to teach the newborn calves to drink from a bucket. Watching each animal grow into an adult, I observed the behavioral and physical differences between individuals. Each animal had its own characteristic temperament, which we sometimes had to account for in our handling and care-giving. Our animals lived a good life by most animal standards. Yet, they were still animals. Every few days, a new calf was born and an older cow was retired and sent to slaughter. The cycle of life I observed in our domestic animals helped me to understand the less predictable life and death struggle affecting the wild deer, rabbits, pheasants, woodchucks and other wild creatures living on our farm.

One day while working in the fields, I heard the cries of an animal in the nearby woods. I quietly approached the sounds and, to my surprise, found a young puppy (dog) trying to keep up with its mother (a feral animal), who raced off when I came in sight. I caught the pup, took it home and we raised it to adulthood. In spite of our care, it never became fully socialized to humans. At that time I knew very little about 'sensitive periods' for socialization and 'imprinting' but it was clear that this dog was different in the way it responded to people. I couldn't help but think its early experience in the woods had something to do with the development of such persistent timidity toward humans. I was impressed with the contrast between this part-wild dog and the behavior of domestic dogs I had known. On another occasion, I captured a young red fox from a woodchuck den and kept it in our basement, where it was free to roam about. I conscientiously fed it meat and tried to coax it to approach me. It remained very fearful. Eventually, I let it go on the farm after accepting the fact that it did not want to share my world.

In retrospect, I can see how these experiences with the feral dog pup and the wild fox cub contributed to my interest in animal behavior and domestication.

My interest in domestication was rekindled in graduate school. Dr John King, my major professor at Michigan State University, maintained a research colony of various species of deermice (*Peromyscus*) and I was hired to assist in his research on behavioral development. I quickly became interested in these animals and chose to study them for my Masters and PhD dissertation research. My research involved a comparison of the behaviors of prairie deermice (*Peromyscus maniculatus bairdii*) from three populations, namely wild-caught, first-generation laboratory-reared and a stock that had been bred in captivity for approximately 15 generations. After finishing my degrees, I took a professorial position in the Department of Zoology at the State University of New York College of Environmental Science and Forestry in Syracuse. My research during the 10 years with SUNY involved the effects of domestication on the behavior of the Norway rat (*Rattus norvegicus*). Much of this work was a comparison of wild-caught and first-generation laboratory-reared wild rats with various strains of domestic rats. My ultimate goal was to take a population of wild rats and monitor the generation-by-generation changes in their behavior and reproductive success. Unfortunately, funds for such a long-term project did not become available. I subsequently took a position with the Department of Animal Science at the University of California, Davis to develop a program of teaching and research on the behavior of domestic livestock. Concentrating on reproductive behaviors, my empirical research on domestication effects came to an end. However, my interest in domestication as a biological phenomenon persisted and I continued to search the literature for studies on this topic. This book is the culmination of a career-long search for information on the process of domestication and its biological effect on captive animals.

The book is conveniently divided into six parts. The first part, titled 'General Aspects', discusses when and reasons why certain species were first domesticated, definitions of domestication, approaches to the study of domestication and pre-adaptations for life in captivity. Historical accounts of domestication are very brief. There are many other writings on this topic (see text for references), which deal with the voluminous and sometimes contradictory literature on this topic in a much more scholarly way than I could have. The second part, 'Genetic Mechanisms Influencing Domestication', provides an overview of the mechanisms influencing changes in the gene pool during domestication. Part III, 'Variation Under Domestication', discusses the effects of domestication on genetic variability and phenotypic variation in behavior, structural traits and physiology. Part IV, 'Adaptation to the Biological Environment', concerns the many ways captive animals adapt to provision of food and water, the lack of predation, infectious diseases and parasites, the presence of humans and other animals. Part V, 'Adaptation to the Physical Environment', discusses the adaptations of animals to selected physical aspects of their captive environment, namely temperature, provision or absence of shelter and their use of space, which is sometimes severely limited. The last part deals with aspects of behavioral development in the captive environment not discussed in Parts IV and V, and how certain developmental processes influence their ability to survive and reproduce, both in captivity and when reintroduced into nature. The book ends with an overview of how the welfare of animals is affected by domestication and selected management techniques associated with captive animal husbandry. Ethical issues associated with the husbandry of captive animals is also discussed.

Since publishing my 'Behavioral aspects of animal domestication' paper in 1984 there has been a noticeable increase in the number of published papers dealing with the husbandry of aquatic species, particularly fish. Much of this work focuses on the nutritional and environmental requirements of rearing larval fishes, factors influencing reproductive infertility in captivity and the genetic implications of farming fish for release in nature. Many species of fish are currently being considered for domestication as part of a global expansion of interest in aquaculture. As a consequence, this book frequently uses studies with fish to illustrate certain points.

The book is basically a review of the literature on the topic of domestication and rearing animals in captivity. Personal opinions are sometimes expressed but usually only after attempting to present the facts surrounding an issue in an objective manner. The title of the book, *Animal Domestication and Behavior*, may seem a bit misleading to some readers, since so much of the book is devoted to the topic of the management of captive animals, whether domesticated or not. This title makes more sense when one considers that the domestication process is difficult to avoid when animals are brought into captivity. Most captive-reared wild animals will express certain aspects of the domestic phenotype simply by being reared in captivity. The application of artificial selection together with the effects of natural selection in captivity can greatly accelerate the domestication process.

Acknowledgments

I am greatly indebted to Per Jensen and an anonymous reviewer who offered comments on the entire manuscript. Rolf Beilharz, Joseph Garner and Katherine Miller reviewed parts of the text and Tim Hardwick, CAB *International* Publisher, made many excellent suggestions. In addition, I would like to thank my students and colleagues (too numerous to name) who have contributed either directly or indirectly to the ideas contained in this paper. I would also like to thank Harold Musselman, my high school biology teacher, who first introduced me to the formal study of zoology and Dr John King, my mentor in graduate school, who gave me the training and encouragement to embark on a career in zoology and who provided me with the necessary resources and inspiration to engage in domestication research. Lastly, I am indebted to my wife, Mabell, for the many sacrifices she has made so I could pursue scholarly endeavors such as writing this book.

Chapter 1

Why Study Domestication?

Animal domestication has been practiced for many centuries and has profoundly affected the course of civilization (Zeuner, 1963; Mason, 1984; Clutton-Brock, 1999). Animal domestication has not only provided a ready supply of food, clothing and companionship but, in many respects, has changed the way people live and view the world (Baenninger, 1995; Tanabe, 2001).

An understanding of the domestication process is important in dealing with issues related to animal welfare (Dawkins, 1980). From a biological standpoint, one of our greatest concerns is the extent to which captive animals are capable of adapting to certain artificial environments or management practices that, from an anthropocentric perspective, might appear to jeopardize their welfare. The process of adaptation to man and the captive environment is a major consideration in nearly all studies dealing with domestication.

Many wild species are bred and raised in captivity. The captive environment can be very different from the natural habitat for these animals. An understanding of the process of domestication and how animals adapt to new environments can provide important insights on how to successfully propagate wild animals in unnatural habitats. The purpose of this chapter is to provide examples of species that have been successfully domesticated and some ways that captive-reared wild and domestic animals benefit mankind.

Examples of Animal Domestication

The domestication of the dog's wolf-like ancestor (*Canis lupus*) sometime between 10,000 and 15,000 years ago is believed to be the first example of animal domestication and the only domestication to have taken place during the hunter–gatherer stage of human cultural development (Brisbin and Risch, 1997). A recent report (Vilà *et al.*, 1997) proposed that dogs emerged from wolves about 135,000 years ago, during the *Homo neanderthalensis* era. Since that paper was published, the reliability of mitochondrial DNA as an evolutionary clock, on which this time estimate was partially based, has been seriously questioned (see Coppinger and Coppinger, 2001, pp. 285–292). The development of large, relatively permanent, agriculture-based societies was the primary event initiating livestock domestication about 10,000 years ago. With the exception of the pig (*Sus scrofa*) and llama families, ungulate domestication (e.g. cattle, sheep and goats) largely began in the Near East (Troy *et al.*, 2001), especially in the eastern Fertile Crescent, a broad arc of land stretching from present-day Turkey to Iran. Pigs were domesticated separately in Europe and Asia about 9000 years ago (Giuffra *et al.*, 2000). Domestication of the llama (*Lama glama*) and alpaca (*Lama pacos*) took place in South America about 6000 years ago (Wheeler, 1995) at about the same time that peoples of the Sredni Stog culture in the Ukraine began domesticating horses (*Equus caballus*) for meat and transportation (Anthony *et al.*, 1991). Domestication of ungulate species hunted for meat can be viewed as an elaboration of the predator–prey relationship whereby herding is used to guarantee a dependable and renewable source of animal protein with minimum risk and expenditure of energy (Meadow, 1989). Rodents were also exploited for meat and clothing. Guinea pigs (*Cavia porcellus*) have been domesticated for 3000–4000 years (Weir, 1974; Kyle, 1987). The indigenous peoples of Peru and surrounding countries in South America bred guinea pigs for food and religious ceremonies. Domesticated guinea pigs arrived in Europe in the late 16th century AD. Rabbits (*Oryctolagus cuniculus*) were domesticated in southern Europe approximately 2500 years ago (Nachtsheim and Stengel, 1977). Domesticated poultry also became an important resource in early agrarian settlements. The house mouse (*Mus musculus domesticus*) was used for scientific research as early as 1664, when Robert Hooke used a mouse to study the properties of air, and particularly oxygen (Festing and Lovell, 1981). The first inbred strain of mice, DBA, was developed by C.C. Little in 1909 (Festing and Lovell, 1981). Once the scientific value of inbred strains became known in the 1920s and 1930s, many other modern strains of domesticated laboratory mice were developed from fancy pet mice and, to a certain extent, wild mice (Festing and Lovell, 1981; Atchley and Fitch, 1991).

Chickens (*Gallus domesticus*) were domesticated from jungle fowl (*Gallus gallus*) in Thailand and its adjacent regions (Wood-Gush, 1959; Zeuner, 1963; Fumihito *et al.*, 1994, 1996) and taken north to China about 6000 BC (West and Zhou, 1988). Remains of domesticated chickens have been found at 16 neolithic sites along the Huang He (Yellow River) in northeast China dating back to that time (West and Zhou, 1988). Wild turkeys (*Meleagris gallopavo*) were first found in North America but were domesticated in Europe starting about 500 years ago (Brant, 1998).

The relationship between man and silkworms (*Bombyx mori*) can be archeologically traced back to about 2500 BC in China (Hamamura, 2001). Rock objects resembling spinning wheels were simultaneously excavated with silkworm cocoons, suggesting that silk thread was being utilized at that time. From the way the cocoons were sharply cut, it appeared that pupae were also used as food.

Clutton-Brock (1992) has listed the approximate dates of domestication, principal region of origin and the presumed wild ancestors of many of our domesticated animals. Kisling (2001) provides an historical account of the practice of keeping wild animals in captivity from ancient collections to present day zoological gardens.

Taxonomic Designations

Linnaeus (1758) gave names to more than 4000 organisms, including all of the common domestic animals of his day. If he was familiar with both wild and domestic forms of a species and they looked alike, they were given the same name. If the wild form of a domesticated species was not known (e.g. sheep and goats), only the domesticated form was named (Clutton-Brock, 1999). For the most part, Linnaeus's nomenclature was retained. However, it is not uncommon to find discrepancies in the literature regarding the taxonomic designations of wild and domestic stocks of a species. For example, wild and domestic forms of the pig are often given the same taxonomic designation, *Sus scrofa*, in spite of the fact that others (e.g. Clutton-Brock, 1999) list the domestic pig as *Sus domesticus*. While wild boar and domestic pigs are typically found in different environments, the phenotypic differences and habitat choices are not as different as, for example, wolves and dogs. In addition, domestic pigs readily establish feral populations and become independent of human provisioning.

In other genera, a taxonomic distinction is commonly made between wild and domestic forms. For example, the scientific name of the domestic chicken is given as either *Gallus domesticus* or *Gallus gallus domesticus*, whereas *Gallus gallus* refers to its wild ancestor, the jungle fowl. The debate is whether the domestic stock represents a separate species or subspecies. Those favoring separate species status typically argue that wild and domestic forms are morphologically, behaviorally and/or ecologically distinct. Those favoring subspecies status acknowledge that wild and domestic stocks are capable of interbreeding and are not genetically distinct. In most of these cases, inter-stock fertility has been demonstrated in captivity without any attempt to confirm interbreeding in nature.

Linnaeus chose the name *Canis familiaris* for the domestic dog to distinguish it from its gray wolf ancestors, *Canis lupus*. This decision makes some sense considering the morphological, behavioral and ecological differences of the two stocks. Nevertheless, dogs and wolves are very capable of interbreeding and producing viable offspring. Honacki *et al.* (1982) suggested that the domestic dog be renamed *Canis lupus familiaris* to reflect its evolutionary descent and its many similarities to wolves, including their inter-fertility. Their proposal has not received much support; the phenotypic differences are too great.

Clutton-Brock (1999) lists four mammals in which the domestic form retains the exact same taxonomic designation as its wild counterpart, namely the rabbit (*O. cuniculus*), brown rat (*Rattus norvegicus*), reindeer (*Rangifer tarandus*) and Bali cattle (*Bos javanicus*). There is no special reason for this practice other than tradition and the rationale that creating separate names would only add to more nomenclatorial confusion.

One could argue that the species name of the wild form should be retained in the domestic stock if the two stocks are capable of interbreeding and they produce fertile offspring. The domestic form could then be given a separate subspecies name (e.g. *domesticus*) to identify their current status. For example, *Sus scrofa domesticus* would immediately inform the reader that the animal concerned is the domestic pig and it produces fertile offspring when interbred with the wild *Sus scrofa*. As it is now, the use of *Sus scrofa* for both wild and domestic stocks does not distinguish between the two and *Sus domesticus* implies that the two stocks are not capable of interbreeding. Although this rationale makes intuitive sense to this author, in the short term, current taxonomic designations for long-standing domesticated species will likely be retained as much to avoid confusion as for any other reason. A few changes in species names may be made as advances in biotechnology provide clarification of the genetic relationship of domestic and captive stock to existing or extinct wild populations.

Reasons for Animal Domestication

There are a number of reasons why populations of animals are bred in captivity and thus potentially subjected to the process of domestication (Table 1.1). The desire for animal products and companionship head the list, with the use of animals in scientific research, and for insect pest control, recreation, species conservation and public display also being important objectives. Of course, reasons for domestication may have changed with time and culture.

The greatest changes in phenotypic traits ascribed to domestication have been attained in species bred for food, fiber, companionship, show and research. Populations bred in captivity for insect pest control, to meet conservation

Table 1.1.   Some objectives of breeding animals in captivity.

1. Domestication
    a. Animal products
    b. Companionship
    c. Scientific study
    d. Recreational uses
2. Conservation
    a. Release in nature
    b. Preservation of species
    c. Germplasm resources
3. Public display
    a. Education
    b. Pleasure

objectives and for supplementing natural populations (restocking) have generally exhibited fewer and less obvious phenotypic changes.

Few species have been domesticated specifically for scientific research (e.g. *Drosophila, Tribolium*). Many species used in research are chosen because of their availability, size, breeding success in captivity and prolificacy, popularity, or the volume of literature on the species in addition to their intrinsic scientific value (Festing and Lovell, 1981). Laboratory mice (*Mus musculus*) were first domesticated as 'fancy' pets (Keeler, 1931; Cooke, 1978) and much later brought into laboratories for research on cancer and immunology (Festing, 1969). The period between 1890 and 1930 saw the rapid development in use of the domesticated albino rat (*Rattus norvegicus albinus*) in laboratory research (Logan, 2001). The Prague/Vienna reproductive physiologist, Eugen Steinach, and the American developmental neurologist, Henry Donaldson of the Wistar Institute of Anatomy and Biology in Philadelphia, are credited with being principal advocates for the early use of the albino rat as a standard laboratory animal in Central Europe and America. The rat was well suited as an experimental animal for Steinach's studies on sexual development because of its altricial nature at birth, relatively slow development and high levels of libido (Logan, 2001). Also, it was a convenient size for handling and experimentation, relatively inexpensive to maintain in the laboratory and a prolific breeder. Donaldson was introduced to albino rats in Vienna during a 1909 trip to Europe (Logan, 2001). His itinerary included stops in various European locations to collect wild European rats to compare brain growth with wild rats caught in the USA. While in Vienna, he was introduced to the zoologist Hans Przibram, who was breeding albino rats for scientific study. Impressed with Przibram's studies, Donaldson promoted the use of albino rats at the Wistar Institute, which by the 1920s had distributed thousands of rats throughout the USA.

Although availability and familiarity are important considerations, non-traditional animal models may be more appropriate for certain kinds of experimentation. For example, cricetidae rodents serve as natural reservoirs or hosts of many human endemic–epidemic diseases in the New World. Several species have been brought into the laboratory and bred specifically to study such diseases. The phyllotine rodent *Calomys* sp. is the neotropical wild rodent most studied and used as a laboratory model (Justines and Johnson, 1970; Mello, 1981). The South American water rat, *Nectomys squamipes*, adapts well to the laboratory (D'Andrea *et al.*, 1996) and is used as an experimental model in schistosomiasis studies.

In recent years there has been increasing concern for the preservation of rare and endangered species. Biologists and conservationists have teamed together to formulate policies for the management of animals in danger of extinction (Gibbons *et al.*, 1995). Artificial breeding colonies (farms) have been established for the purpose of re-establishing populations of these species in their natural habitat (e.g. Beck *et al.*, 1994; Kleiman, 1996) and providing stock for display in zoos and wildlife parks. The success of such programs will depend on a thorough knowledge of the domestication process and its effects on the animals involved, as well as an understanding of the development of behaviors deemed important for survival and reproduction in the wild. In general, domestication reduces the probability of successful reintroductions.

A Symbiotic Relationship

Animal domestication has been suggested to involve the formation of a symbiotic, give-and-take relationship between man and animals (Rindos, 1980; Budiansky, 1994). Man, the domesticator, benefits from the products and services provided by animals under his care and the animals benefit from the care and protection provided by man. As human populations increase in modern-day societies and land areas devoted to agriculture decrease, people must become more efficient in utilizing existing wild and domestic animal and plant resources. A case can be made to domesticate species representative of the native fauna endemically adapted to local regions (Jewell, 1969; Crawford, 1974; Stanley-Price, 1985). For example, the recent domestication of the musk ox (*Ovibos moschatus*) in North America (Wilkinson, 1974), eland (*Taurotragus oryx*) in Africa (Carles *et al.*, 1981) and in the USSR (Treus and Kravchenko, 1968), the giant pouched rat (*Cricetomys gambianus*) in Africa (Ajayi, 1974) and the capybara (*Hydrochaeris hydrochaeris*) in Venezuela and Brazil represent efforts to utilize indigenous animal species for food and fiber.

Rindos (1980) has pointed out that the development of new domesticates is impeded by a tendency to breed and subsist on but a few preferred and competitive species, especially as the yield of these species increases worldwide. The use of certain traditional domesticated species in areas of the world only marginally suited for their survival and well-being is a questionable practice but they fit into current patterns of trade and have clearly understood market values.

Recent Surge in Aquaculture

Aquaculture is the world's fastest-growing agribusiness and is believed necessary to meet the rapidly increasing global demands for fish and other aquatic food species (FAO, 1997). Certain aquatic species such as the carp have been domesticated for many centuries (Balon, 1995). In recent years, there has been a rapid proliferation in the captive propagation of many aquatic vertebrates adapted to fresh or salt water (Bardach *et al.*, 1972; Doroshov *et al.*, 1983; Heslinga and Fitt, 1987; Anonymous, 2001). The first public salmon hatchery in the USA was constructed in 1871 in the state of Maine (Moring, 2000). Currently, over 90% of all adult Atlantic salmon are in aquaculture facilities (Gross, 1998). The culture of the freshwater prawn, *Macrobrachium rosenbergii*, either for commercial purposes or research activities has been reported in at least 77 countries (New, 1995). A number of other promising candidates for aquaculture have been identified (Hassin *et al.*, 1997; Falk-Petersen *et al.*, 1999; Hickman and Tait, 2001).

Invertebrate Domestication

In ancient Greece and Rome, epitaphs indicate that insects such as cicadas and crickets were often kept as pets, although not on the same level as dogs and cats (Bodson, 1997). In modern-day Japan, insects such as crickets, grasshoppers and beetles are frequently maintained as pets by children (Laurent, 1997). Insect rearing was first confined to certain easy-to-rear species such as fruit flies (*Drosophila*), house flies (*Musca domestica*), cockroaches and certain insects typically found in stored food products (Knipling, 1984). In 1936, colonies of the screwworm (*Cochliomyia hominivorax*) were established to make possible the screening of hundreds of candidate chemicals and other formulas for treating livestock wounds to prevent screwworm infestation. Laboratory populations of body lice (*Pediculus humanus humanus*) and malaria-transmitting mosquitoes (*Anopheles*) were established during World War II to screen thousands of candidate insecticides and repellents to control these and other human pest species in many parts of the world. After World War II, the US Department of Agriculture established research colonies of various crop pests such as the boll weevil (*Anthonomus grandis grandis*) and pink bollworm (*Pectinophora gossypiella*) to study their biology and methods of control. In short, the laboratory rearing (i.e. colonization) of various insect species during the last century has made it possible to study their genetics, physiology and behavior in a controlled manner. These colonies have facilitated the identification and synthesis of insect sex pheromones and research on plant resistance. The field of biological control has been greatly advanced by having methods of rearing insect parasites and predators in captivity for introduction in nature or augmentation of natural populations and by the propagation of hosts for insect pathogens (Knipling, 1984).

Gon and Price (1984) provide a partial list (Table 1.2) of domesticated and beneficial insects and other invertebrates, reflecting the diversity of species used. To meet the need for new animal domesticates, methods should be developed to determine which species are best suited for domestication and how modern techniques of animal breeding and management can be applied to efficiently produce the most desirable and useful laboratory populations (King and Leppla, 1984).

Conclusions

Whether the objective is to domesticate new species or to minimize negative impacts of domestication on captive populations of wild animals, it is important to understand the process of domestication and its effects on both genotype and phenotypic traits. Many species have been domesticated to benefit mankind. Food, fiber, work, transportation, research and recreation are the primary reasons animals are maintained in captivity. While mammals and birds receive the most attention, it is important to note the contributions made by many aquatic vertebrates and invertebrates and a host of insect species. The ensuing chapters

Table 1.2. A partial list of domesticated and beneficial invertebrates, showing the diversity of species currently used and domesticated. (From Gon and Price, 1984.)

Taxon	Common name	Food/product	Biocontrol	Research	Status[a]
Cnidaria					
Hydra spp.				X	1
Platyhelminthes					
Dugesia spp.	Planaria		X	X	1
Trematode spp.	Flukes			X	1
Annelida					
Lumbricus spp.	Earthworms	X		X	1,2
Tubifex spp.		X			1
Palola sp.	Palolo worm	X			1
Mollusca					
Helix spp.	Escargot	X			1
Murex spp.	Tyrian snail	X			1
Nucella spp.		X			1
Loligo spp.	Squid	X		X	1,2
Sepia spp.	Cuttlefish	X		X	1,2
Octopus spp.		X		X	1,2
Tridacna spp.	Giant clam	X			1
Ostrea spp.	Oyster	X			1,2,3
Euglandina sp.	Cannibal snail		X		1,2
Arthropoda					
Scarabeid spp.	Scarab beetle grubs	X			1
Schistocera spp.	Grasshoppers	X			1
Agrotis sp.	Bugong moth grub	X			1
Melophorus spp.	Honey ants	X			1
Macrotermes spp.	Termites	X			1
Eurycantha sp.	Walking stick	X			1
Spaerodema spp.	Waterbugs	X			1
Daphnia spp.	Water fleas	X		X	1
Artemia sp.	Brine shrimp	X		X	1,2
Macrobrachium spp.	Prawn	X			3
Panaeus spp.	Shrimp	X			1,2
Procambarus spp.	Crayfish	X		X	1,2
Paralithodes sp.	King crab	X			1
Homarus sp.	Lobster	X			1,2
Araneid spp.	Spiders	X	X	X	1
Periplaneta spp.	Cockroach			X	1,2
Gomphadorina sp.	Giant roach			X	3
Apis sp.	Honeybee	X		X	3

Table 1.2. *Continued.*

Taxon	Common name	Food/ product	Biocontrol	Research	Status[a]
Bombyx spp.	Silk moth	X		X	3
Laccifer sp.	Lac insect	X			3
Dactylopius sp.	Cochineal insect	X			3
Aphytis spp.	Parasitic wasp		X		3
Eucelatoria sp.	Parasitic fly		X		3
Chrysolina sp.	Predaceous beetle		X		3
Orthophagus spp.	Dung beetle		X		3
Dermestes spp.	Carrion beetle			X	3
Drosophila spp.	Pomace fly			X	3
Echinodermata					
	Sea urchins	X		X	1
	Sea cucumbers	X		X	1

[a]1, Wild-caught; not domesticated; 2, Reared or mass-reared without conscious artificial selection; 3, Domesticated; both reared and bred for specific purposes.

of this book have been written to provide a review of conceptual issues and factual information relating to the process of domestication and the management of captive wild and domestic animals. The emphasis is on animal behavior.

Chapter 2

Domestication Defined

Since this book is about the captive rearing and domestication of animals, it is important at the outset to seek an acceptable definition for domestication. A review of the literature reveals some common denominators and some differences in the way domestication is viewed.

Approaches to Defining Domestication

Animal domestication is best viewed as a process, more specifically, the process by which captive animals adapt to man and the environment he provides. Since domestication implies change, it is expected that the phenotype of the domesticated animal will differ from the phenotype of its wild counterparts. Adaptation to the captive environment is achieved through genetic changes occurring over generations and environmental stimulation and experiences during an animal's lifetime (Price, 1984). In this sense, domestication can be viewed as both an evolutionary process and a developmental phenomenon.

Darwin (1859, 1868) suggested that domestication is more than taming, that it includes breeding animals in captivity, is goal-oriented, may occur without conscious effort on the part of man, increases fecundity, may bring about the atrophy of certain body organs, enables animals to achieve greater plasticity and is facilitated by subjugation to man, the domesticator. Some contemporary definitions postulate that domestication is a condition in which the breeding, care and feeding of animals are more or less controlled by humans (Bökönyi, 1969; Hale, 1969; Clutton-Brock, 1977). This definition implies that a population of animals is rendered domestic by exposure to the captive environment and by the institution of certain management practices. Ochieng'-Odero (1994) proposes that domestication consists of habituation and conditioning to environmental stimuli associated with the captive environment. Whereas many important aspects of the domestic phenotype are environmentally induced or can be linked to

©CAB *International* 2002. *Animal Domestication and Behavior*
(E.O. Price)

certain experiences, there are many adaptations to the captive environment that can be best explained by genetic changes accompanying the domestication process.

In response to claims that animal domestication was solely an experiential phenomenon, Price and King (1968) and Bartlett (1993) proposed that domestication is an evolutionary process involving the genotypic adaptation of animals to the captive environment. Ratner and Boice (1975) took a more ontogenetic approach by acknowledging the contributions of both genetic change and experience in the development of the domestic phenotype. More recently, Lickliter and Ness (1990) proposed a 'developmental systems' approach to domestication. In their view, domestic phenotypes are not transmitted in the genes nor contained in features of captive environments but are constructed by the 'coaction of organic, organismic, and environmental factors during ontogeny'.

Domestication Defined

It is difficult to formulate a definition of domestication that is general enough to account for the many factors contributing to the domestication process yet specific enough to be meaningful in terms of the evolutionary and biological processes involved. For the purpose of this book, domestication is defined as 'that process by which a population of animals becomes adapted to man and to the captive environment by some combination of genetic changes occurring over generations and environmentally induced developmental events recurring during each generation' (Price, 1984). This rather simplified definition of domestication does not assume that genes and the environment operate as independent factors which additively combine to produce the domestic phenotype. Neither does it assume that one can be understood in isolation from the other. As Lickliter and Ness (1990) point out, development of the domestic phenotype can only be understood in terms of the complex interplay of organic, organismic and environmental factors during ontogeny.

This definition is meant to apply to domestication in general. It assumes that the captive environment is different from the ancestral wild environment of the species. It also assumes that certain general animal management and housing practices are consistently applied over time in rearing and maintaining each species in captivity. These are reasonable assumptions even for species which are highly pre-adapted for life in captivity. Differences between the ancestral wild and captive environments of a population allow evolutionary mechanisms to bring about directed changes in the gene pool (i.e. mostly through changes in gene frequencies) as the population adapts to captivity over generations. Consistently applied animal management and housing practices allow certain environmentally induced developmental events to contribute to the domestic phenotype in a reoccurring manner in each generation. For example, research will be cited later in this book showing that reduced fear of humans, an important component of the domestic phenotype, is attained in certain captive rodents by rearing in traditional open laboratory cages. Conspecifics (members of one's own species) reared in cages with burrows or other forms of shelter where they could visually (and physically) escape from the presence of humans were found to exhibit stronger avoidance behaviors toward human-like stimuli. Of course, genetic selection for ease of handling also contributes heavily to the evolution, and thus development, of this behavioral trait.

Adaptation to the Captive Environment

The domestic phenotype refers to that cadre of phenotypic traits that facilitates the adaptation of captive animals to their environment. The range of environmental conditions typically provided captive populations of some species (e.g. livestock) on a global basis will be greater than for other species (e.g. laboratory rodents). Hence, for any given captive population, attainment of the domestic phenotype must be evaluated on a relative scale, based on the degree of adaptation to the range of environmental circumstances in which the animals are most likely to be found. Since degree of adaptation forms a continuum and is difficult to measure, it is sometimes difficult to determine the extent to which a population has become domesticated. Such decisions will necessarily be somewhat subjective. Assuming a stable environment over time, the domestication process is complete only when adaptation to the captive environment, including man, has reached some maximal steady state (i.e. terminal plateau). Of course, phenotypic traits may continue to evolve post-domestication as illustrated so well by the various breeds of domestic dogs, cats and farm animal species.

Conclusions

For many persons, animals are domesticated when they come under man's control. While this is a necessary prerequisite for domestication, there is much more. First, domestication is a process involving both genetic changes occurring over generations and experiences associated with living in captivity which typically recur in each generation. Domestication is about adaptation to man and the environment he provides. Phenotypic adaptations to the captive environment will occur based on the same evolutionary processes that enable free-living populations to adapt to changes in their environment. The major difference is that in captivity, man can accelerate phenotypic changes, that would otherwise not appear or persist in nature, through artificial selection (and now gene transfer). Many of these latter changes are initiated post-domestication.

Chapter 3

Approaches to the Study of Domestication

Before getting into the biological mechanisms associated with the process of domestication, it is important to note how current knowledge of domestication has been derived. This chapter introduces and discusses the four major approaches that have been used to gain a better understanding of the domestication process: (i) comparison of wild and domestic stocks (of a species); (ii) monitoring the generation-by-generation changes in a population that has recently been brought into captivity; (iii) searching for 'domestication' genes; and (iv) the study of wild × domestic hybrids.

Comparing Wild and Domestic Stocks

The most commonly used approach in studying the domestication process is to compare wild and domestic stocks of a given species (e.g. Price, 1973; Desforges and Wood-Gush, 1976; Huck and Price, 1976; Boice, 1980; Price, 1980; Wong *et al.*, 1982). This approach assumes that the wild stock used is representative of the wild ancestors of the domesticated population investigated, a requirement that is difficult if not impossible to attain, even for our more recently domesticated animals. First, the ancestral origin of many of our domesticated species is uncertain (Zeuner, 1963; Hyams, 1972). For decades the common laboratory mouse was believed to have been derived from wild *M. musculus* (Schwarz and Schwarz, 1943). Then, Marshall (1981) argued that the morphology of the laboratory mouse resembled *M. domesticus* more closely. Now, Blank *et al.* (1986) have provided information suggesting that the common inbred strains and substrains of laboratory mice have been derived from several *Mus* species including *M. musculus*, *M. castaneus*, *M. molossinus* and *M. domesticus*. This conclusion is based on a survey of protein and DNA polymorphism data suggesting extensive fixation of alleles from various loci in the above species. An additional dilemma is that the wild counterparts of some of our common domesticated animals have become extinct or are unavailable for scientific experimentation (Isaac, 1970). Even when ancestral populations are available, one cannot assume these populations have remained static since the domestication process was initiated.

Selection of an experimental wild population may present an additional problem. Because of geographical variation, no single population selected for study may adequately represent the genetic diversity present in the species (Mitchell *et al.*, 1977; Berry, 1978; Van Oorschot *et al.*, 1992). Leopold (1944) reported differences in the escape patterns of certain populations of wild and domestic turkey poults (*M. gallopavo*) in response to alarm calls. However, Hale (1969) noted that differences among wild subspecies are even greater, leaving open the question whether the differences observed by Leopold were due to domestication. Garcia-Marin *et al.* (1991) found that only 3% of the total genetic variation among hatchery stocks of brown trout (*Salmo trutta*) could be explained by stock differences whereas 60% of the total genetic variation among wild trout was due to differences between populations.

Once a suitable wild population is identified, capture bias may confound efforts to obtain a representative cross-section of individuals from that population. Baker *et al.* (2001) reported that free-living subordinate red foxes (*Vulpes vulpes*) were trapped more frequently than dominant individuals. This concern is especially important if wild-caught individuals rather than subsequent, captive-reared generations are used for study.

A related problem lies in identifying a representative domestic population (Box, 1973, p. 122). The many breeds or strains of our common domestic animals and the geographical variation associated with most of these stocks re-emphasize the problem of choosing representative populations for study. Boreman and Price (1972) systematically hybridized four strains of domestic rats (*R. norvegicus*) to produce a single, more representative domestic stock for study.

In addition, most comparisons of wild and domestic populations are conducted on animals reared either in nature or captivity but not both environments. If we assume that wild animals are best adapted to wild environments and domestic animals are best adapted to domestic environments (Garcia-Marin *et al.*, 1991), such comparisons are likely to be skewed in favor of one of the test populations, usually the domestic population since most domestication research is conducted in the captive environment. Even when wild and domestic populations are reared in both environments (Wecker, 1963; Price, 1969), the comparative approach provides little or no information on how the traits actually develop or how rapidly phenotypic changes occur during the domestication process. The comparative approach only provides a description of differences between specific populations of wild and domestic animals at a single point in time. Consequently, its major usefulness is in generating hypotheses regarding the effects of domestication upon behavior. Gross (1998) has summarized some of the genetic- and developmentally based differences between farmed and wild Atlantic salmon (*Salmo salar*) (Table 3.1). This information is especially useful since many of the studies compared both

Table 3.1. Some genetic and developmental differences between cultured (farmed) and wild salmon. (From Gross, 1998.)

Genetic		Developmental	
Confirmed	Suspected	Confirmed	Suspected
Increased growth rate	No ejaculation	Lower stamina	Gut length
Increased age of maturity	Reduced male courtship	Smaller eggs	Diet preference
Increased weight	Higher hatchery survival	More eggs	Stream knowledge
Increased disease resistance	Higher temperature tolerance	More fat	Body odor
Decreased stress response	Shallower depth preference	Smaller rayed fins (parr and adult)	
Lower genetic diversity	Larger testes	Smaller heads on parr	
Allele frequency change	Juvenile color	Bulkier body on adult	
Malic enzyme allele change	Adult color	Distorted jaw on adult	
Reduced predator response		Longer head on adult	
Increased juvenile aggression		Longer adipose fin on adult	
Increased tameness		Smaller hearts in females	
Lower survival in wild		Decreased juvenile color	
		Decreased adult color	

stocks reared in either captive or wild environments. Studies comparing wild animals reared in nature and their domesticated counterparts reared in captivity (Diefenbach and Goldberg, 1990; Mesa, 1991) are necessarily confounded by rearing environment.

The Longitudinal Approach

In contrast to the comparative approach just described, the longitudinal approach to domestication research monitors phenotypic changes in a population of animals over generations of breeding in captivity. With this information, one can better estimate the rate at which phenotypic changes occur during domestication under different regimens of breeding and husbandry. The problem of species representativeness is less restrictive when using the longitudinal approach since the emphasis is on how individual populations change over time (generations).

An example of the longitudinal approach is a study by King and Donaldson (1929) and King (1939), who monitored changes in the growth rate and reproductive success of a population of wild Norway rats (*R. norvegicus*) over the first 25 generations in captivity. During this period, positive gains were noted for body weight, percentage of mated pairs that produced offspring, number of litters born and length of the reproductive lifespan. The investigators also reported that the tendency to escape and the resistance to handling declined over generations, although these responses were not quantified.

Connor (1975) also used the longitudinal approach in an attempt to re-enact the domestication process in a population of wild house mice (*M. musculus*). Three subpopulations of genetically wild mice were established: (i) a randomly bred wild population maintained in standard laboratory cages; (ii) a randomly bred wild population maintained in an enriched naturalistic environment (located in the laboratory); and (iii) an inbred population of wild mice (brother/sister matings) maintained in standard laboratory cages. In addition, Connor established three populations of inbred domestic mice for comparative purposes. A battery of six behavioral tests (19 variables that clearly differentiated wild and domestic mice) was administered to the various populations over ten generations of breeding in captivity. No behavioral differences were found between tenth-generation wild mice reared in standard laboratory cages and wild mice maintained in the naturalistic environment. Yet both of these stocks still differed significantly from domestic mice. Inbreeding reduced intermale aggression and resistance to capture by humans but failed to affect body weight, handling-elicited vocalizations, investigation of intruders and escape from intruders. These results suggest that ten generations of breeding in captivity is insufficient to produce the domestic phenotype in this species, with or without systematic inbreeding.

The Search for 'Domestication' Genes

In addition to the comparative and longitudinal approaches to domestication research, one may look for so-called 'domestication genes' in animal populations. Keeler *et al.* (1968) and Keeler (1975) reported that tameness in captive red foxes (*V. vulpes*) is linked in a correlated (pleiotropic) manner to alleles for pelage color-ation. Avoidance of humans by captive foxes was inversely related to the number of mutant coat-color alleles in the genotype. Keeler had much earlier (1942) reported that docility in the domestic Norway rat (*R. norvegicus*) is associated with the nonagouti (black) pelage-color allele. The nonagouti allele is masked by the albino locus in this species (Robinson, 1965). Cottle and Price (1987) confirmed Keeler's hypothesis. They found that captive-reared wild-genotype Norway rats homozygous for the recessive nonagouti allele, which confers black pelage coloration, could be handled with greater ease than animals heterozygous or homozygous for the agouti allele, whose pelage coloration is brown (Table 3.2). Interestingly, the black and agouti subjects did not differ in three other tests that clearly differentiate wild and domestic stocks of rats (open-field test, platform

Table 3.2.   Percentage of agouti (*n* = 62) and black (*n* = 28) Norway rats
exhibiting the behaviors listed on the first day of handling tests. (From Cottle
and Price, 1987.)

Behavior	Agouti (*AA*)	Black (*aa*)
Could be touched	64%	96%
Could be stroked	32%	89%
Attacked–attempted to bite	59%	25%
Jumped	91%	57%

jumping and response to a novel food). Hughes *et al.* (1981) also reported no differences between black and agouti wild Norway rats in open-field (novel arena)
behavior and activity-wheel running.

A decade later, Hayssen (1997) replicated the Cottle and Price (1987) study
with deermice (*Peromyscus maniculatus*) some 20 generations removed from the
wild. The results were very similar; nonagouti deermice were more touchable,
strokeable, catchable and less likely to attack the experimenter's hand than the
agouti subjects. While nonagouti and agouti rats did not differ in the platform
jump measures, nonagouti deermice were less likely than agouti subjects to
jump off the platform. In addition, nonagouti mice spent significantly more
time grooming.

Biochemical studies with recombinant agouti protein (Lu *et al.*, 1994; Willard
et al., 1995) indicate that agouti is an antagonist of melanocyte-stimulating
hormone (MSH) at its receptor on pigment cells and potentially at other melanocortin receptors in neural tissues. Melanocortins are potent neuromodulators with
diverse effects on mammalian behavior and physiology (e.g. O'Donohue and
Dorsa, 1982). Thus, the behavioral differences between agouti and nonagouti rats
and mice are probably related to the regulation and effects of MSH. In addition,
Hayssen *et al.* (1994) reported that agouti and nonagouti deermice differ in their
neural catecholamine profiles, especially in the mid- and hindbrain. Neural MSH
is located primarily in the hypothalamus, and pituitary MSH is regulated by
dopamine neurons (see Hayssen, 1997, for references). Interestingly, selection for
tameness (toward humans) in foxes, mink and rats is accompanied by changes in
brain biochemistry (see Chapter 14).

Although most behaviors are polygenic (i.e. influenced by many genes),
certain alleles have a relatively large impact on the development of certain
behavioral characters specific to domestic animals. The nonagouti allele in rats
and mice may be such a major 'domestication' gene. It is not surprising, then, that
over 70% of approximately 140 inbred mouse strains and over 80% of approximately 50 inbred rat strains that have been characterized at the agouti locus are
homozygous for the nonagouti allele (Festing, 1979a,b, 1989; Staats, 1981, cited
in Hayssen, 1997).

Hybridization of Wild and Domestic Stocks

Hybridization of wild and domestic stocks has also been used to provide information on behavioral development and the process of domestication (Leopold, 1944;
Boreman and Price, 1972; Rood, 1972; Smith, 1972; Cheng *et al.*, 1979b). Maternal effects can be assessed by comparing reciprocal hybrids with each other and
with the parental strains. Hybridization can also be used to provide information
on the mode of inheritance of the trait or traits in question, namely directional
dominance, over-dominance (heterosis) or intermediate inheritance (Bruell,
1967). Of course, the outcome of hybridization experiments may be affected by
the choice of parent stocks (i.e. representativeness) as in the comparative approach
to domestication research. In addition, hybridization offers no direct information
regarding the rate of phenotypic changes during the domestication process.

When wild and domestic stocks are interbred, the behavior of the hybrids
usually falls intermediate to the two parent populations. Price and Loomis (1973)
examined the behavior of wild and domestic Norway rats (*R. norvegicus*) and their
reciprocal hybrid crosses in response to being placed in a $122 \times 30.5 \times 28$ cm
novel enclosure. Wild rats initiated activity in the enclosure more quickly, were
more active during the 5-min test and spent more time grooming than domestic
rats. Hybrids were intermediate (Table 3.3). Boreman and Price (1972) observed
the relative social dominance of wild, domestic and hybrid strains of Norway rats
(*R. norvegicus*) in mixed groups of 12 animals (four animals per strain) housed in
a $7.6 \times 3.6 \times 2.7$ m room for 13 consecutive days. Domestic rats were dominant
to wild rats in both spontaneous and competitive interactions and hybrids were
intermediate (see p. 176 for data and additional explanation). These results are
not surprising, inasmuch as most behaviors are influenced by many genes acting
synergistically or in an additive manner. Intermediate inheritance is characteristic
of traits with an intermediate optimum; selection has removed dominant genes
from the gene pool (Bruell, 1967).

Price and Loomis (1971) found little support for a maternal effect in explaining the differences between wild and domestic rats (*R. norvegicus*) in response to a

Table 3.3. Mean (± SE) behavioral responses of wild and domestic Norway
rats and their reciprocal hybrids when exposed for 5 min to a novel
$122 \times 30.5 \times 28$ cm enclosure. (From Price and Loomis, 1973.)

Behavior	Strain[a]			
	DD	DW	WD	WW
Number of subjects	23	21	22	17
Latency to be active (s)	22.1 ± 3.0	8.7 ± 1.2	6.7 ± 0.9	5.2 ± 1.0
Total activity units	25.1 ± 6.8	32.4 ± 5.5	47.5 ± 6.5	63.6 ± 7.8
Time grooming (s)	25.4 ± 4.4	72.2 ± 10.6	65.1 ± 9.7	119.4 ± 25.5

[a]D, domestic; W, wild; first letter indicates maternal strain.

novel environment. Conversely, Richardson *et al.* (1994) discovered a maternal effect in comparing the basal metabolic rate (BMR) of wild and domestic house mice (*M. domesticus*) and their reciprocal hybrids. Hybrid mice from wild dams had significantly lower BMRs than hybrid mice from domestic dams, and both hybrid groups had lower BMRs than pure laboratory mice (−23.1 and −11.2%, respectively). First-generation laboratory reared wild mice also exhibited lower BMRs than the domestic stock (−17.2%).

Cases where the hybrids score significantly higher (or lower) on a quantitative scale than either parent stock (over-dominance) usually indicate that the parent stocks have experienced a loss in vigor due to inbreeding (Bruell, 1967). Cheng *et al.* (1979b) reported that a greater percentage of hybrid mallard ducklings (*Anas platyrhynchos*) exhibited the 'following' (imprinting) response than did young from pure wild or game-farm parent stocks. However, the hybrids exhibited an intermediate level of arousal, which was believed to facilitate the expression of following behavior. Wild ducklings were over-aroused and game-farm ducklings were under-aroused for maximum following under the experimental conditions employed.

Lickliter and Ness (1990) argued that our understanding of the domestication process has been hampered by the lack of a coherent framework of theories and concepts that can be empirically tested. Lickliter and Ness are correct in assuming that a developmental systems approach to domestication research is likely to be more productive than attempts to separate the relative contributions of genetic and environmental factors. All behaviors develop through an interaction of genes and environment, and studies comparing the effects of this interaction in nature and captivity will contribute the most to our understanding of the domestication process.

Conclusions

Of the four approaches to domestication research, the longitudinal approach provides very useful information. By monitoring changes in populations of animals over generations in captivity, the longitudinal approach can provide information on the rate and magnitude of phenotypic changes over generations without concern about stock or sample representativeness. Unfortunately, longitudinal studies on domestication have been few and far between, and the long-term nature of such work creates logistical problems that discourage would-be researchers. The use of invertebrates with relatively short generation intervals (e.g. *Drosophila*) would partially alleviate this problem if variables relevant to the domestication process can be identified and effectively tracked over time.

The search for 'domestication genes' is intriguing but success is limited by the polygenic nature of most behavioral traits. The one exception to date is the research linking coat-color genes in mammals to brain biochemistry and behaviors which facilitate domestication and ease of handling. Further work in this area could provide a major contribution in explaining differences between wild and domestic animals in their responses to people.

Of the four approaches, studies comparing wild and domestic populations of a species are most common, primarily because such studies are relatively easy to initiate once stocks of animals become available and investigations are typically short term. The usefulness of comparative studies is limited by the perceived representativeness of the animal stocks employed. It is difficult, if not impossible, to find contemporary wild populations that match the wild ancestors of many of our common domesticated species. Biological differences found between wild and domestic stocks will to some degree apply only to the specific populations studied.

Hybridization studies can provide useful information on the inheritance of traits important to domestication, but little more.

Chapter 4

Pre-adaptations for Domestication

Some species readily adapt to life in captivity; others do not. It is the purpose of this chapter to discuss those factors which predispose certain species to not only become candidates for domestication, but to survive and reproduce in captivity.

Captive Niches to be Filled

Many cultural and pragmatic factors have determined which animal species have become domesticated and when and where their domestication has occurred (Isaac, 1970; Reed, 1980). With the possible exception of the dog (see later discussion in this chapter), the first requirement for the successful domestication of animals is that man, the domesticator, has a recognized need or desire that can only be satisfied by controlling, protecting and breeding a certain population of animals (Downs, 1960). It is generally accepted that no important animal domestications associated primarily with food production have been made in tropical regions, where food supplies are generally available through hunting, fishing and gathering (Brisbin, 1974). The chicken (*G. domesticus*) and turkey (*M. gallopavo*) were originally domesticated for religious reasons, and the water buffalo (*Bubalus bubalis*) primarily as a beast of burden (Zeuner, 1963). It is likely that the house cat (*Felis silvestris catus*) and the ferret (*Mustela furo*) were domesticated to protect stored grains and other food materials from rodents (Zeuner, 1963). Other species were domesticated as sources of food, clothing, labor, transportation, adornments or sacrificial offerings. In contrast, the domestic value of many wild species has been largely ignored or, if recognized, social and technological barriers have often prevented their domestication. The quagga (*Equus quagga*), a small, horse-like animal from Central and South Africa, was treated as vermin by the Boer and English farmers and was hunted to extinction despite its docile nature, 'tameability', sturdiness and resistance to the endemic diseases that affected imported horses of European descent (Ridgeway, 1905). The plains Indians of North America had a recognized need for domesticated animals as a food source. However, their nomadic lifestyle and their inability to gather enough food to support such large animals as the bison or the pronghorn antelope in captivity precluded the domestication of these species (Downs, 1960).

Developmental Plasticity

The degree to which a wild population of animals is pre-adapted for domestication largely depends on the degree of developmental plasticity of the species and the extent to which the captive environment allows for the development and expression of species-typical behavioral patterns compatible with husbandry techniques. The degree of pre-adaptation is relative to the specific conditions under which a group of animals is maintained. Just as there is geographical variation in the environments of free-living animal populations, variation exists among captive environments (Hediger, 1964; Box, 1973). Hence, the degree of pre-adaptation of a species for domestication is dependent on the capacity of species members to adapt through developmental and evolutionary processes to a variety of environmental and husbandry conditions (Balon, 1995).

The pre-adaptation of a species for domestication should, in theory, vary inversely with the number and extent of differences between the natural and captive environments. In addition, populations with the fewest pre-adaptations to captivity will experience the greatest number of changes in selective pressures in terms of: (i) the number of traits affected; (ii) the direction of selection; and (iii) the intensity or severity of selection.

Few Species Domesticated

Man has domesticated relatively few animal species, either by choice or because of failure to provide a captive environment that meets the minimal requirements for successful reproduction. He has exploited a few species that are relatively convenient and economical to breed and to maintain in captivity (e.g. many ungulates and gallinaceous birds). Hemmer (1988) discusses the pre-adaptations of several European deer species for farming/domestication. He and Rammelsberg (1986a, cited in Hemmer, 1988) rated roe deer (*Capreolus capreolus*), fallow deer (*Dama dama*), sika deer (*Cervus nippon*), red deer (*Cervus elaphus*) and elk (*Alces alces*) on several pre-adaptation criteria, including 'level of social activity', 'frequency of aggressive behavior', 'social intolerance', 'influence of the captive environment', 'influence of the seasons' and 'behavioral intensity'. Fallow deer emerged as having the best pre-adaptations for farming/domestication while roe deer and elk were rated the poorest. The most serious drawback of fallow deer is their tendency to panic when startled.

Pre-adaptations for Vertebrate Domestication

Table 4.1 lists a number of behavioral traits that are considered favorable and unfavorable for the domestication of vertebrates. In general, pre-adaptations for vertebrate domestication include gregariousness, nonaggressive nature, promiscuous sexual behaviors, readily breed in captivity, precocious young, readily

Table 4.1. Behavioral characteristics considered favorable and unfavorable for the domestication of vertebrate animals. (Modified from Hale, 1969.)

Favorable characteristics	Unfavorable characteristics
Social structure of populations	
Social organization – dominance hierarchy	Social organization – territoriality
Large gregarious social groups	Family groups important
Males affiliated with social group	Males typically live in separate groups
Intra- and interspecies aggressive behavior	
Nonaggressive	Naturally aggressive
Sexual behavior	
Promiscuous matings	Form pair bonds prior to mating
Males dominate females	Females dominate males/males appease females
Male initiated	Female initiated
Sexual signals provided by movements or posture	Sexual signals provided by color markings or morphology
Parental behavior	
Precocial young	Altricial young
Young easily separated from parents	Prolonged period of parental care
Response to humans	
Tameable/readily habituated	Difficult to tame
Short flight distance from man	Long flight distance from man
Nonaggressive toward humans	Aggressive toward humans
Readily controlled	Difficult to control
May solicit attention	Independent/avoids attention
Temperament	
Limited sensitivity to changes in environment	Highly sensitive to changes in environment
Locomotor activity and habitat choice	
Limited agility	Highly agile/difficult to contain or restrain
Small home range	Requires large home range
Wide environmental tolerance	Narrow environmental tolerance
Nonshelter seeking	Shelter seeking
Feeding behavior	
Generalist feeder or omnivorous	Specialized dietary preferences/requirements

tamed, ease in handling, limited sensitivity to environmental change, limited agility, wide environmental tolerance and generalized feeding behaviors. Ease of handling and favorable reproductive success in captivity have encouraged the recent successful captive propagation of several wild rodent species (Mello, 1981, 1986; Delany and Monro, 1985; Murphy, 1985; Kyle, 1987; Smythe, 1987; Ohno *et al.*, 1992; D'Andrea *et al.*, 1996). Most of these traits allow for the efficient exploitation of animals by man under a variety of prevailing economic and eco-logical conditions (Tennessen and Hudson, 1981). With the advent of modern-day husbandry practices such as improved diets and assistance in reproduction, many formerly important pre-adaptations (e.g. social and sexual behaviors of the male) have become less critical (Siegel, 1975). Artificial insemination, cryopreservation of semen and embryos, embryo recovery and transfer, *in vitro* production of embryos, and micromanipulation techniques (including sperm injection, cloning, artificial incubation, assisted hatching and artificial rearing of young) are being used to improve the reproductive performance of animals in captivity (Cseh and Solti, 2000). Consequently, an increasing number of species ranging from primates to bacteria (Davis, 1987) can be rapidly and economically propagated in captivity. Tennessen and Hudson (1981) have maintained that if domestication of a species is warranted economically and ecologically, success will depend largely on the suitability and flexibility of the management system employed.

A few species have become domesticated in spite of lacking important pre-adaptations, the domestic cat, *F. silvestris catus*, being our best example. First, the domestic cat is basically territorial and most individuals are not well adapted to living in large social groups (Leyhausen, 1988). More importantly, domestic cats can be relatively aloof in the company of people. Cameron-Beaumont *et al.* (2002) compared 16 species and subspecies of small cats (Felids) kept in zoos for affiliative behavior toward people (i.e. tameness) and found that the ocelot lineage had the highest proportion of individuals showing affiliative behavior. Interest-ingly, this is the group most distantly related to the domestic cat. This result suggests that the domestication of *Felis silvestris lybica* alone among felids is likely to have been the result of a specific set of human cultural events and requirements in the Egyptian New Kingdom (e.g. rodent control), rather than the consequence of a unique tendency for tameness in this subspecies. Considering these facts, it is not surprising that the breeding of domestic cats has not been subjected to the same degree of control as most domesticated species, nor has their freedom of movement been as restricted. In this sense, domestic cats are pre-adapted for a rather unique commensal relationship with humans.

Pre-adaptations of the Wolf for Domestication

Theories on the evolution of the dog are particularly interesting and deserve special consideration. It is unlikely that the wolf-like ancestors of the dog were captured, tamed and selectively bred from the very start of their domestication. Wolves are very difficult to contain, do not make good pets even when tamed and would have had little utilitarian value to people living in Mesolithic times. Morey (1994) and Coppinger and Coppinger (2001) argue that the domestication of the wolf probably did not proceed as a well-organized plan. The wolf-like ancestors of the domestic dog probably first associated with man during the Mesolithic period as scavengers around permanent human settlements, eating discarded food and human excrement as it became available. The naturally tamer individuals were more reproductively successful in this niche and tended to mate with other individuals willing to risk close proximity to humans. In time, populations of these 'village wolves', which through natural selection for tameness had become well adapted to living in such a symbiotic relationship with humans, arose in various parts of the globe. Some of these progenitors of the domestic dog were nurtured and received special attention because of their unusual coloration, companion-ship, hunting ability, protection, etc., without any attempts by man to achieve long-term domestication (Manwell and Baker, 1984). In more recent times, systematic artificial selection for specific phenotypic characters was applied, resulting in unique populations of relatively tame and morphologically distinct dogs. There was no one point when 'village wolves' suddenly became 'village dogs'; the process was very gradual.

The observations of Coppinger and Coppinger (2001) of populations of 'village dogs' living commensally with present-day hunter–gatherer peoples conjure images of what the dog's ancestors might have experienced in Mesolithic times. Relative to the myriad of dog types found in modern-day society, dogs on the tiny island of Pemba, off the East African Coast, were remarkably uniform in appearance suggesting rather intense selection for some optimum phenotype. They were very uniform in size, about 14 kg, and slender in body type. They had short, smooth coats of variegated colors, some with large spots, some with markings on their heads, ears, legs or tails. Their ears were pendent or erect but bent over slightly at the tips (tulip ears). Interestingly, each house on the island had its own refuse dump and a latrine. Dogs tended to adopt certain houses and essentially waited for food they could scavenge. They could be approached by humans but rarely allowed direct contact. There were no dog packs; groups were seldom larger than three individuals, who were probably related to one another. The people of Pemba were basically hunter–gatherers, living off the sea (coral reefs), perhaps not that unlike the supposed villagers in late Mesolithic times when the domestication of the dog's wolf-like ancestor began. The inhabitants of Pemba did not consciously tame the village dogs or invite them into their homes since they were generally considered unclean. Nevertheless, they tolerated the presence of the dogs and inadvertently provided an appropriate niche for their coexistence. From this scenario, it is easy to extrapolate how such populations of tame animals were eventually exploited by humans and how artificial selection may have been deliberately or unconsciously applied to establish unique populations. Artificial selection and subsequent hybridization of animals from these various populations in more modern times could have easily resulted in many of the recognized dog breeds today. If this theory is correct, it was the decision of humans to adopt a life-style that led to the development of relatively permanent settlements in late

Mesolithic times that paved the way for the domestication of the dog. The morphological and behavioral pre-adaptations of the wolf-like ancestors of dogs to feed and reproduce in the company of man and their capacity for socialization to man predisposed them to be adopted by man and vice versa (Messent and Serpell, 1981; Coppinger and Coppinger, 2001). Thus, it seems most plausible that the wolf-like ancestors of the dog initially domesticated themselves and that humans became their domesticators only later in the process.

While the process described above is the most likely explanation for the domestication of the wolf, it is not a likely scenario for the establishment of most of our domestic animal stocks. Most domesticated species were probably first captured or lured into confinement and soon bred for their utilitarian value, temperament and desirable morphological traits (e.g. pelage coloration). The advent of stable agrarian societies and the ability to provide food for captive herbivores set the stage for these important domestications.

Reproductive Failure

Some species are pre-adapted to survive in captivity but consistently fail to reproduce (Medina *et al.*, 1996; Hassin *et al.*, 1997; Sato *et al.*, 2000). Almost all fish reared in captivity exhibit some form of reproductive dysfunction (Zohar and Mylonas, 2001). The most commonly observed reproductive dysfunctions in cultured fish are the unpredictability of final oocyte maturation in females and the diminished volume and quality of sperm in males (Mylonas and Zohar, 2001). This infertility is largely due to failure to simulate the natural conditions of spawning; consequently, the pituitary fails to release the maturational gonadotropin, luteinizing hormone. Manipulations of various environmental parameters, such as temperature, photoperiod, salinity, tank volume and depth, substrate vegetation, etc., can often facilitate spawning but in some species hormonal treatment is the only reliable means of controlling reproduction (Zohar and Mylonas, 2001). A case in point is the white grouper (*Epinephelus aeneus*), a fish native to the Mediterranean. In captivity, groupers demonstrate a fast growth rate, hardiness, disease resistance and have high market value. Although the oocytes of adult females reach the final stages of vitellogenesis in captivity, final oocyte maturation, ovulation and spawning in captivity do not occur spontaneously (Hassin *et al.*, 1997). Injections or implants of gonadotropin-releasing hormone (GnRH) induce ovulation but not natural spawning. However, successful fertilization of eggs can be attained artificially. A second problem related to captive white grouper propagation is that all young fish are females. Adult females eventually spontaneously invert to males when they get older, but because most fish used/caught are younger females there is often a lack of males for establishing grouper broodstocks. Females can be induced to show sex inversion by injections of 17-α-methyltestosterone, but males produced in this fashion will often revert back to females following hormone removal.

The neotropical anostomid fish *Leporinus elongatus* matures but does not spawn in captivity. Sato *et al.* (2000) have shown how spawning in this fish can be induced with carp pituitary extract. This same treatment can be used to induce northern pike (*Esox lucius*) to spawn in captivity (Szabó, 2001). Van Eenennaam *et al.* (2001) reported the first successful artificial spawning of captive green sturgeon (*Acipenser medirostris*) by injections of gonadotropin-releasing hormone analog (GnRHa) and domperidone. Medina *et al.* (1996) demonstrated that female *Penaeus kerathurus* shrimp can be successfully reared in captivity but fail to reproduce because they do not produce mature oocytes. In contrast, Primavera (1985) reported that some 14 species of marine penaeid shrimp that successfully mature and spawn in captivity have been identified. Browdy *et al.* (1986) have added *Penaeus semisulcatus* to that list.

Control of Sex

Ability to control sex may constitute an important pre-adaptation for the captive propagation of species in which one sex is valued more highly than another. Technology recently developed with domestic ungulates to identify and separate X and Y chromosome-bearing sperm allows animal breeders to selectively produce female or male offspring (Hohenboken, 1999; Seidel and Johnson, 1999). In recent years there has been rapid progress in the development and application of sex control technologies in finfish aquaculture (Donaldson, 1996). Schutz and Harrell (1999) noted that 47 different species of fish have been sex-reversed by the dietary administration or direct injection of exogenous hormones such as 17-α-methyltestosterone, which reverses the sex of genetic females to phenotypic males. Sex reversal can be useful in the domestication process by increasing the ratio of females to males in broodstock populations (Schutz and Harrell, 1999). In many species, one sex grows faster, matures later or has a higher market value than the other sex. In tilapia, males are preferred. In flatfish and salmonids, females are preferred for culture. Sex control is also important for reproductive containment (Donaldson, 1996; Schutz and Harrell, 1999). The culture of monosex or sterile populations can reduce or eliminate reproductive interaction between escaped farmed fish and wild conspecifics and prevent cultured fish from forming self-sustaining feral populations. Sex control can also be used to reproductively contain genetically modified (e.g. transgenic) animals.

Pre-adaptations of Invertebrate Species for Domestication

The domestication of invertebrates has not received the attention given to vertebrates. Most invertebrates breed exceptionally well in captivity (Balmford *et al.*, 1996) suggesting that their basic needs can be readily met by the captive environments typically provided. One can debate whether the western honeybee

(*Apis mellifera*) is a true captive domesticate or lives as a symbiont with man. It is tolerant of human management, it is adaptable to a broad range of climatic conditions and it provides honey (Delaplane and Mayer, 2000). Its ability to

Table 4.2. Behavioral characteristics considered favorable and unfavorable for the domestication of invertebrate animals. (From Gon and Price, 1984.)

Favorable characteristics	Unfavorable characteristics
Social structure of populations	
Gregarious	Large territories
Small territories	Groups monosexual
Males affiliated with female groups	
Intra- and interspecies aggressive behavior	
Nonaggressive in intra- and interspecies interactions	Aggressive in intra- and interspecies interactions
Altruistic	
Sexual behavior	
Male initiated	Requires long, correct behavioral sequence, with no guarantee of mating success
Sex signals via movement or posture	
Pheromonally induced	
Promiscuous	Requires death of one or both mates
	Pair bonding
Parental behavior	
Egg guarding	No parental care
Precocial young	Altricial young
Nonplanktonic young	Alloparental cannibalism
Young easily separated from adults	
Nest building, shelter building	
Reaction toward humans	
Readily habituated	Wary
Little disturbed	Easily disturbed
Nonantagonistic	Antagonistic, toxic or dangerous
Feeding behavior	
Generalist feeder	Requires specific, difficult items
Feeds on common, if specific, items	Cannibalistic
Noncannibalistic	Requires live food
Accepts artificial diet	Maximal feeding with no endogenous satiation mechanism
Endogenous feeding satiation	
Locomotor activity and habitat choice	
Nonmigratory	Migratory (obligate)
Sessile or small home range	Nonhoming, free ranging
Limited agility	Extreme agility
Wide environmental tolerance	Specific environmental requirements
Nonshelter seeking	Shelter required
Ecological versatility	Ecological specialization

pollinate plant crops has been critical for modern-day agriculture. About 130 agricultural plants in the USA are pollinated by bees (McGregor, 1976), at an annual value to US agriculture of many billions of dollars (Robinson *et al.*, 1989).

Many invertebrates are propagated in captivity for release in nature to achieve biological control of important insect pests. Other species are studied in the laboratory to better understand biological mechanisms which can be exploited in control measures. The fruit fly (*Drosophila*) has been used for decades to scientifically study genetic mechanisms important in population dynamics. Gon and Price (1984) offer a list of behaviors (Table 4.2) that predispose invertebrates to domestication and captive propagation.

Conclusions

The choice of certain species for domestication is ultimately based on some perceived benefit to man. Among those species selected for captive propagation, some will readily survive and reproduce, others will survive but not reproduce, while still others will not survive in captivity under existing management practices. Certain traits confer fitness on captive animals and predispose them to domestication. These pre-adaptations may be shaped in nature (prior to being brought into captivity) or may simply reflect the developmental plasticity of the species. In either case, species successfully domesticated will likely possess many pre-adaptations for living in captivity. A case can be made for the wolf-like ancestors of the dog choosing humans in a kind of self-domestication, but it is doubtful that other species were likewise attracted to human settlements and domesticated in a similar manner.

Chapter 5

Inbreeding

Genetic adaptation to captivity probably occurs to some degree in all wild populations brought into captivity. The rate and extent of genetic change over time will differ with the species and the specific housing and management conditions employed (Bartlett, 1984; Frankham and Loebel, 1992). The genetic processes with the greatest potential impact on the domestication process are inbreeding, genetic drift and selection. Whereas inbreeding and genetic drift produce random changes in gene frequencies, the changes resulting from selection are directional. There are three primary selective phenomena that influence populations of animals undergoing domestication (Price and King, 1968): (i) artificial selection; (ii) natural selection in captivity; and (iii) relaxation of natural selection. As a result of these three processes acting individually or in combination, selection with respect to specific traits may change in intensity or direction, or both.

The present chapter examines the incidence and consequences of inbreeding within populations of captive animals. Small population sizes with limited numbers of breeding animals increase the chance breeding of related individuals. In addition, man's desire to preserve favored traits has often resulted in the conscious interbreeding of genetically similar animals. Inbreeding depression (loss of vigor or fitness) is a likely consequence of close inbreeding. It is noted that crossbreeding animals from different populations may serve to restore vigor and fitness, but outbreeding may have detrimental consequences when co-adapted gene complexes are disturbed.

Genetic Variability

Inbreeding may be difficult to avoid in relatively small, closed (or relatively closed) captive populations (Fuller and Thompson, 1960; Lacy et al., 1993). Close inbreeding such as brother–sister or father–daughter matings will typically result in a reduction in genetic variability within a population (Fig. 5.1). Genetic

©CAB *International* 2002. *Animal Domestication and Behavior*
(E.O. Price)

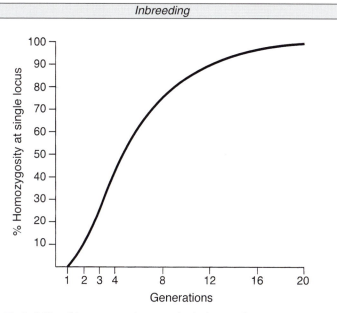

Fig. 5.1. Probability of homozygosity at a single locus after 20 generations of brother–sister matings (inbreeding). Genetic variability is reduced as homozygosity is increased (Green, 1966).

variability consists of both allelic diversity and heterozygosity and can be measured in both individuals and populations (Ballou and Foose, 1996). Allelic diversity in a population refers to the number of different alleles at various gene loci on the chromosomes. Heterozygosity is determined by the percentage of gene loci that are heterozygous (carry different alleles) for an individual or population.

Inbreeding 'Depression'

Inbreeding may result in inbreeding 'depression', which is the lowering of vigor or fitness brought about by the expression of deleterious genes previously masked by dominant genes at the same or a different (epistatic) locus (Shields, 1982; Thornhill, 1993; Falconer and MacKay, 1996). Inbreeding depression is most evident in characters with the greatest influence on fitness (Lynch and Walsh, 1998; DeRose and Roff, 1999). Slate *et al.* (2000) demonstrated that the lifetime reproductive success of wild male and female red deer (*C. elaphus*) is negatively correlated with degree of inbreeding and positively correlated with outbreeding (heterosis) and multilocus heterozygosity. Shikano *et al.* (2001) reported that inbreeding in guppies (*Poecilia reticulata*) resulted in a linear decrease in salinity tolerance (survival time in 35-ppt seawater) at a mean rate of 9.1% per 10% increase in the inbreeding coefficient.

Among domestic animals, inbreeding depression has been demonstrated for such quantitative traits as juvenile mortality, egg hatchability, clutch size, milk yield, litter size and growth rate (Falconer and MacKay, 1996). Ralls *et al.* (1988) reported that full-sib mating increased juvenile mortality by an average of one-third in 40 captive wild animal populations representing 38 species. Laikre and Ryman (1991) describe inbreeding depression in a population of captive wolves in Scandinavian zoos. Inbreeding reduced genetic variability, reproduction, longevity and juvenile weight and increased the incidence of blindness. Mäki *et al.* (2001) describe the positive association between degree of inbreeding and the incidence of elbow and hip dysplasia in Labrador retrievers (*C. familiaris*), a growth disorder of the bone which causes arthritis. Hedrick and Kalinowski (2000) note that evidence of inbreeding depression in captive or laboratory environments may typically be less than, or at least be different from, the detrimental effects of inbreeding depression in more harsh natural environments. Husbandry and management conditions in captivity may buffer animals against the negative effects of inbreeding depression.

Effect on aggression and reproductive behaviors

Inbreeding is known to reduce the success of animals in aggressive and sexual interactions. Eklund (1996) studied the aggressiveness and competitive ability of male wild house mice (*M. musculus*) produced under three levels of inbreeding: (i) full-sib matings; (ii) first-cousin matings; and (iii) unrelated individual matings. Paired encounters were staged between males of the three inbreeding levels when they reached sexual maturity. The offspring of full-sib matings (highest level of inbreeding) won the smallest percentage of encounters and had the lowest scores for aggression (Fig. 5.2). Sharp (1984) pointed out that inbreeding affects the competitive mating ability of male fruit flies (*Drosophila*). Aspi (2000) found that courtship singing in inbred *Drosophila montana* males after 20 generations of brother–sister matings was 14% less frequent than for outbred males. Male courtship song frequency is closely associated with male mating success and offspring survival. Female fruit flies can increase the viability of their offspring by about 24% by choosing a male breeding partner whose courtship song frequency is one standard deviation higher than average in the population (Hoikkala *et al.*, 1998). Miller and Hedrick (1993) created 18 experimental lines of *Drosophila melanogaster* derived from wild stock which were homozygous for chromosome 2. (These lines were comparable to inbred lines subjected to two generations of full-sib mating.) The homozygous flies in six lines which exhibited reduced fitness were allowed to compete with heterozygous subjects for females for 2 h as a measure of male mating success. Males in the homozygous lines showed markedly reduced mating success. Homozygous males in the worst mating line inseminated only two of 87 stock females while males in the best-performing inbred line only inseminated 34 of 89 females.

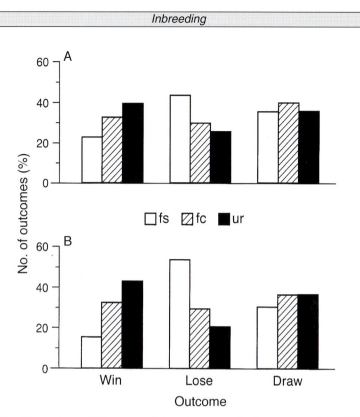

Fig. 5.2. Percentage of three possible outcomes (win, lose or draw) in all trials (A) and first trials (B) in paired encounters between male house mice produced by full-sib matings (fs), first-cousin matings (fc) and matings between unrelated parents (ur) (Eklund, 1996).

Population size and inbreeding depression

A study by Bryant *et al.* (1999) demonstrated how inbreeding depression can be exacerbated in small populations. Populations of the house fly (*M. domestica*) were either initiated and maintained at constant sizes of 40, 200 or 2000 individuals, or were initiated with two pairs of flies and allowed to grow to 40 individuals. The latter low-founder-number populations exhibited low larval viability (22%) after 24 generations compared to the 200 individual populations (49%) or the 2000 individual populations (69%). In addition, 56% of the low-founder-number populations became extinct during the 24 generations. Populations maintained at 40 individuals from the start exhibited somewhat higher larval viability (35%) and somewhat lower population extinction rates (40%) than the low-founder-number populations after 24 generations.

Margan *et al.* (1998) demonstrated with *Drosophila* that several small populations (pooled), when compared to single large populations of equivalent size, were found to have lower average inbreeding coefficients and higher genetic diversity.

Although inbreeding may be more prevalent in small than in large populations and genes may be more likely lost or fixed, the outcomes will vary in different small populations, thus preserving some genetic diversity among populations.

Inbreeding depression is not inevitable

Inbreeding depression is not an inevitable consequence of close inbreeding or small populations (Coulson *et al.*, 1999). Kalinowski *et al.* (1999) did not find any evidence for a reduction in juvenile mortality or litter size in captive Mexican wolf (*Canis lupus baileyi*) or red wolf (*Canis rufus*) populations in spite of being founded by only three and 13 individuals, respectively. Likewise, heterosis (i.e. increase in size, vigor or other characteristics) is not inevitable following outbreeding. Shields (1982) proposed that many natural populations are inherently philopatric and thus are adapted to local inbreeding. In such populations and in populations that are frequently downsized to relatively small numbers (i.e. 'bottlenecks'), natural selection will remove deleterious alleles from the gene pool. Relative to these conditions, an argument can be made that inbreeding depression may be more common in large populations that retain deleterious alleles than in small isolated groups of animals in which deleterious genes are constantly purged from the gene pool (Brewer *et al.*, 1990).

Inbreeding depression may be more pronounced when animals are maintained under suboptimal conditions (Kalinowski *et al.*, 2000). There is increasing evidence that inbreeding depression is greater for stressed animals living in either controlled (Miller, 1994; Pray *et al.*, 1994) or natural environments (Jimenez *et al.*, 1994; Keller *et al.*, 1994). Barlow (1981) reviews evidence supporting this thesis by demonstrating that heterosis is more pronounced in stressful environments. Heterosis (i.e. hybrid vigor) is often observed when inbred lines suffering from inbreeding depression are mated together, because heterozygosity is restored.

Specificity of stressors

Dahlgaard and Hoffmann (2000) addressed the question of whether the deleterious effects of inbreeding are general to all stressors as opposed to being stress-specific. More specifically, they looked at the resistance of *D. melanogaster* to acetone, desiccation and heat stress as well as productivity (viability and fecundity). Inbreeding reduced resistance to acetone and desiccation but did not affect response to heat stress. However, following heat stress, the productivity of inbred flies was poorer than that of outbred flies. It was concluded that the expression of recessive deleterious genes through inbreeding did not confer a generalized reduction in resistance to stressors. Rather, inbreeding within a specific environment and selection for resistant genotypes may purge a population of deleterious genes specific to a limited number of stressors.

Genes with large vs. small effects and family representation

Willis (1999) discussed the relative impact of deleterious genes of large vs. small effect on inbreeding depression. Theoretical models show that if inbreeding depression is due to recessive deleterious genes of large effect, then inbreeding depression may be rapidly purged following moderate inbreeding because of the relatively intense natural selection on the traits involved. In contrast, if inbreeding depression is caused by many deleterious alleles each with relatively small effects, little purging may occur in the short term. In fact, such mildly deleterious alleles can become fixed (i.e. alternative alleles are lost) by genetic drift in effectively small populations.

Inbreeding depression can be minimized and genetic diversity maintained by the occasional introduction of outbred or genetically distinct individuals into the population (Bryant *et al.*, 1999; Ball *et al.*, 2000). These same objectives can be achieved by developing a breeding system in which founder/family representation is equalized in each generation (Loebel *et al.*, 1992). When mating is random, some families will inevitably be represented less than others or lost, resulting in increased inbreeding in the poorly represented families and reduced genetic variation. Severe inbreeding and inbreeding depression are less likely for the population as a whole with equal family representation in each generation.

Inbreeding and artificial selection

Inbreeding is sometimes practiced together with artificial selection in order to reach a specific phenotypic goal (e.g. body conformation, trainability). Through systematic inbreeding, one can attain a degree of homogeneity and constancy of characteristics that is normally not seen in wild populations. The loss of variability due to inbreeding is less costly to a population being maintained by humans, who often attempt to preserve variants that would normally not survive in nature. Crnokrak and Roff (1999) compared the inbreeding depression of free-living animals with the inbreeding depression found by Ralls *et al.* (1988) in captive-bred populations of wild species. Estimates of the cost of inbreeding on the juvenile mortality of *wild* mammal populations were substantially higher than in *captive* populations of wild mammals. The authors could not identify the environmental conditions responsible for the higher inbreeding depression in free-living animals. They hypothesized that relatively harsh living conditions in the wild can reduce the survival of unfit animals, whereas in captivity, management techniques and health care are available to promote and prolong their survival.

It has been proposed that inbreeding depression may be reduced by selecting as breeding stock individuals who have prospered under inbreeding (i.e. controlled inbreeding). Templeton and Read (1984) used this hypothesis to explain the increased viability noted in a relatively small captive population of Speke's gazelle (*Gazella spekei*) during its first few generations in captivity. However, closer examination of the data suggested that the increased viability may be better explained by improvement in the management system under which they were maintained (Kalinowski *et al.*, 2000). Bryant *et al.* (1999) found with the house fly (*M. domestica*) that selecting the two top-performing pairs of ten full-sib matings as founders did little to improve fitness over random selection of (two) pairs. At present, experimental support for reducing inbreeding depression through controlled inbreeding is weak (Lacy and Ballou, 1998).

Restoration of heterozygosity with crossbreeding

If inbreeding and any accompanying 'inbreeding depression' can account, in part, for the biological changes that occur during domestication, then the crossbreeding of individuals from different inbred lines (i.e. outbreeding) should theoretically restore heterozygosity and result in a more wild-like and vigorous phenotype through the process of heterosis. Smith (1972) tested this hypothesis by comparing the behavior of a population of wild house mice (*M. musculus*) with three inbred domestic strains and all possible hybrid combinations of the latter. Stocks were compared in five behavioral tests, including open-field behavior, wheel-running activity, home cage emergence, avoidance conditioning and underwater escape conditioning. Although behavioral heterosis was observed for the hybrids in all but the open-field measures, crossbreeding the three inbred stocks did not restore the wild phenotype.

Henderson (1970) likewise discovered hybrid vigor when comparing the behavior of inbred and hybrid lines of domesticated laboratory mice (*M. musculus*). Six inbred strains and their 15 possible F1 crosses were reared in either standard laboratory cages ($14 \times 20 \times 9$ cm) or much roomier ($55 \times 25 \times 15$ cm) enriched cages containing a hollow log, small maze, steel tube, rocks, and wood and wire ramps. At 6 weeks of age, the animals were food-deprived for 24 h and then individually placed in a large enclosure ($75 \times 75 \times 60$ cm) with a wire basket containing food attached to one of the walls 50 cm above the floor. The mice could only reach the food by climbing and tightrope-walking a wire-mesh ladder, running along a narrow ramp and entering a series of elevated steel water pipes which opened onto a small platform from which the food could be reached. This task was chosen because it required skills that could be acquired in the enriched rearing environment and because it simulated conditions mice are likely to encounter in the wild when searching for food. The experimenters recorded the time from introduction into the test enclosure until the mice reached the food. Hybrid animals reached the food significantly faster than their inbred parents in 11 of 15 (73%) comparisons and, in both rearing conditions, hybrid means were significantly greater than means of their highest-scoring parent strain, indicating heterosis and suggesting that the inbred lines were experiencing inbreeding depression.

Launey and Hedgecock (2001) crossbred inbred lines of Pacific oysters (*Crassostrea gigas*) derived from a natural population and compared the segregation ratios of microsatellite DNA markers in F2 or F3 families. Evidence was found suggesting that the wild founders of the inbred lines carried a minimum of 8–14 highly deleterious recessive mutations (i.e. a high genetic load), which were manifested by inbreeding depression in the inbred lines and restoration of yield and vigor in the hybrids (heterosis). The authors maintain that the Pacific oyster is a particularly good model for understanding the genetic and physiological causes of inbreeding depression and hybrid vigor.

Outbreeding depression

Lynch (1991) discusses how the mating of individuals from distant populations can either increase fitness by reversing the negative effects of inbreeding depression or decrease fitness by the breakup of favorable (co-adapted) gene combinations (i.e. genes that work well together). The latter phenomenon is sometimes referred to as 'outbreeding depression' (Shields, 1982; Thornhill, 1993; Lynch and Walsh, 1998). Aspi (2000) found evidence for outbreeding depression in the frequency of courtship song in male fruit fly (*D. montana*). When strains from two different geographical locations were interbred, the hybrids had lower fitness than either parent strain. de Boer (1982, cited in Lacy *et al.*, 1993) reported that crosses between chromosomal races of the owl monkey (*Aotus trivirgatus*) often produced hybrids with little or no fertility. Moreno (1994) proposed that favorable inter-actions within sets of functionally related gene loci may be very common. In some cases, outbreeding depression may be misdiagnosed as the more familiar inbreeding depression (Crnokrak and Roff, 1999).

Conclusions

It is commonly believed that some inbreeding is inevitable in small, closed populations of captive animals. However, because the effects of inbreeding on gene frequencies are random, there is no consistent relationship between inbreed-ing and the process of domestication *per se*. Inbreeding is not a prerequisite for domestication nor does it necessarily accompany the process of domestication. However, it can be assumed that a greater degree of inbreeding is tolerated in captive populations than in nature because of man's ability to preserve genetic variants that would otherwise not survive in nature. From that standpoint, inbreeding is probably more common in captive-bred populations than in nature. To the extent that inbreeding is applied in captive breeding programs and has a negative effect on fitness, captive-bred animals may be less vigorous than their free-living counterparts. Attempts to restore fitness by crossbreeding inbred lines are often successful. Outbreeding may not always have positive outcomes. It has been hypothesized that breeding animals from different populations may some-times result in the breakup of co-adapted gene complexes (genes that work well together) with negative consequences on fitness.

Chapter 6

Genetic Drift

The last chapter described how inbreeding can cause random fluctuations in gene frequencies in captive animal populations. The present chapter describes another mechanism which can bring about random changes in gene frequencies, namely genetic drift. First, genetic drift is defined. Then, methods of measuring drift and ways to avoid drift are discussed. Lastly, the point is made that drift may lead to increased genetic variability between groups of animals even when variability is eroded within populations.

Genetic Drift Defined

Genetic drift is the random fixation or loss of genes in the population gene pool. It is typically manifested by the loss of genetic variability in relatively small populations of animals. Alleles are found in pairs on chromosomes and, by chance, certain alleles may be randomly 'fixed' or lost to the gene pool when there is only a small number of breeding individuals in the population, in much the same manner that one is likely to get all 'heads' or all 'tails' in just a few tosses of a coin. Drift is common when populations are founded with relatively few individuals ('founder effect'). For example, it is believed the first captive colony of Syrian golden hamsters (*Mesocricetus auratus*) was established by a single littermate pair (Murphy, 1985). Mongolian gerbils (*Meriones unguiculatus*) distributed in laboratories worldwide are believed to have descended from a founder population of 20 pairs of animals trapped in Manchuria in 1935 and brought to Japan for breeding (Rich, 1968; Neumann *et al.*, 2001). Eleven pairs were imported into the USA in 1954 and of these animals, five females and four males were used to start the breeding colony at Tumblebrook Farms Ltd, which has provided the breeding stock for laboratories in the USA and Europe (Schwentker, 1963; Marston, 1972).

Drift may also occur when an existing population experiences a severe reduction in numbers (bottleneck). Likewise, every time a small daughter colony is established, the phenomenon of genetic drift may be repeated because of the limited number of breeding animals.

Measuring the Impact of Genetic Drift

In this context, genetic variation is typically estimated by: (i) measuring the degree of heterozygosity or the proportion of individuals in a population which are heterozygous at a given gene locus; (ii) measuring the number of alleles present at a locus; and (iii) determining the percentage of loci which are polymorphic (Allendorf, 1986; Agnese *et al.*, 1995; Richards and Leberg, 1996). The possible loss of heterozygosity and allelic diversity should be considered in the design of breeding systems for captive animals. In general, heterozygosity provides a good measure of the ability of a population to respond to selection immediately following a reduction in numbers (i.e. bottleneck). The number of alleles remaining after a bottleneck is important for the long-term response to selection and for the survival of the population (Allendorf, 1986). Heterozygosity affects the ability of a population to adapt to short-term changes in its environment, whereas allelic diversity is more important in the long-term adaptation of populations. Allelic diversity provides the flexibility for populations to make major shifts in phenotypic traits over generations. Allendorf (1986) and Fuerst and Maruyama (1986) point out that allelic diversity is lost much more rapidly than heterozygosity during bottleneck and founding events. Hence, a larger founder population is needed to protect allellic diversity than heterozygosity. Of course, 'founders' can be added to the population periodically to protect it against the loss of genetic variation by genetic drift (Odum, 1994; Ballou and Foose, 1996).

Since founder populations of captive wild and domestic animals often consist of relatively few individuals, it is not unusual for the genetic variability of these populations to soon decline due to drift and inbreeding (Verspoor, 1988). Neumann *et al.* (2001), working with captive-reared wild and domestic Mongolian gerbils (*M. unguiculatus*), found a marked reduction in heterozygosity and allele number at nine polymorphic dinucleotide repeat loci in the laboratory stock. Mean (± SE) heterozygosity was only 0.14 (± 0.06) in the domestic gerbils compared to 0.76 (± 0.02) in the wild animals. Mean (± SE) allele number was 1.78 (± 0.28) at the nine loci in the domestic population and 9.20 (± 0.57) in wild stock. The 45 domestic gerbils used in the study represented five different strains from the USA and Europe and the 40 wild animals used were the second-generation descendents of gerbils trapped in their native habitat (Mongolia) at six different locations. Ryman and Stahl (1980) compared gene frequencies at two polymorphic loci in three 'domestic' stocks of brook trout (*S. trutta*) with gene frequencies at the same loci in two natural populations of trout from the point of origin of the hatchery stock. The polymorphism present at one locus in one of the natural stocks was lost in the corresponding hatchery population. At the other locus, there was a marked shift in gene frequencies. Verspoor (1988) reported that mean heterozygosity in a relatively small population of Atlantic salmon (*S. salar*) was reduced 26% in one generation of captive breeding. Mason *et al.* (1987) found that a population of soybean loopers (*Pseudoplusia includens*) maintained in the laboratory for 24 generations over a 2-year period showed a 62% reduction in heterozygosity at six enzyme loci.

In a computer simulation program used to examine interacting effects of genetic drift, mutation, immigration, selection and population subdivision on the loss of genetic variability from small, managed populations, genetic drift was the overriding factor controlling the loss of genetic variation (Lacy, 1987). As in inbreeding, the loss in genetic variability due to genetic drift can negatively impact fitness-determining traits. Ringo *et al.* (1986) reported that drift in two captive populations of fruit flies, *Drosophila simulans*, reduced the propensity of flies to mate and the number of offspring produced relative to an outcrossed base-line population.

Avoiding Genetic Drift

Genetic variability can be maintained when founder populations are relatively large and efforts are made to avoid inbreeding (Ferguson *et al.*, 1991). Lengthening the generation interval (time between birth of individuals and birth of their offspring) will also help to preserve genetic diversity. The longevity of animals living in captivity (Conway, 1978) allows animal breeders to preserve genetic variability in this manner. Of course, longer generation intervals slow down (in real time) the genetic adaptation of populations to the captive environment.

The traditional doctrine that founder events reduce genetic variation may not always be correct. A founder event (population bottleneck) followed by a rapid increase in population size ('founder-flush') may actually increase the genetic variation available to selection due to significant shifts in the organization of the gene pool, such as in the rapid increase in frequencies of previously rare alleles (Carson, 1990; Meffert, 1999). Meffert and Bryant (1991) demonstrated no loss in mating activity in four of six lines of houseflies (*M. domestica*) that experienced five successive founder-flush bottlenecks to a single male–female pair. Reductions in genetic variance within populations due to genetic drift are most likely to result from a repeated succession of bottlenecks without recovery of population size.

Genetic Variation Between Populations

While genetic variance within populations may decrease due to drift, variability between populations may not be affected or even increase (Dobzhansky and Pavlovsky, 1957; Wang *et al.*, 1990). Dobzhansky and Pavlovsky (1957) tested the hypothesis that the outcome of natural selection should be more variable among small than among large populations. They established ten populations of fruit flies (*Drosophila pseudoobscura*) consisting of 4000 flies each and ten populations founded by 20 flies each. At the start of the experiment, the PP gene arrangement was present on 50% of the third chromosomes examined from the source population. At the end of the study, 19 generations later, the PP gene arrangement was present on 27.4% of the third chromosomes examined from the ten large populations and 32.7% of the third chromosomes from the ten small populations. This difference was not statistically significant. However, the percentage of PP chromosomes found among the ten large populations ranged from 20.3 to 34.7%, while among the small populations, the percentage varied from 16.3 to 47.3% (Fig. 6.1). This difference in variation was statistically significant and evidently due to the different size of the foundation stocks.

One way to preserve genetic diversity in captive populations is to establish a number of subpopulations under the assumption that if drift occurs, the genes fixed or lost will not be the same in the various groups. In a similar manner, genetic diversity may be preserved in natural populations by geographical and/or behavioral isolation of subpopulations (Danielsdottir *et al.*, 1997), an important consideration when restocking captive-reared animals in nature. Fiumera *et al.* (2000) report on the loss of genetic diversity in captive-bred populations of a Lake Victoria cichlid (*Prognathochromis perrieri*) and the importance of captive subpopulations located at a number of institutions in maintaining genetic diversity in the population as a whole. The original captive population was founded by

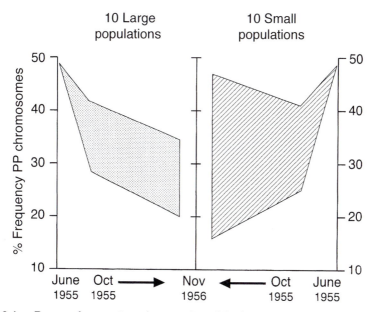

Fig. 6.1. Range of percentage frequencies of the homozygous *PP* gene arrangement on the third chromosome of individuals in ten populations of *Drosophila pseudoobscura* founded by 4000 individuals (left) and ten populations founded by 20 individuals (right) over 19 successive generations of laboratory breeding (Dobzhansky and Pavlovsky, 1957).

10–20 adult fish. By the fourth generation of captive breeding, approximately 19% of the original alleles were lost and heterozygosity was reduced by 25%. At the same time, approximately 33% of the total genetic diversity was maintained solely between subpopulations. The authors speculate that the polygynous mating system of the fish studied may have contributed to the relatively rapid loss in genetic diversity. Rather than all fish contributing offspring for the next generation, reproductive success was highly variable. A breeding program that attempted to equalize family size in each generation would have helped to preserve genetic diversity. Of course, equalizing family size to reduce genetic change under captive rearing is only effective if it correlates with equal reproductive success within and between families (Waples, 1999).

Populations of social animals in captivity typically possess a number of demographic characteristics that naturally increase the magnitude of genetic drift and inbreeding relative to other evolutionary forces. Small populations, polygynous mating, female philopatry and constraints on gene flow between populations tend to reduce genetic variation within subpopulations and increase genetic variation between populations (Storz, 1999).

Conclusions

Like inbreeding, genetic drift is most likely to occur in relatively small populations of animals, whether in nature or captivity. A large majority of captive populations of animals founded by humans are relatively small in numbers and thus are subject to genetic drift. Genetic heterozygosity and allelic diversity within populations of a species are typically reduced when drift occurs, thus potentially compromising the ability of each population to adapt to changes in its environment. This may be most important to populations in the early generations in captivity, particularly if the species is not well pre-adapted to the captive environment. Since the effects of genetic drift are random, genetic variability between populations of a species may persist, offering animal breeders the opportunity to maintain genetic variation by periodically crossbreeding animals from different populations. Unlike inbreeding, genetic drift is not typically associated with a loss of physical vigor or vitality.

Chapter 7

Artificial Selection

Artificial selection is one of the best-understood aspects of the domestication process (Price and King, 1968) and is the only genetically based mechanism that is unique to domestication. Artificial selection may be applied either inadvertently (unconsciously) or intentionally (consciously). This chapter provides examples of inadvertent and conscious artificial selection and discusses the differences between artificial and natural selection. It ends by describing how artificial selection experiments can provide insights regarding mechanisms of gene expression and the evolution of quantitative traits.

Inadvertent Artificial Selection

Personal biases and preferences often influence the selection of breeding stock (Muntzing, 1959) and such biases may be very subtle. Inadvertent artificial selection by animal caretakers is commonplace. Frankham *et al.* (1986) proposed that unconscious selection for tameness is inevitable in captive populations of animals. Marliave *et al.* (1993) reported that ten generations of laboratory rearing of the coonstripe shrimp (*Pandalus danae*) resulted in inadvertent selection for reduced intensity of escape responses. Regular handling of the shrimp in a study of protandric hermaphroditism would often injure individuals with intense tail-flip escape responses. By the tenth generation, the intensity of escape responses had become greatly reduced and dopamine levels (frequently associated with emotional reactivity) were reduced to only 5.5% of the level in wild stock. Hybrids between wild and 'domestic' shrimp were intermediate for both variables.

Waples (1991) claimed that some salmon hatcheries in the Pacific Northwest (USA) have inadvertently selected for early-season spawning by collecting eggs only from early-spawning wild fish. (The number of eggs taken per year is frequently on a quota basis.) In addition, conscious selection for early spawning is often applied to hatchery-raised broodstock, since they produce young that are larger at traditional release times or can be released earlier in the spring. These early-released fish may have higher survival rates providing unpredictable (undesirable) springtime water conditions do not occur.

Inadvertent artificial selection may also account for an acceleration in sexual development and improved sexual vigor in a long-standing laboratory population of oriental fruit flies (*Dacus dorsalis*). In a comparison of a domestic stock of *D. dorsalis* maintained in the laboratory for about 330 generations with wild *D. dorsalis* collected from three islands in Hawaii, Wong *et al.* (1982) found that 100% of the domestic females had mated by day 12 after eclosion whereas 90% of the wild females were not mated until day 25. In addition, the mating speed (proportion of flies mating per hour) of male domestic flies was about four times that of the wild males and the domestic females mated about twice as fast as the wild females. The authors concluded that the laboratory rearing system employed over the 28 years of propagating the laboratory population had selected for early sexual development and rapid mating behavior (i.e. sexual vigor). It should be noted that the wild flies used in this study were older than the laboratory flies (28–30 days vs. 9–10 days), since it took them longer to reach sexual maturity. Could mating speed have been influenced by the different ages of the two stocks at testing? To what extent are mating speed and early sexual maturation related in this species? It is also possible that light intensity could have confounded the results of the mating speed experiment. Copulations were monitored for 4 h starting at 10.00 and 11.00. Wild fruit flies mate more frequently under low light conditions, which may correspond with the cooler and more humid times of the day. Selection for sensitivity to light intensity may have been relaxed in long-standing laboratory populations of fruit flies because of the constant temperature and humidity of the laboratory environment. Hence, domestic stocks may be more active in the middle of the day. Koyama *et al.* (1986) reported that sexual activity in a laboratory strain of melon flies (*Dacus cucurbitae*) was initiated earlier in the afternoon (i.e. at higher light intensity) than in a comparable wild population.

Conscious Artificial Selection

Artificial selection has been demonstrated for a plethora of phenotypic traits. The average rate of phenotypic change in response to (conscious) artificial selection stands in sharp contrast to the relatively slow rate of evolutionary change in free-living populations of wild animals (Lush, 1945; Haldane, 1949). In but a few generations of intense artificial selection, populations of animals have been altered with respect to a wide variety of traits (Siegel, 1975; Plomin *et al.*, 1980; Grandin, 1998; Rauw *et al.*, 1998).

Artificial selection of dog breeds

The variety of physical and behavioral traits seen in the many breeds of domestic dogs (*C. familiaris*; Scott and Fuller, 1965) reflect the genetic changes that can be realized in a species through conscious artificial selection and hybridization. Dog breeds are more diverse in shape and conformation than any other domestic animal species.

It is estimated that selective breeding of dogs for specific functions and appearance began about 5000 years ago (Case, 1999). Dogs resembling today's greyhounds represent one of the oldest breeds and were often shown on paintings and pottery in Egypt and western Asia as early as 2900 BC. The ancient Romans were also heavily involved in the systematic breeding of dogs. Written records from the fifth century BC describe dogs used for herding, sport, war, arena fighting, scent hunting and sight hunting. The aristocracy also kept smaller 'house' dogs. Mastiff breeds were developed primarily in the area of Tibet and were later used in war by the Babylonians, Assyrians, Persians and Greeks. Wolf-like breeds similar to today's spitz were developed early and are the likely ancestors of modern-day huskies, Keeshond and other arctic breeds. Pointers were developed early on to hunt small game and most of the sheepdog breeds probably originated in Europe and were selected to herd livestock.

The majority of exisiting purebred dog breeds have their origins in some type of working animal (Case, 1999). The greatest era for the proliferation of dog breeds was in the Middle Ages in western Europe, spanning the 13th to 15th centuries AD. Notable during this period were the many breeds (e.g. deerhounds, wolfhounds, otterhounds, bloodhounds, etc.) established to hunt different species of game (Menache, 1997).

Artificial selection in other species

Artificial selection has not only been used to establish various breeds within domestic animal species but it has been used to establish populations differing in very specific characteristics. For example, Bernon and Siegel (1983) demonstrated the extent to which sexual performance in male chickens (*G. domesticus*) can be artificially selected in 20 generations (Table 7.1). Koteja *et al.* (1999) selected laboratory mice (*M. domesticus*) for activity wheel use and, after 13 generations, reported a 2.2-fold increase in number of revolutions per day. Interestingly, the increase was in running speed not time spent in the wheels and the selection-line mice 'traversed' 12.1 km per day. Especially noteworthy is the work on selecting silver foxes (*V. vulpes*) for nonaggressive behavior towards man (i.e. tameness) which began in 1959 at the Institute of Cytology and Genetics in Novosibirsk,

Siberia (Belyaev, 1979). This selection has resulted in a line of foxes that show little fear of people and resemble domestic dogs in their behavior (Trut, 1999; Fig. 7.1). The unselected control population still exhibits wild-type behavior, including strong defensive responses toward humans.

Domestic farm animals artificially selected for production traits have shown remarkable gains in recent decades (see review by Rauw *et al.*, 1998). Growth rate of broiler chickens (*G. domesticus*) has increased from 10 g day^{-1} in 1960 to 45 g day^{-1} in 1996. Growth in Norwegian meat-type pigs (*S. scrofa*) has increased from 629 g day^{-1} in 1960 to 962 g day^{-1} in 1996 due to selection. Milk yield of dairy cows (*Bos taurus*) has typically doubled since the early 1950s.

Table 7.1. Mean (± SE) total number of mounts and completed matings exhibited by male chickens after 20 generations of artificial selection for high and low levels of sexual performance.[a] (From Bernon and Siegel, 1983.)

Selection line	Number of subjects	Number of mounts	Number of completed matings
High	44	31.2 ± 1.9	24.4 ± 1.4
Control	63	7.2 ± 0.8	5.2 ± 0.6
Low	77	0.5 ± 0.1	0.3 ± 0.1

[a]During eight 10-min tests with females.

Fig. 7.1. Silver foxes selected for nonaggressive behavior toward humans show little fear of people and resemble domestic dogs in their behavior. (Photo courtesy of Institute of Cytology and Genetics, Novosibirsk, Siberia.)

How Artificial and Natural Selection Differ

Certain theoretical differences exist between artificial and natural selection. Artificial selection is goal-oriented from the standpoint that man selects those individuals for breeding whom he believes will produce (transmit) a desired phenotype. Thus, all artificial selection occurs prior to reproduction. In natural selection, some selection occurs prior to reproduction because some individuals do not live long enough to breed. Differential reproduction is the essence of the natural selection process. In the latter, selective advantage can only be measured '*ex post facto*', after an animal's contribution to future generations is determined. Natural selection is goal-oriented only from the standpoint that it is directed towards maximizing individual fitness (i.e. the number of reproducing offspring left for future generations). Artificial selection may reduce fitness by providing a selective advantage for nonadaptive traits. In artificial selection, the breeder not only decides which animals will serve as breeding stock but also the number of breeding opportunities for each individual. In both artificial and natural selection, an animal producing only a few offspring may contribute as many or more repro-ducing offspring for the next generation than one that produced many young. However, this is more likely under artificial selection because it is typically a more intense selection process.

Darwin, in an 1859 letter to Lyell (cited in Richards, 1998), contrasts artificial and natural selection. First, Darwin noted that artificial selection brings about changes more rapidly than natural selection. Animal scientists agree that the intensity of artificial selection is usually much greater than the intensity of selection occurring in nature (Lush, 1945). Secondly, Darwin maintained that man can only select for the external and visible phenotypic characteristics of the organism. While this was true when Darwin was alive, successful selection has now been demonstrated for a host of physiological traits (e.g. nerve conduction velocity, Hegmann, 1975; adrenocortical response, Jones *et al.*, 1994) thanks to modern-day technological advances. Thirdly, Darwin believed that artificial selection is pursued to meet man's desires and whims rather than necessarily for the good of the individuals comprising the population. It is difficult to argue this point when so much of artificial selection is directed at very specific phenotypic traits with little impact on general fitness. Coppinger and Coppinger (2001) have noted that in domestic dogs, hair color strongly influences who gets mated to whom. Current interest in longevity and lifetime reproductive success in economically important species (i.e. animals reared for food and fiber) constitutes a rare exception to Darwin's observation. Reed and Bryant (2000) and others have demonstrated that experimental populations of insects maintained in the laboratory under a shortened generation-interval breeding program exhibit reduced longevity, presumably because of the random accumulation of deleteri-ous alleles negatively affecting this trait. There is an apparent trade-off between longevity and early progeny production. This result is interesting, considering that shortened generation intervals are recommended to attain the most rapid response to artificial selection. Youngson and Verspoor (1998) cite several reports supporting the hypothesis that the lifetime reproductive success of hatchery-reared Atlantic salmon (*S. salar*) is reduced relative to their wild counterparts because of selection for production traits and the effects of domestication.

Lessons Learned from Studying Artificial Selection

Darwin proposed that the breeding of domestic animals constituted a 'grand experiment' from which the laws of organic nature and the existence of natural selection could be established (Richards, 1998) and one could gain a better understanding of individual inheritance (Bartley, 1992). He argued that since artificial selection could produce relatively large changes in organisms, natural selection could also. Darwin believed that artificial selection could be used to test the importance of specific characters on survival and reproduction (i.e. how natural selection works). This hypothesis has been supported in several ways. Artificial selection should proceed more rapidly in the direction of more favorable genetically homeostatic conditions (Lerner, 1954). Hetzer and Miller (1972) artificially selected for backfat thickness in Duroc pigs over 16 generations. Selection in both directions was successful but the rate of genetic change was 1.8 times greater for the high backfat line (Fig. 7.2). Increased fat deposition is considered beneficial for many traits associated with fitness. Ricker and Hirsch (1985) provided evidence for the existence of favorable, co-adapted gene complexes (genes that work together well) in their study selecting for geotaxis (gravity-oriented behavior) in *D. melanogaster*. Their work represents one of the longest selection experiments ever conducted, with more than 600 generations of intermittent artificial selection recorded over a 26-year period. A case was made for the possible loss of previously established co-adapted gene complexes and the establishment of new co-adapted gene complexes (utilizing genes associated with extreme geotaxis expression) in the later generations of the geotaxis lines, particularly the low (negative response) line. In the earlier generations of selection (up to generation 450), the geotaxis scores of the flies in each line would regress (reverse) if selection was relaxed, as if some previously optimal genetic state was being sought. Relaxed selection in the later generations did not have this effect, even though genetic variability still existed at the loci affecting geotaxis as shown by both lines responding to reverse selection. Belyaev (1979) strongly believed that during domestication, previously existing co-adapted gene complexes are lost and new co-adapted gene complexes become established after many generations of artificial selection for specific traits. Differences in response to relaxed and reversed selection in artificially selected lines can prove useful in identifying the existence of such favorable combinations of heterozygous gene complexes at loci affecting the selected traits.

Daniels and Bekoff (1990) have proposed that unconscious passive artificial selection is occurring on a grand scale in nature as people alter the environment through such practices as pesticide use, construction of dams on waterways and industrial emissions. While such environmental perturbations are, in fact, unnatural (i.e. artificial), the mechanism of action is still natural selection. Humans

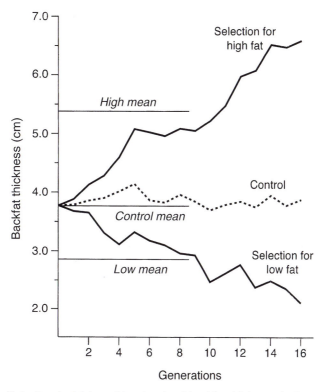

Fig. 7.2. Selection for high and low levels of backfat thickness in Duroc pigs over 16 generations (Hetzer and Miller, 1972).

are not selecting breeding stock prior to reproduction. Similarly, the environment of animals is also artificially altered when translocating them from nature to captivity. Except when man chooses potential breeding partners, differential reproduction in captivity is the hallmark of a natural selection process.

Conclusions

Artificial selection is the only genetic mechanism unique to the domestication process. It is almost always effective in producing phenotypic change when consciously applied. In some respects, artificial selection serves as a showcase for how natural selection works. It has provided insights on how evolution proceeds and how genes are expressed. Phenotypic changes can be rapid when artificial selection is intensely applied. Artificial selection differs from natural selection by the fact that it is applied prior to reproduction whereas natural selection is measured after reproduction, by an individual's contributions to the next generation. Artificial selection can be a double-edged sword, since it is typically applied to one or a few specific traits rather than the entire phenotype. Selection for specific traits may result in unexpected effects on correlated characters, some of which may counter breeding objectives or jeopardize animal welfare. It has also been hypothesized that artificial selection for specific characters may sometimes break up co-adapted gene complexes (i.e. groups of genes that work well together). No matter what the outcome, artificial selection studies on captive wild and domestic animals have proved to be the 'grand experiment' that has greatly enhanced our understanding of individual inheritance.

Chapter 8

Natural Selection in Captivity

In the absence of artificial selection, natural selection provides the basic selective mechanism for genetic change in captive populations. When artificial selection is administered, man does not necessarily select those individuals with the greatest fitness (for captivity) as breeding stock for his selection programs. That is the role of natural selection. Differential mortality and reproduction, including reproductive failure, among artificially selected populations is one way that natural selection in captivity is manifested. This chapter focuses on the various ways that natural selection in captivity is expressed, namely mortality and reproductive failure, and changes in these parameters over generations in captivity. The underlying objective is to demonstrate that natural selection does not cease once a population of animals is brought into captivity, but rather continues to operate regardless of whether or not artificial selection is applied.

Mortality in Captive Populations

Mortality in the early generations in captivity can be severe. Blus (1971) established a breeding colony of short-tailed shrews (*Blarina brevicauda*) in captivity and found that only 11% of his 117 wild-caught shrews and 9% of his 46 captive-born shrews survived for 12 months in the laboratory. Mean age at death for the captive-born stock was approximately 5 months.

Reproductive Failure in Captive Populations

Natural selection typically accompanies artificial selection in captive breeding populations. Selection imposed on captive populations that cannot be ascribed to artificial selection must be 'natural' (Price and King, 1968; Hale, 1969; Wright, 1977). Artificial and natural selection often work in opposition. Animals selected to be parents of the next generation do not always reproduce (Mellen, 1991; Roest, 1991; Van Oorschot *et al.*, 1992). When captive animals reproduce, they do not always provide an expected number of offspring. Reproductive failure may result from physiological or psychological stress caused by such factors as sensory and locomotor deprivation (Hediger, 1964), social incompatibility (Bluhm, 1988), dietary deficiencies (Johnson and Boyce, 1991), parasitism (Hughes and Sokolowski, 1996) and human disturbance (Blus, 1971). In short, reproductive failure in the early generations of captive breeding probably reflects the inability of many individuals to reach a sufficient level of adaptation to the captive environment for successful reproduction to occur. Some examples are cited below.

Price (1980) reported that only 18 of 37 (49%) first-generation captive-reared male Norway rats (*R. norvegicus*) copulated when individually housed overnight with a hormone-induced estrous female (domestic strain). Trut (1999) noted that only 14% of field-trapped Norway rats produced offspring surviving to adulthood. Price (1967) found that 30 of 50 (60%) female wild-caught prairie deermice (*Peromyscus maniculatus bairdii*) produced offspring after being paired with wild-caught males for 4 months in laboratory cages. By comparison, 68 of 75 (91%) female prairie deermice approximately 25 generations removed from the wild produced litters during a comparable period of time. Dohm *et al.* (1994) reported that only seven of 17 (41%) pairs of wild-caught house mice (*M. musculus*) successfully bred in captivity (produced litters) compared to 19 of 20 (95%) pairs of domestic laboratory mice (ICR strain) maintained under identical conditions. Reproductive success was better for wild-caught mice when they were mated to domestic partners. Thirteen of 17 (76%) wild dam × domestic sire pairs produced litters and 12 of 13 (92%) wild sire × domestic dam pairs bred successfully. Thomas and Oommen (1999) reported that about 50% of male wild-caught South Indian gerbils (*Tatera indica cuvieri*) were reproductively inactive when placed with females in the laboratory. Likewise, about 50% of their first-generation laboratory-born male offspring were reproductively inactive. In contrast, 90% of wild-caught females gave birth and successfully reared offspring. Moscarella and Aguilera (1999) brought wild-caught *Oryzomys albigularis*, a sigmodontine rodent, into the laboratory and obtained successful reproduction with only five of 11 pairs (45%). Blus (1971) obtained young from seven of 11 (64%) female wild-caught short-tailed shrews (*B. brevicauda*) after their first pairing in the laboratory, while only 12 of 57 (21%) captive-born females produced offspring from their first pairing. Thirty per cent of the captive-reared females produced young when the first pairing was made after they had reached 100 days of age. Hattori *et al.* (1986) reported that only five of 23 pairs (22%) of wild-caught Watase's shrews (*Crocidura horsfieldi watasei*) bred in captivity. Female shrews that were pregnant when captured frequently cannibalized their young soon after parturition. Wolf *et al.* (2000) noted that over 50% of the prime-breeding-age male black-footed ferrets (*Mustela nigripes*) managed under the Black-footed Ferret Species Survival Plan failed to sire offspring in each of 3 years in the mid-1990s (60–73 males per year) and four of 19 males (21%) failed to produce offspring in all 3 years. Causes of reproductive failure are listed in Fig. 8.1. Improper breeding posture was the major cause of reproductive failure. Some male ferrets were unable to successfully restrain the female in a correct breeding position to allow copulation (Williams *et al.*, 1991). Semen characteristics were similar in the successful and unsuccessful ferrets. Sharpe *et al.* (1998), working with ruffed grouse (*Bonasa umbellus*), found that of 63 eggs collected from wild nests, 59 (94%) were fertile compared to only 29 of 157 eggs (18%) collected from 11 captive-reared birds. Only three of the 11 captive-reared birds produced fertile eggs. In addition, of the 59 fertile eggs collected in the field, 36 (61%) hatched whereas only 11 of the 29 (38%) fertile eggs from captive birds hatched.

Clark and Price (1981) found that of 21 first-generation captive-reared female wild Norway rats (*R. norvegicus*) that gave birth for the first time, only nine (43%) reared at least some of their offspring to weaning age. One litter was cannibalized at birth and 11 litters were abandoned at various stages before weaning. In contrast, 100% of 49 domestic first-time mothers (Long–Evans and Sprague–Dawley strains) reared all or part of their litters to weaning age under the same laboratory conditions. Litter sizes at parturition for these captive-reared wild females averaged about two-thirds the number of pups as for the domestic females $(6.4 \pm 1.0, 9.7 \pm 1.8$ and 10.4 ± 2.8 for the wild, Long–Evans and Sprague–Dawley domestic females, respectively). Interestingly, litter sizes of wild-caught female Norway rats are comparable to those of domestic females whether they are

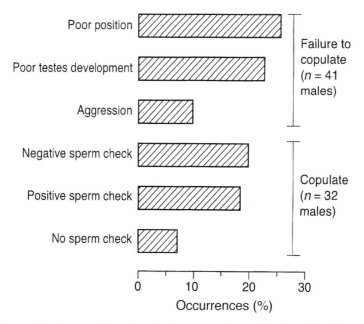

Fig. 8.1. Behavioral and physiological causes of reproductive failure in adult male black-footed ferrets (Wolf *et al.*, 2000).

living in nature or housed in the laboratory (Miller, 1911; Davis, 1951; Boice, 1972). This observation suggests that captive rearing reduces the litter sizes of wild female rats and that the constraints on litter size gradually disappear through the process of domestication (i.e. natural selection in the laboratory). Wyban and Sweeney (1991) reported that seven of 74 (9.5%) captive female shrimp broodstock (*Penaeus vannamei*) were responsible for more than 50% of the total nauplii production in a lifetime reproduction trial. Nineteen females (26%) produced approximately 75% of the nauplii. The offspring of the highest-producing female were reared to adult size and the reproductive success of 18 of her daughters was compared to the reproductive success of 18 female control shrimp. After 85 days, the high-producing female's daughters had produced 4.5 million nauplii from 81 spawns, with an average hatch rate of 47% while the control females produced only 1.2 million nauplii from 46 mates with an average hatch rate of 30%. Similarly, Bray *et al.* (1990) demonstrated that less than 25% of first-generation captive-reared female *Penaeus stylirostris* shrimp produced nearly 70% of the larvae.

Behavioral Factors in Reproductive Failure

Reproductive failure in captivity is often behaviorally based. Wielebnowski (1999) reported that captive cheetahs (*Acinonyx jubatus*) assessed as 'tense', 'insecure' and 'fearful' were less likely to breed in captivity than individuals scoring lower on these traits. Sloan (1973) tested the hypothesis that wild-caught Norway rats (*R. norvegicus*) that behaved most like domestic rats would be most likely to successfully breed in captivity. Some 280 wild Norway rats (140 females and 140 males), trapped as juveniles and raised to adulthood in the laboratory, were subjected to a battery of nine behavioral tests that clearly differentiated wild and domestic Norway rats. They were then randomly paired (mated). An equal number of wild rats caught as adults were treated in the same fashion, except that they were not exposed to the battery of behavioral tests (control for testing effects). Domestic rats, both tested and untested, and of approximately the same age, provided baseline data for both behavioral and reproductive performance. On the basis of individual scores in the behavioral tests, Sloan was unable to predict the reproductive success of the wild rats. Nevertheless, reproducing wild females, as a group, behaved more like domestic rats than did wild females that did not reproduce, for ten of 11 behavioral variables in which reproducing and nonreproducing wild rats were significantly different. The hypothesis that reproductive success in captivity is greatest for those individuals that are most domestic-like in their behavior was partially supported.

Infant mortality in captivity is often based on animals' failure to exhibit appropriate parental behaviors. Bardi *et al.* (2001) noted that of 1093 live births of cotton-top tamarins (*Saguinus oedipus*) at the New England Regional Primate Research Center over a 10-year period, 546 (50%) were rejected and 134 (12%) were killed by their parents. Infant rejection was more common for inexperienced mothers and in family groups without siblings. Abuse rates were higher with experienced mothers and when mothers were sick.

Founder Events, Inbreeding Depression and Natural Selection in Captivity

Population bottlenecks created by founder events are typically associated with reduced genetic variation for subsequent selection. Founder-flush events (see p. 40), which often occur when relatively small populations of wild animals are brought into captivity, do not necessarily reduce genetic variation for complex traits. In fact, they have been known to increase genetic variation (Meffert, 1999) as when previously rare alleles become more common in the gene pool. Increased genetic variation potentially facilitates adaptation to the captive environment via artificial and natural selection in captivity.

Inbreeding depression may increase the intensity of natural selection in captivity. Mortality rates and reproductive failure often increase as inbreeding enhances the phenotypic expression of deleterious alleles. Interestingly, Miller and Hedrick (2001) demonstrated with *D. melanogaster* that under certain circumstances a series of founder events (population bottlenecks) can purge a population of inbreeding depression and its related decline in fitness. 'It appears that a population's response to bottlenecks is dependent not only on the population's initial genetic constitution and the traits involved, but also on the chance genetic processes associated with bottlenecks' (Hedrick, 2001, p. 600). Constraints on fitness caused by the chance fixation of deleterious genes through genetic drift are balanced by the effects of selection, recombination and chance mutation, which increase fitness.

Attempts to satisfy human whims and desires in artificial selection programs are often accompanied by unwanted or detrimental side-effects (i.e. factors reducing fitness). Selection for the Rex hair color in rabbits has resulted in certain metabolic and endocrine disturbances that increase mortality and susceptibility to specific diseases (Muntzing, 1959). Natural selection in captivity acts to remove such traits that reduce fitness.

Changes in Intensity of Natural Selection over Generations in Captivity

The intensity of natural selection on captive populations undergoing domestication depends on: (i) the extent to which the captive environment allows for the development and expression of species-typical biological characteristics (Spurway, 1955); and (ii) the rate of adaptation to the captive environment or number of generations in captivity. In theory, species that possess relatively few pre-adaptations for their respective captive environments will experience rather intense natural selection, and thus show relatively poor survival or reproductive success.

Adaptation and reproductive success

In general, natural selection in captivity is most intense during the first few generations following the transition from field to captive environments. Evolution and adaptation to the captive environment occur rapidly during this time because of the change in direction and intensity of natural selection on so many different traits and the relatively large number of correlated characteristics affected (Price *et al.*, 1993). The degree of adaptation to the captive environment will increase as the frequencies of 'favorable' genes increase in response to selective pressure.

An improvement in reproduction (i.e. fitness) over the initial generations in captivity can reflect the climb to a new adaptive peak as individuals become increasing well-adapted to the captive environment over successive generations (Fig. 8.2). King and Donaldson (1929) reported a marked increase in the reproductive success of wild genotype female Norway rats (*R. norvegicus*) during the first nine generations in captivity (Table 8.1). During this period the animals were maintained under the same conditions and were fed the same diet. Breeding stock for each generation was selected only from litters in which all animals survived to 60 days of age. Unfortunately, the investigators failed to maintain an unselected control population. Consequently, it was not possible to determine the effect of this selection on the traits observed. Although conscious (and, perhaps, unconscious) selection for 'vigor' was practiced, it seems plausible that much of the marked improvement in reproductive success seen during the first few generations of breeding in captivity was due to natural rather than

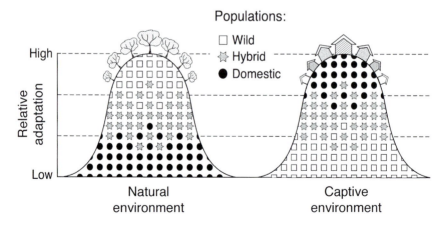

Fig. 8.2.   Relative adaptation (i.e. fitness) is reversed for wild and domestic stocks in natural and captive environments represented by separate adaptive peaks. Adaptation is typically intermediate for wild × domestic hybrids in both environments.

artificial selection. Litter sizes of the captive wild rats reached levels comparable to domestic and free-living wild rats after about 20 generations of breeding in captivity and the length of the reproductive period (between first and last litters) eventually doubled from 204 to 440 days (King, 1929, 1939). Wallace (1981) reported an increase in the percentage of female wild house mice (*M. musculus*) that successfully bred in her laboratory between the wild-caught and first-generation laboratory-reared generations (88.2% of 120 females and 97% of 98 females, respectively). Fertility in subsequent generations was confounded by inbreeding. Blottner *et al.* (2000), in a study measuring changes in testicular activity of gerbils (*M. unguiculatus*) over generations of laboratory rearing, reported that the F2 generation of gerbils exhibited significantly greater testis weights, testis/body weight ratios and testicular testosterone concentrations than the F1 generation offspring of gerbils trapped in their native Mongolia. They postulated that the intensified spermatogenesis and hormone production seen in the F2 generation reflects the relatively rapid changes in reproductive parameters that frequently occur under relatively constant laboratory conditions. Kawahara (1972) reported on the breeding success of an unselected population of Japanese quail (*Coturnix japonica*) following the transition from nature to captivity (cages). Starting with 268 wild-caught individuals, the percentage of birds that had layed eggs by 20 weeks of age was 50, 61 and 67% in the first three generations, respectively. In a domestic control strain, all females were in lay by 20 weeks of age. Searle (1984) noted that wild-caught common shrews (*Sorex araneus*) successfully weaned only eight of 15 litters (53%) conceived in nature but born in the laboratory. In contrast, 12 of 14 litters (86%) conceived and born in captivity to wild-caught females were successfully reared to weaning age. Females pregnant at the time of capture not only did not have as long to adapt to their new environment prior to giving birth but some of them might not have been otherwise predisposed to reproduce in captivity.

Table 8.1. Percentage of female wild-genotype Norway rats that gave birth to offspring during the first nine generations in captivity. (From King and Donaldson, 1929.)

Generation	Number of females mated	Percentage that gave birth
Wild-caught	20	30
2	59	63
3	60	70
4	54	81
5	56	89
6	52	88
7	54	89
8	53	91
9	51	94

Reproductive success of insects over generations in captivity

Laboratory colonization of insects is generally accompanied by an improvement in reproductive success over generations as the colonies become adapted to their new environment. Leppla *et al.* (1983) reported a progressive increase in reproduction of colonies of the Mediterranean fruit fly (*Ceratitis capitata*) established from field-collected pupae and maintained in captivity for 12 generations. The average number of viable pupae per female increased from less than two in the first generation to more than six in the 12th generation (Fig. 8.3), with pupal weight and viability remaining constant. Improved reproduction was due to an increase in the proportion of reproducing females from about 30% to 60% over the 12 generations. Switching the diet and oviposition substrate of the E colony (to the F regimen) after five generations was as detrimental as initial colonization. Leppla *et al.* (1976) monitored a number of reproductive parameters during five generations of laboratory propagation of Caribbean fruit flies (*Anastrepha suspensa*) and compared those changes with a laboratory population some 35 generations removed from the wild. Oviposition rate (eggs deposited) of the wild females gradually increased from less than 5% in the first generation to 43% of the rate of

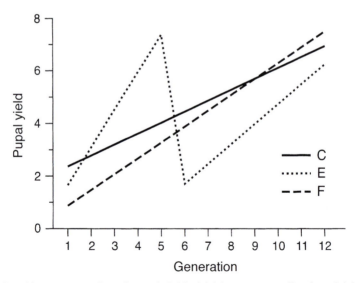

Fig. 8.3. Linear regression of pupal yields (viable pupae per live female) for colonies of Mediterranean fruit flies reared in the laboratory for 12 generations using three different husbandry techniques. The C flies were given fruitless guava branches and papaya fruit (for oviposition and larval development), the E strain was reared on papaya fruit but without guava branches for five generations and then switched to an artificial diet and oviposition substrate. The F flies were reared on the artificial diet and oviposition substrate starting with generation 1 (Leppla *et al.*, 1983).

the domestic stock in the fifth generation (138 vs. 320 eggs per mated female, respectively). In addition, the wild stock required 42–45 days to complete each generation while the domestic stock cycled every 35–38 days. Egg development in the two stocks did not differ but larval development was about 2 days longer and the pre-oviposition period was about 4–5 days longer in the wild flies. Leppla *et al.* (1980) compared the development, survival and reproduction of wild and laboratory strains of the cabbage looper (*Trichoplusia ni*) when exposed to a new rearing regimen. The wild stock was derived from field-trapped animals and the laboratory stock had been maintained in captivity for about 75 generations (6 years). The two stocks had similar development and survival rates during the first ten generations but differed in reproduction. A higher proportion of laboratory females mated, mated more frequently and produced a greater average number of viable eggs. Successive generations of both strains showed an initial decline in fecundity, with population recovery by the fifth generation for the laboratory stock and the eighth generation for the wild strain. Proshold and Bartell (1972) reported that it took nearly nine generations of laboratory rearing for a wild-type population of tobacco budworms (*Heliothis virescens*) to mate as frequently as a comparative laboratory population.

Frankham and Loebel (1992) compared the reproductive fitness of a wild-caught population of *D. melanogaster* with a population obtained from the same site in nature and bred in captivity for eight generations. Both populations were tested at the same time under the same conditions. The reproductive fitness index of the captive population was twice that of the wild population (60.0 vs. 30.7, respectively). Increased insect density and a dietary change were cited as the two most important factors responsible for the changes in fitness observed. Sgrò and Partridge (2000) investigated changes in life-history traits of *D. melanogaster* associated with using two relatively common laboratory rearing techniques. Populations of fruit flies were maintained in the laboratory for 2 years either in bottles, where adults could produce offspring only until day 14 (discrete, short generations), or in population cages, where there was no restriction on offspring production (overlapping, unconstrained generations). After 2 years, bottle-reared flies showed a marked increase in early adult fecundity (eggs laid) and remating, and a decrease in late adult fecundity compared to cage populations and wild flies freshly caught in the field from the origin site of the laboratory populations. This result was not unexpected considering the intensity of natural selection for reproduction in the early-adult stage of the bottle-reared flies. Observations such as these highlight the need to account for the genetic adaptation of wild populations to captivity (i.e. natural selection in captivity) when studying evolutionary theory with captive animals and in programs dealing with conservation and biological control.

Dietary influences on natural selection in captivity

Economopoulos and Loukas (1986) demonstrated a link between the food available to captive olive fruit flies (*Dacus oleae*) and genetic changes over generations in the laboratory. The study focused on changes in allele frequencies at the alcohol dehydrogenase (ADH) locus when field-collected olive fruit flies were brought into the laboratory and fed an artificial larval medium. After four generations in captivity, the experimental populations showed a marked increase in the *I* allele and a decrease in the frequency of allele *S*. Control flies kept on olives in captivity showed no allele frequency changes at the ADH locus, suggesting that the larval medium may have been directly linked to the genetic changes observed. This was confirmed by providing olives rather than the artificial medium to the experimental population after 24 generations of captive breeding. In only one generation, the *I* allele decreased from 28.3% to 8.1% and the *S* allele increased from 21.7% to 48.2%. Alcohol dehydrogenase is involved in the metabolism of a variety of alcohols and serves as a detoxifying agent. Thus, it is important during larval growth and development in different types of fruits. It was apparent that the *I* and *S* alleles have somewhat different functions in the metabolism of alcohols found in ripe olives and that the frequencies of *I* and *S* alleles in the population reflect the nature of alcohols found in the larval medium.

Traits resistant to natural selection

It cannot be assumed that captive rearing will necessarily cause changes in important life-history traits. Gustafsson *et al.* (1999b) and Špinka *et al.* (2000) found only a few minor quantitative differences between the maternal behaviors of domestic and domestic × wild hybrid sows (*S. scrofa*). Price and Belanger (1977) compared the maternal behavior of wild and domestic stocks of Norway rats (*R. norvegicus*) in a laboratory setting. Domestic females showed shorter latencies for nest construction than wild mothers, and both field-trapped and captive-reared wild females tended to be more efficient in pup retrieval than domestic females. No strain differences were found in nursing behaviors. Both domestic and field-trapped wild mothers exhibited relatively high levels of aggression toward female intruders placed in their cage; captive-reared wild females were not as aggressive. Maternal behavior represents a composite of behaviors that may be relatively resistant to modification during domestication because of its high impact on fitness and importance in animal production systems. Relaxed selection on maternal behaviors may be more common in species where animal caretakers frequently serve as surrogate mothers or assist females in rearing their young.

Cohen (2000) compared the predatory capabilities of a population of predatory big-eyed bugs (*Geocoris punctipes*), which had been reared in the laboratory for 90 generations (starting with a founder population of about 1000–3000 field-collected insects), with those of the F1 offspring of wild-caught counterparts. The domesticated population had been fed an artificial beef-product diet for more than 60 generations and the F1 wild stock was maintained on a natural diet of insect eggs, larvae and green beans. To test their predatory capabilities, individuals of both populations were exposed to either tobacco budworm larvae, *H. virescens*, or pea aphids, *Acyrthosiphon pisum*. Domestication did not significantly influence predatory feeding efficiency as measured by prey handling times (total period of contact with prey including attack, prey preparation by extraoral digestion and ingestion of prey biomass), amounts of prey extracted and extraction rates. This result is interesting considering that the domesticated *G. punctipes* were fed a nonmoving diet of noninsect origin for so many generations. However, this species may be pre-adapted for such an artificial diet, since in nature they normally feed on plant materials or nonmoving prey such as insect eggs or aphids. In addition, *G. punctipes* may have been pre-adapted to using the beef-product diet because the latter was designed to mimic the nutritional composition and texture of the contents of lepidopteran eggs.

Matos *et al.* (2000) studied a population of *Drosophila subobscura* over 29 generations in the laboratory and found no significant changes in time of development, age at first reproduction, fecundity, starvation resistance and longevity. However, a second captive population established when the first population had reached 24 generations in captivity was different from the first population in age of first reproduction and fecundity during its first nine to ten generations in captivity. These differences gradually declined with time in captivity of the second population. This latter study illustrates the need for replication and control when investigating animal adaptation to novel environments.

Connor (1975) and Smith and Connor (1978) both failed to find emerging genetic differences between wild and domestic house mice (*M. musculus*) over ten generations of laboratory breeding. Smith and Connor (1978) compared the behaviors of wild mice and two stocks of laboratory mice (DBA and C57BL) in an activity wheel and open field (freezing, wall-seeking and activity) from the third to 13th generations of laboratory rearing of the wild mice. The wild stock did not become more similar to the domestic strains during this period. It is clear that natural selection in captivity does not necessarily produce rapid changes in phenotypic characteristics.

Conclusions

Natural selection in captivity is ultimately manifested in differential mortality and reproductive success. Those animals that come into captivity with the fewest pre-adaptations for their new environment should theoretically experience the most intense natural selection. A number of studies were cited in this chapter that showed enhanced reproductive success of long-standing captive populations of animals relative to stocks of wild counterparts recently introduced into captivity. These comparative studies supported the concept of natural selection in captivity presented here but did not account for possible sampling error (see Chapter 3) and the potential confounding influence of other genetic mechanisms such as artificial selection, inbreeding and genetic drift. An ideal experiment to demonstrate natural selection in captivity is to show an improvement in reproductive success over the first few generations in captivity in the absence of artificial selection and any changes in housing and management systems. Although there is some evidence in the literature that appears to meet these criteria, several care-fully controlled studies with different taxa are needed to positively confirm the merit of this hypothesis.

The nature, timing and intensity of natural selection in captivity may be dependent on housing and management conditions employed and the trait in question. For example, the study by Economopoulos and Loukas (1986) demon-strated how a change in diet can have a relatively rapid effect on gene frequencies in captive olive fruit flies (*D. oleae*). Other traits (e.g. maternal and predatory behaviors, etc.) may be more conservative and resistant to the effects of natural selection in the captive state.

Chapter 9

Relaxation of Natural Selection

Relaxation of selection on certain traits is inevitable when animals experience a relatively major change in their environment (see review by Coss, 1999), such as the transition from nature to captivity. Environmental change is accompanied by changes in selective pressures. The purpose of this chapter is to discuss those traits that might be most subject to relaxed selection during the process of domestication and factors responsible for changes in selective pressures on those traits. The chapter also addresses the relaxation of artificially selected traits and problems in determining the relative impact of relaxed selection.

Changes in Selection Pressures in Captivity

It is a reasonably safe assumption that some relaxed selection will accompany the transition from field to captive environments (Price, 1976a; Waples, 1991). Certain behaviors important for survival in nature (e.g. food finding, predator avoidance) lose much of their adaptive significance in captivity. Hence, one would expect natural selection in captivity on such behaviors to lose its intensity. As a result, changes in the gene pool of the population are likely to occur and genetic and phenotypic variability for many traits are likely to increase. For example, behaviors of free-living prey species toward predators may be changed after relatively long periods of freedom from predators (Coss and Biardi, 1997). Caution in accepting novel foods may decline over time in captivity. Free-living herbivores are sometimes exposed to toxic plants. Wild animals living close to human habitation may be exposed to poison baits. In contrast, captive animals are generally protected from toxic food items. Hence, it seems reasonable to expect relaxed selection for food neophobia in captive animal populations. Kronenberger and Medioni (1985) provided some support for this hypothesis by demonstrating that domestic house mice (*M. domesticus*) were more accepting of novel saccharin-flavored water (0.1% solution) than their wild counterparts (*M. musculus*). Hoffmann *et al.* (2001) reported that starvation resistance (50% mortality) in *D. melanogaster* decreased from 50.1 to 35.9 h over the first 4 years in the laboratory. Similarly, desiccation resistance (50% mortality) declined from 14.3 to 9.8 h in 4 years. In nature, the availability of food and humidity can be highly variable. Fruit flies are selected for their ability to survive such changes. In the laboratory, flies are maintained on an *ad libitum* artificial diet and under relatively constant humidity. Consequently, selection is relaxed on their ability to survive starvation and desiccation events.

Sensitivity to Changes in the Environment

In nature, predator detection and avoidance is aided by acute sensitivity to changes in the environment and appropriate behavioral responses to either avoid detection or capture by predators. In captivity, traits reducing the likelihood of detection, such as immobility or 'freezing', and traits facilitating escape, such as speed and agility, are not as important and may be subject to relaxed selection. Lack of 'freezing' behavior by domestic rodents in response to environmental change is one of the more consistent differences observed between wild and domestic rodents (Price and Huck, 1976; Smith, 1978).

Physical Condition

In nature, locating sources of food, water and shelter, mating activities and avoiding predators can require relatively high levels of physical fitness. Physical stamina and agility are less important in captivity due to the absence of predators and provisioning of basic necessities of life by man. For these very reasons, Dohm *et al.* (1994) hypothesized that relaxed selection has reduced the locomotor performance of laboratory mice (*M. domesticus*). They tested their hypothesis by comparing various measures of exercise physiology and voluntary wheel-running activity of a random-bred strain of domestic mice (ICR strain) with first-generation captive-reared wild mice (*M. musculus*) and their reciprocal crosses. After effects of body mass and other appropriate covariates were accounted for, wild and hybrid mice exhibited forced maximal sprint running speeds that averaged about 50% higher than the laboratory mice (Table 9.1). Wild and hybrid mice also had significantly higher (+22%) mass-corrected maximal rates of oxygen consumption (VO_{2max}) during forced exercise and greater (+12%) relative heart ventricle masses than lab mice. Differences in swimming endurance were equivocal. Although laboratory mice spent the same amount of time in the running wheels as wild and hybrid mice, the wild stock and hybrids ran significantly more total revolutions (+101%) and at a higher average velocity (+69%) when they were active. The results were consistent with the hypothesis

Table 9.1. Mean (± SD) forced racetrack sprint speed and voluntary running-wheel activity in wild and domestic mice and their reciprocal hybrids. (From Dohm *et al.*, 1994.)

Behavior	WW	DD	WD	DW
Forced sprint speed (m s^{-1})	2.88 ± 0.24	1.38 ± 0.25	2.68 ± 0.35	2.47 ± 0.49
Total revolutions in 24 h	11,404 ± 4,441	7,100 ± 3,309	11,475 ± 5,609	12,029 ± 5,998
Average running rpm	24.3 ± 6.34	15.0 ± 6.10	23.0 ± 7.88	25.0 ± 8.45

that behavioral traits are more subject to relaxed selection than associated traits at lower levels of biological organization. The magnitude of the differences between wild and laboratory mice were greater for behavioral measures than physiological traits.

Cognitive Abilities

There is also reason to suspect that natural selection for cognitive abilities may be relaxed in captive animal populations. The evolution of cognitive abilities in animals involves the evolution of more sophisticated anticipatory mechanisms, reaching further back in time away from the expected event (Dawkins, 2000). Frank (1980) reviewed evidence that domestic dogs are inferior to wolves in observational learning. In nature, fitness is enhanced by the ability of individuals to quickly learn the consequences of their behavior or the behavior of other animals. In captivity, humans typically provide animals with the basic necessities for survival and may buffer the negative consequences of their mistakes. Opportunities to exercise cognitive abilities are reduced when the animals' environment limits physical activity and social interactions.

Changes in Animal Management Practices

Traditionally applied management systems may facilitate the relaxation of natural selection in captivity. Dairy calves are typically taken from their mothers at birth and hand-reared, whereas beef calves remain with their dams. Hence, it was not surprising when Buchenauer (1999) reported that dairy cows generally exhibit less intense maternal behavior than beef cows. Selman *et al.* (1970) noted that beef cows initiated licking their calves immediately after parturition, while dairy cows were sometimes slow or failed to initiate grooming. Furthermore, beef cows stood immobile quickly to the teat-seeking behavior of their calves, whereas a few of the dairy cows in their study remained lying. Liimatainen *et al.* (1997) compared the mating success of 'wild' Mediterranean fruit fly males (*C. capitata*) from a mass-reared population maintained in the laboratory for approximately 40 years with a population of *C. capitata* reared in the laboratory for three generations. Flies were sexually inexperienced when tested and were used only once. The proportion of courtships that led to mating was nearly twice as great for the 'wild' males as for the laboratory stock (47 vs. 23%, respectively). The reduced mating success of the mass-reared laboratory stock reflected a reduction in the sequential structure of courtship resulting in a less integrated pattern of signal–response between the sexes. In unsuccessful matings involving laboratory males, both sexes performed all of the different courtship acts but did so largely independently of one another without an orderly signal–response sequence. Mass-reared (laboratory) males often failed to respond appropriately to females who approached or stood nearby. These findings have important implications for sterile insect technique programs, since mating success of the mass-reared males may decline over generations of laboratory breeding. Over time, mass-reared males may lose the more precise, orderly pattern of responding to female signals necessary for successful mating. The lack of orderliness in the courtship behavior of the laboratory males may be a consequence of breeding fruit flies under high-density conditions where the need to attract mates even over short distances is less important, resulting in relaxed selection for integrated signal–responses between the sexes during courtship. Rössler (1975a) also found that male Mediterranean fruit flies from a stock bred in captivity for 13 years had relatively low mating success compared to males from a stock held in captivity for less than a year.

Changes in Mating Systems

Captive breeding systems which eliminate male–male competition and female choice may also weaken reproductive behaviors. There are four basic mating systems used in captivity: (i) multi-sire mating; (ii) single-sire mating; (iii) controlled (hand) breeding in which sexually receptive males and females are introduced to one another in pairs just long enough for mating to occur; and (iv) artificial insemination. Natural selection for ability to compete with other males for access to females is eliminated in all but the multi-male mating system. In addition, female choice is largely eliminated in the single-sire and controlled breeding systems, and in artificial insemination programs, male and female do not meet at all. Only a few of our common domesticated species are typically bred using the multi-sire mating system (Table 9.2). One of those exceptions is the commercial breeding of broiler chickens (*G. domesticus*); breeders recommend eight or nine males per 100 females at 20–30 weeks of age (Hazary *et al.*, 2001). In the absence of male–male competition and female choice, males with relatively poor libido or mating technique may leave as many offspring for the next generation as males with relatively high libido and good technique. It just takes the former longer to impregnate females; breeding is prolonged. The considerable variability in sexual performance observed among male farm animals, with many males exhibiting substandard libido and mating ability (see review by Price, 1987), is probably a consequence of relaxed selection for sexual performance in captive populations.

Price (1980) compared the copulatory behavior and reproductive success of male wild and domestic Norway rats (*R. norvegicus*) when tested alone (with an estrous-induced female) and with a male competitor. Wild rats achieved ejaculations with fewer intromissions and at a faster rate than their domestic counterparts (Fig. 9.1). In addition, fewer wild rats exhibited mounts without vaginal penetration, and ejaculatory clasps were longer in duration than for domestic males. Interestingly, reproductive success of wild and domestic males was nearly equal in both the laboratory and an outdoor enclosure. Similar differences in copulatory behavior between male wild and domestic house mice (*M. musculus*) were reported by Estep *et al.* (1975). Price (1967) reported that 20 of

Table 9.2. Occurrence of male–male competition in mating systems typically employed in breeding common domestic animals.

Species	Common	Somewhat common	Rare
Dairy cattle			X
Beef cattle		X	
Sheep		X	
Dairy goats			X
Swine			X
Horses			X
Laboratory rodents			X
Chickens	X		
Turkeys			X

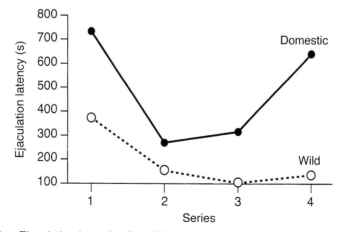

Fig. 9.1. Ejaculation latencies for wild and domestic male Norway rats during the first four ejaculatory series when individually exposed to a single estrous female (Price, 1980).

70 (29%) females of a semi-domestic stock of prairie deermice (*P. m. bairdii*), approximately 25 generations removed from the wild, cannibalized or deserted at least one of their first four litters of offspring. In contrast, only two of 35 (6%) wild-caught females destroyed at least one litter. It was hypothesized that the intensity of natural selection against factors controlling cannibalism in nature is more severe than in captivity. In nature, most female deermice that live to sexual maturity will seldom produce more than three or four litters in their life-span (Howard, 1949), while female deermice in the laboratory are likely to produce 12–20 litters in their reproductive lifespan. Thus, a wild female in nature that cannibalizes 75% of her litters will probably leave no more than five descendants, assuming an average litter size of five young, while a comparable female in the laboratory can cannibalize 75% of her litters and still leave 15–25 descendants. The latter's offspring will probably have as good a chance of being used for breeding stock in the next generation as the offspring of a non-cannibalizing female, particularly when an effort is being made to avoid inbreeding. In captivity, artificial breeding (selection) programs allow a female who destroys all but one litter to be as 'fit' as one that successfully raises all of her offspring. The odds are against such an occurrence in nature.

Lifetime reproductive success

In nature, fitness is measured by the lifetime reproductive success of animals. Natural selection may favor animals who conserve resources to ensure future reproductive success. Gustafsson *et al.* (1999b) speculate that during domestication animal breeders have consciously or unconsciously selected females who place more emphasis on the nurturing of present offspring, since access to the resources needed for future reproduction (e.g. food, shelter) are more or less ensured by humans. Natural selection pressures to conserve resources for future reproduction are relaxed. If this is so, the survival, growth and well-being of the offspring of domestic animals should be enhanced relative to their wild ancestors without sacrificing long-term reproductive success. This hypothesis remains to be tested.

Social dominance becomes less important for reproductive success

Another consequence of removing competition between males in captive animal mating systems is that selection is relaxed on behaviors which confer dominance. When access to females is no longer contingent on attaining dominance, males are able to breed at a younger age and selection for early sexual maturation may occur in cases where young males have access to females. Relaxed selection for sexual dimorphism may occur in the absence of male–male competition and when females no longer have a choice of mating partners. Male traits normally impor-tant for establishing dominance and gaining access to females, such as body size, aggressive potential, and horn size and shape (Zohary *et al.*, 1998), may be reduced or modified. On the other hand, one may argue that the nurturing effect of the captive environment minimizes the expenditure of energy by animals in procuring basic needs, thus allowing more resources to be directed toward body and horn growth (Zohary *et al.*, 1998).

Survival Rates Greater in Captivity

Less intense selection can be also be inferred when populations experience greater survival rates in captivity than their wild counterparts experience in nature. At the New York Zoological Park, annual mortality of adult gallinaceous birds and ducks is less than 7%, whereas in nature, 40–60% die each year (Conway, 1978). Howell *et al.* (1985) estimate that survival of salmon in hatcheries from egg to smolt stage is approximately 50%, whereas survival in wild populations during this period is generally under 10%. Jonsson and Fleming (1993) suggest an even greater difference in survival (20-fold) between cultured and wild salmon to the smolt stage. The nurturing effect of the hatchery environment is confirmed when farmed smolts are released into natural waterways. Survival to the adult stage is about half that of wild fish (Jonsson *et al.*, 1991). Skaala *et al.* (1996) reported that when first-generation hatchery-reared brown trout (*S. trutta*) were released into two natural waterways in Norway, their survival rates were approximately one-third those of wild trout indigenous to the area.

Problems in Measuring Relaxation of Natural Selection

To demonstrate relaxed selection experimentally, one must differentiate its effect from the effect of a reversal in the direction of natural (or artificial) selection. For example, assume that heightened emotional reactivity confers an advantage in avoiding predators in nature but impairs reproductive success in captivity. If a reduction in emotional responsiveness is exhibited by the individuals of a population undergoing domestication, how does one differentiate the relative contributions of relaxed selection due to the absence of predators and of directed natural selection for reproductive success in captivity? Such an assessment is difficult since quantitative characters are influenced by many genes interacting, often in very subtle ways (Falconer and Mackay, 1996). Haase (2000) compared a number of reproductive parameters in male wolves (*C. lupus*) and dogs (*Canis domesticus*) living under similar conditions in captivity. Wolves exhibited maximal levels of testosterone and 5-α-dihydrotestosterone in the peripheral plasma and maximal testicular weights during the winter months (which is their natural breeding season), whereas the dogs exhibited relatively high levels of these parameters during all seasons. Has the loss of reproductive seasonality in the dog been due to relaxed selection, since they are normally protected from harsh climatic factors and their basic needs are provided by human caretakers? Has natural selection in captivity favored dogs that breed more than once a year? Or have dogs been artificially selected for greater fecundity? In reality, any combination of these (or other) factors might have been involved.

Relaxation of Artificially Selected Traits

It is not as difficult to demonstrate relaxed selection on an artificially selected trait, particularly if the trait has been subject to intense selection, and other selective forces and environmental factors are held constant for several generations before and after the selection pressure is removed (Barria and Bradford, 1981a,b). However, phenotypic changes do not always accompany relaxed selection, as was demonstrated when artificial selection for geotaxis (gravity-oriented behavior) was relaxed in populations of domesticated *D. melanogaster* (Ricker and Hirsch, 1985).

Conclusions

Environmental changes such as those typically associated with the transition from nature to captivity will inevitably lead to changes in selection pressures and relaxed selection on some traits. The relative safety of most captive environments due to the absence of predators and animal poisons should theoretically result in a reduction in sensitivity to novel and unfamiliar stimuli. The ready availability of food, shelter and other needed resources may relax selection on physical agility and stamina and the conservation of energy to prolong lifetime reproductive success. Cognitive abilities may become less important when the necessities of life are provided by man and social groups are relatively stable. Mating systems that eliminate male–male competition and female choice remove much of the adaptive significance of male aggressive potential and the need to attain dominance for access to mates.

Relaxed selection on artificially selected traits is relatively easy to document, but measuring the extent to which naturally selected traits have been relaxed is more problematic. It can be difficult to separate the effects of relaxed selection from a reversal in the direction of natural selection.

Further Reflection on Genetic Mechanisms Influencing Domestication

It is clear that domestication, like any evolutionary phenomenon, involves the complex interplay of both random and nonrandom genetic mechanisms, the effects of which are difficult to separate even in the laboratory. Using the longitudinal approach, one can set up experiments with somewhat equivalent populations to demonstrate the effects of inbreeding, genetic drift and selection over generations of breeding in captivity. Whereas inbreeding, drift and artificial selection are subject to a reasonable degree of control by the experimenter, natural and relaxed selection in captivity cannot be easily differentiated and often must be lumped together. Study of animal adaptation under domestication thus has many of the advantages of the controlled selection experiment, yet it retains some of the realism of natural selection (Frankel, 1959). Consequently, efforts to determine the relative contribution of different genetic mechanisms to the domestication process under different circumstances will be impeded. Even so, it is not difficult to understand why Haldane (1954, p. 119) wrote, 'One of the most hopeful fields for the study of evolution is the domestication of animals and perhaps also of plants'.

Chapter 10

Genetic Variability and Behavior

There is no consistent pattern to the effects of domestication and rearing animals in captivity on phenotypic variation. Variation has increased for certain traits in some populations and decreased in others. Not surprisingly, studies on genetic variation show this same pattern. This chapter provides examples of both increased and decreased genetic variation in captive populations and how it has affected variation in phenotypic traits, especially animal behavior. The chapter also continues our discussion of how artificial selection for one trait results in correlated effects on other characters.

General Factors Affecting Genetic Variability

It is difficult to generalize about the effects of domestication on genetic variability. The proportion of phenotypic variation due to genetic factors in captive populations is reduced as the result of fixation of genes by selection (both artificial and natural) and genetic drift and barriers to gene flow between populations. Conversely, genetic variability will increase as a result of the preservation of new or novel phenotypes that arise from mutation or recombination (Frankel, 1959) and the crossbreeding of animals from different populations. The increased longevity of animals in captivity provides captive parents the opportunity of passing on their genes in more combinations to more young for a greater number of breedings than would typically occur in nature (Conway, 1978). Conversely, the disproportionate breeding of a small percentage of captive individuals in the early stages of domestication can reduce genetic variability in the population. If captive environments are typically more variable than those of wild populations, as Darwin suggested (Richards, 1998), one would expect greater genetic and phenotypic variability between populations of domestic animals than between

©CAB *International* 2002. *Animal Domestication and Behavior*
(E.O. Price)

populations of their wild counterparts. The current ease with which animals and semen can be transported between populations has also facilitated efforts to increase and maintain genetic variation both within and between captive animal populations. These factors affecting genetic variability must be understood before genetic improvement of economically important traits can be made (Hedgecock *et al.*, 1976). Documentation of genetic variation within rare breeds is important in developing and administering conservation programs (Kantanen *et al.*, 2000; Giovambattista *et al.*, 2001).

Considerable Genetic Variability in Captive Populations

There is increasing evidence that considerable genetic diversity remains among various domestic stocks of animals in spite of intensive selection for economically important traits and other characteristics. Dunnington *et al.* (1994) examined the genetic diversity among commercial lines of domestic chickens (*G. domesticus*) compared with random-bred control lines and red jungle fowl (*G. gallus*), the wild ancestor of domestic chickens, using DNA fingerprinting techniques. It was concluded that the majority of commercial breeder populations available in the USA contains a considerable reservoir of genetic diversity. They speculate that the large number of chickens used for breeding in the USA and worldwide has been instrumental in maintaining existing diversity. Baccus *et al.* (1983) estimate that the average genetic heterozygosity at various protein loci in populations of European cattle breeds (*B. taurus*) varies between 4 and 8%, which compares well with the mammalian mean of $4.14 \pm 0.25\%$ (Fig. 10.1). Bradley and Cunningham (1999) suggest that the retention of genetic variability in domestic cattle may have resulted from the rapid expansion of cattle populations once bred in captivity and early repeated breeding with wild aurochs (*Bos primigenius*), a successful, wide-ranging species. Vilà *et al.* (2001) used mitochondrial and microsatellite markers to examine the genetic diversity of 191 present-day domestic horses (*E. caballus*) and wild horse remains from archeological sites 12,000–28,000 years old and found a high level of genetic diversity in domestic stocks relative to the ancient population sampled. They concluded that horses were probably domesticated from multiple populations of wild horses. Belliveau *et al.* (1999) compared the genetic variability in American mink (*Mustela vison*) from four mink ranches in Nova Scotia with wild mink trapped in eastern Canada using seven microsatellite loci. Estimates of genetic heterozygosity in the ranch mink populations ranged from 0.53 to 0.61 and the average number of alleles per locus ranged from 4.14 to 5.14. These values were comparable to or greater than those found in the wild mink sampled (0.50 and 4.43, respectively). In spite of many years of intense selection for fur quality traits, linebreeding and positive assortative mating, genetic variation in the captive black mink was relatively high. Some populations have shown very little reduction in genetic variability due to domestication. Three hatchery and four naturalized populations of Pacific oysters (*C. gigas*) in Australia were compared with one another and with two endemic Japanese populations at 17 allozyme loci

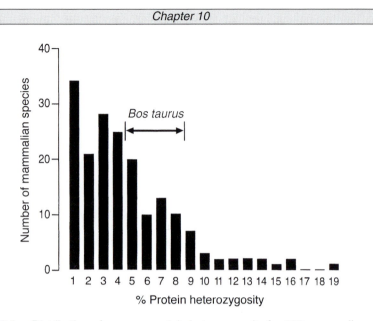

Fig. 10.1. Distribution of average protein heterozygosity for 183 mammalian species or subspecies. The range of a series of estimates calculated for domestic cattle populations is denoted by an arrow on the horizontal axis. (Adapted from Baccus *et al.*, 1983, and Bancroft *et al.*, 1995, by Bradley and Cunningham, 1999.)

(English *et al.*, 2000). The naturalized populations had been introduced to Tasmania from Japan approximately 50 years prior to the study. Overall, all populations showed a high degree of genetic variability. For example, the percentage of polymorphic loci ranged from an average of 70.6% (hatcheries) to 73.5% (naturalized and Japan). The mean number of alleles per locus ranged from 3.0 (hatcheries) to 3.3 (naturalized) to 3.5 (Japan). In contrast, losses of genetic variability in hatchery-reared Pacific oysters have been documented in Britain and the USA (Gosling, 1982; Hedgecock and Sly, 1990).

The amount of genetic variability present in captive populations relative to their wild ancestors may depend on such factors as the specific traits being considered, the distribution, size and genetic diversity of the populations sampled, the size and representativeness of the founder population and the conditions under which individuals have been maintained and bred (Slater and Clayton, 1991; Peterka and Hartl, 1992; Cronin *et al.*, 1995; Hedrick *et al.*, 2000). The diversity in morphology, physiology and behavior of the various breeds of domestic dog (*C. domesticus*) illustrates the extent to which phenotypic changes may be attained in domesticated species. The variability in the domestic dog seen today is due to: (i) the loss of phenotypic buffering mechanisms that tend to preserve the relatively uniform physical characteristics of wolves; (ii) artificial selection of characteristics of value to humans; and (iii) human preferences for exotic phenotypes, which has preserved mutations in the gene pool that would otherwise have been lost or selected against in nature (Ginsburg and Hiestand, 1992). The fact that selection is still applied to preserve breed characteristics suggests that sufficient genetic variability exists to permit further modification, including reversion to behaviors previously selected against.

The retention of genetic variability, *per se*, during the process of domestication is no guarantee that a population has retained its adaptive fitness. Domestication may alter the distribution of such variability as well as the arrangement of genotypes (Bartlett, 1985). Genetic variation has been found in wild house mice (*M. musculus*) that is not present in laboratory populations. In contrast, there are hundreds of mutant strains of mice held in laboratories (e.g. nude, obese, muscular dystrophy strains) that could not survive in the wild. It could be argued that, taken as a whole, laboratory mice are substantially more variable than their wild ancestors in much the same way that the domesticated dog is now more variable than its wild wolf-like ancestors (Festing and Lovell, 1981).

Genetic Variability Often Reflected in Ease of Artificial Selection

Increased genetic variability may be reflected in ease of selection for various traits. Clayton and Paietta (1972) artificially selected for early and late eclosion (i.e. hatching) times in wild and domestic populations of *D. melanogaster*. They found that the domesticated population (in captivity since 1926) exhibited significantly greater selectability (i.e. greater genetic variability) than the wild population, presumably because of relaxed selection for eclosion time in the laboratory environment. In nature, flies must eclose in the early morning hours when humidity is relatively high; in captivity, flies are reared in bottles with continuous high humidity.

Loss of Genetic Variability in Captivity

In some cases, domestic stocks are less variable than their wild progenitors. Berlocher and Friedman (1981) examined the genetic variability within one wild and three laboratory populations of the black blowfly, *Phormia regina*. Variability was determined by starch gel electrophoresis performed on six enzymes which gave satisfactory results. Genetic variation decreased in the order: field-collected stock, first-generation laboratory-reared and then laboratory stock. Wild flies had an average heterozygosity (0.16) typical of insects, in general. Heterozygosity of the first-generation laboratory-reared stock was slightly lower (0.14), while heterozygosity of the laboratory stock was considerably lower (0.02). Only six genotypes were possible in the laboratory population, while the wild population supported 900 possible genotypes. Clearly this laboratory stock was not representative of *P. regina* as a whole. This result was not surprising considering that the laboratory population had been maintained under uniform laboratory conditions for at least 300 generations (15 years). Gibert *et al.* (1998) measured wing length and body pigmentation in female wild-living fruit flies (*D. melanogaster*) and their ninth-generation laboratory-reared descendants. Phenotypic variability for wing length was 8.8 times greater in wild-living females than in the laboratory-reared stock. Poor feeding conditions and higher developmental temperature were likely explanations for the greater variability in wing length of the wild-living flies. Phenotypic variability for pigmentation was similar for the two stocks. Shikano *et al.* (2000) studied the relationship between genetic variation and salinity tolerance (survival time after transfer from fresh water to 35 ppt artificial seawater) in six wild populations and 14 domestic strains of the guppy (*P. reticulata*), a fish native to northeastern South America. The domestic strains had been maintained in captivity for 20–30 generations (10 years). Salinity tolerance and mean genetic heterozygosity, estimated from five polymorphic and 23 monomorphic allozyme loci, were greater in the wild populations than in the domestic strains. A positive correlation was found between mean heterozygosity and salinity tolerance among wild and domestic populations but not within populations (individual scores). A loss in heterozygosity due to inbreeding and consequent inbreeding depression (Chiyokubo *et al.*, 1998) may have been responsible for the reduced salinity tolerance of the domestic stocks. Dupont-Nevit *et al.* (2001) studied changes in genetic variability in a population of the edible snail, *Helix aspersa*, over six successive generations of laboratory breeding. The wild-caught generation consisted of 500 snails sampled from 20 different geographical areas. No artificial selection was applied during the first three generations but in generations four through six, snails were selected for increased adult weight or they remained in an unselected control line. Genetic variability was described using genealogical parameters such as inbreeding, effective number of founders and ancestors, effective number of remaining genomes and additive genetic variance. Large decreases in all parameters of genetic variability were observed over the six generations consistent with predictions of relatively strong natural selection. For example, the effective number of ancestors (animals which contributed highly to the population studied) decreased from 97.4 in the first laboratory-reared generation to 34.1 in the sixth generation. Similarly, the effective number of remaining founder genomes, which takes into account the random losses of genes during reproduction, decreased from 96.2 to 34.1 between the first and sixth generations. The authors concluded that natural selection was most intense on families that did not adapt well to artificial rearing or that for other reasons experienced poor reproduction. Some snails failed to mate under the laboratory conditions provided. Wimmers *et al.* (2000) evaluated the genetic variability of local chicken populations (*Gallus*) from Bolivia, India, Nigeria and Tanzania with 22 microsatellites. Between two and 11 alleles per locus were detected and all populations showed relatively high levels of heterozygosity ranging between 45 and 67%. Levels of heterozygosity are generally lower in commercial strains of domestic chickens (*G. domesticus*; Crooijmans *et al.*, 1996).

Norris *et al.* (2001) compared the microsatellite polymorphism between two laboratory colonies and a large field population of *Anopheles gambiae*, the primary mosquito vector of human malaria in West Africa. Nine loci on chromosome 3 were examined. One of the two laboratory colonies had been in culture for more than 15 years and the other for about 3 years. Genetic variability was markedly lower in the laboratory populations. Number of alleles per locus was many times lower in the laboratory populations than in the wild sample (2.3 and 3.7 vs. 22.1, respectively). Similarly, mean heterozygosity (number of loci with more than one allele) was significantly lower in the laboratory stocks (0.19 and 0.39 vs. 0.68, respectively). Genetic variability was lowest in the laboratory stock that had been cultured for the longest time. The field population carried many rare alleles presumably absent in the laboratory stocks when they were captured (founder effect) or lost in the first few generations in the laboratory. Mean heterozygosity in the single wild population used was similar to that of another wild population captured in the same area 4 years earlier (0.68 vs. 0.73, respectively).

Xu *et al.* (2001) used six microsatellites to compare the genetic diversity of four wild and two cultured populations of giant tiger prawns (*Penaeus monodon*) in the Philippines. The two cultured populations showed less genetic diversity than the wild populations, based on pairwise comparisons of allelic and genotypic frequencies. Palma *et al.* (2001) found evidence for the loss of rare alleles (found in wild populations) in cultured stocks of gilthead sea bream (*Sparus aurata*). Norris *et al.* (1999) used microsatellite markers to differentiate between farmed and wild strains of Atlantic salmon (*S. salar*). Fifteen microsatellite loci were analyzed within and between three farmed and four wild populations of European Atlantic salmon. The oldest farmed population was established in Norway in the late 1960s from a number of wild Norwegian populations. The Irish farmed population was derived from this population during the period 1982–1986. A third Norwegian farmed population was established in 1971 from wild fish obtained from 41 rivers in Norway. Three of the wild populations were from Ireland and one from Norway. All seven populations contained relatively high levels of genetic diversity. Allelic diversity (number of alleles per locus) was 20–48% lower for the farmed populations than for three of the wild populations. The fourth wild population also showed reduced allelic diversity. The loss of allelic diversity in the farmed populations was possibly due to the loss of low-frequency (rare) alleles when the populations were established since they had gone through a population bottleneck. Genetic heterozygosity (percentage of loci that hold more than one allele) did not differ between two of the farmed populations and three of the wild populations. Rare alleles contribute little to heterozygosity, which could explain why heterozygosity was not appreciably reduced in the captive-reared populations. Crozier and Moffett (1989) and Mjolnerod *et al.* (1997) also found no significant difference in heterozygosity between wild and farmed strains of Atlantic salmon (*S. salar*). However, 20–30% lower heterozygosity in farmed salmon has been reported by Cross and King (1983), Stahl (1983) and Verspoor (1988). In general, allelic diversity is a more sensitive measure of differences in genetic variation between wild and captive populations of animals than overall heterozygosity (Norris *et al.*, 1999). Losses in heterozygosity have usually been attributed to inbreeding resulting from poor breeding-management programs.

Small population size does not necessarily result in a loss of heterozygosity if an effort is made to avoid inbreeding. Some inbreeding is inevitable in very small populations.

Variation Between and Within Wild and Captive Populations

Tsakas and Zouros (1980) compared the electrophoretic profiles of laboratory and geographically diverse wild populations of the olive fly (*D. oleae*) and found that the genetic differences between laboratory and wild populations were greater than between the various wild populations sampled. This finding is not particularly surprising considering the relative differences between laboratory and natural environments compared to various natural habitats in which this species is found. Crozier (1998) monitored allozyme variation at five polymorphic loci in populations of hatchery and wild Atlantic salmon (*S. salar*) derived from a single wild broodstock to investigate the genetic consequences of hatchery rearing. Genetic composition of the wild fish population did not change over time. However, in the hatchery environment, fish that developed more rapidly and smolted early were found to be more genetically heterogeneous than fish that smolted later in life, whether in the hatchery or in the wild. As a consequence, preferential selection of early-smolting fish for hatchery broodstock or the predominant use of early-smolting fish for stocking in the wild could alter existing gene pools. The relatively rapid development of early-developing smolts in the hatchery was believed to be due to hatchery management and conditions such as *ad libitum* feeding, grading by size and absence of predation during the first year. It was concluded that hatchery rearing of salmon can alter the genetic composition of populations even within a single generation. Could a similar phenomenon be operating in livestock production systems? In meat production enterprises, conscious or unconscious selection for the most rapidly developing individuals could be tantamount to selecting the most genetically heterogeneous individuals within the population. In nature, rapid development may not carry the same advantages as in captivity, particularly if survival and reproduction (i.e. fitness) are determined by such factors as social status, avoidance of predators, seasonality of breeding, etc.

Guttinger (1985) reported that the most striking difference between the songs of wild and domesticated canaries is in the degree of variability of song architecture (greater for wild birds) and the size of the individual repertoire (twice as many syllable types in wild breeds). Artificial selection among domestic canaries for long rhythmical repetitions of identical notes, especially low frequencies, can explain the reduction in song variability and repertoire size. Slater and Clayton (1991) noted that domesticated European zebra finches are prone to lack the 'distance call' commonly heard in wild populations. In captivity, with many birds housed in close proximity, distance calls may be used less and therefore the young may be exposed to them less than in the wild. On the other hand, the tempo of the song of domestic birds was faster and phrase length was shorter than in the wild stock. Slater and Clayton hypothesize that breeders may have selected individuals with more complex songs (e.g. more elements per second) for their breeding stock and thus reduced the proportion of call notes in the song.

Correlated Effects of Artificial Selection for Behavioral Traits

Selection for one phenotypic trait may affect the expression of other traits. This phenomenon is often referred to as 'pleiotropy' (Lerner, 1954). Pleiotropy reflects the interconnectedness of genes and combinations of genes and the structural, physiological and behavioral traits they support. Beilharz *et al.* (1993) have proposed an interesting alternative explanation for correlated characters, which involves the differential allocation of environmental resources. They hypothesize that as one trait is selected, the utilization of environmental resources affecting that and other phenotypic traits are re-apportioned so that some traits are affected positively and some negatively. For example, the 'resource allocation theory' supports the hypothesis that in captivity, environmental resources formerly required for survival in nature are shifted to traits directly affecting fitness, such as reproduction, or to artificially selected traits. The theory is supported by the observation that in captive animals not selected for increased body size, domestication has resulted in a rather consistent reduction in the size of body structures considered less important in captivity than in nature, while structures supporting reproduction have increased in size (see Chapter 11). Schütz *et al.* (2001) tested the hypothesis that behaviors with high energetic costs should decrease in frequency in domesticated breeds artificially selected to invest a relatively high proportion of energy in reproduction and production traits. A strain of commercial White Leghorn layer hens (*G. gallus domesticus*) artificially selected for feed conversion efficiency and egg production was compared with red junglefowl (*G. gallus*), their wild counterparts, in four tests of behavior. In one of the more interesting tests, the birds were given a choice between feeding on sunflower seeds hidden in wood-shavings vs. freely visible and available commercial layer-hen feed. The commercial hens preferred to feed on the easy-to-obtain layer-hen feed, while the junglefowl spent more time in the arm of the maze engaged in the more energetically demanding task of locating hidden seeds. In short, the selected layer hens chose the more energy-conserving foraging process, thus supporting the resource allocation theory. Schütz and Jensen (2001) extended these findings by conducting a similar investigation with White Leghorn hens, junglefowl and Swedish bantam chickens, also an unselected breed. In this study, commercial layer-hen feed rather than sunflower seeds was hidden among the wood-shavings. Whereas the layer hens preferred to feed from the site where food was freely available, the junglefowl and bantam stock obtained a higher proportion of their feed from the site that required effort (e.g. ground-scratching, exploring the ground). In addition, the layer hens were more inactive and engaged in fewer social interactions than the junglefowl and bantams. Overall, layer hens spent proportionately more of their time engaged in energetically low-cost behaviors than the unselected stocks. The authors hypothesized that a reduction in energy spent feeding and in other activities left more energy available for layer hens to devote to production traits. Is the reduction in energy-demanding activities in laying hens a correlated response to selection for feed conversion efficiency and egg production, as the resource allocation theory proposes, or has natural selection for expression of energy-demanding behaviors coincidentally been relaxed, since in captivity they are no longer so important for survival and reproduction? If captive animals were required to work for their food and engage in other energy-demanding behaviors like their free-living counterparts, would response to artificial selection for production traits be as rapid? Response to artificial selection for one or more production traits utilizing captive stocks with high and low energy requirements for survival and reproduction would be an interesting test of the resource allocation theory.

The resource allocation theory also proposes that intense artificial selection for a specific trait (e.g. growth rate in broiler chickens or milk production in dairy cattle) can utilize available resources to the extent that little is left for other demands. Under these circumstances, undesirable side-effects of selection may be manifested (Beilharz *et al.*, 1993; Rauw *et al.*, 1998; see also Chapters 11 and 20).

While artificial selection for production traits can have correlated side-effects on behavior, selection for behavioral traits can have correlated side-effects on fitness-determining factors such as maturation rate and reproduction. Klotchkov *et al.* (1998) found that mink (*M. vison*) selectively bred for aggressiveness or lack of aggressiveness towards humans over seven to ten generations showed correlated changes in the functional state of the ovary at 7 months of age and thus the rate of sexual development. At 7 months of age, 76% (76/100) of the females selectively bred for docility were in pro-estrus, the transition from pro-estrus to estrus, or estrus, in contrast to 32% (44/136) of those bred for aggressiveness toward humans and 53% (743/1411) of females from an unselected control line. These changes progressively increased during selection. For example, after three to four generations of selection, 14% of the docile mink were in estrus or in the transition from pro-estrus to estrus at 7 months of age compared to 43% after seven to ten generations. Selection for aggressiveness toward humans led to a reduction in the number of follicles, and folliculogenesis was completely inhibited in some aggressive females. In a similar selection study, Malmkvist *et al.* (1997, cited in Malmkvist and Hansen, 2001) reported that mink in a line selected for 'confident' behaviors toward humans bred earlier in life than individuals from a strain selected for fearfulness toward humans. Klotchkov *et al.* (1998) cite a number of experiments conducted in the Soviet Union with mink (*M. vison*), foxes (*V. vulpes*) and rats (*R. norvegicus*) showing that selection for tameness or docility results in changes in serotonin and monoamine levels in the brain. These systems are known to be involved in the regulation of pituitary hormone secretion and neuroendocrine rhythms and will be discussed in greater detail in Chapter 14.

Eysenck and Broadhurst (1964) found that following artificial selection for high and low rates of defecation in a novel environment, the Maudsley 'reactive' and 'nonreactive' strains of domestic Norway rats (*R. norvegicus*) differed significantly in 24 of 32 different behavioral tests and in 19 of 24 physiological measures. Unfortunately, causal relationships between these variables were not investigated. Trut (1996) reported that selection for tameability in silver foxes (*V. vulpes*) resulted in a correlated increase in the ratio of males to females born into the population (0.54 ± 0.005 and 0.51 ± 0.007 in the selected and unselected lines, respectively). The male-to-female sex ratio was even higher (0.58 ± 0.015) in the offspring of selected (docile) female foxes heterozygous for the 'star' mutation.

Correlated Effects of Selection for Growth on Behavior

Selection for growth may have unexpected effects on behavioral traits. Selection for the large double-breasted phenotype in domestic turkeys (*M. gallopavo*) has greatly hampered the physical ability of males to mate naturally (Etches, 1996, p. 234). Consequently, most commercial turkeys are now bred by artificial insemination. Selection for increased mean performance (e.g. growth rate) is often accompanied by increased variability (Falconer and Mackay, 1996). Yet, relatively high variation in growth rate within commercially important captive animal populations is generally considered undesirable. It is often a consequence of intense competition for feed resources. The activity of subordinate animals, including their feeding behaviors, may be inhibited by the mere presence of dominant individuals (Symons, 1971). In aquatic species, waterborne chemical compounds can inhibit the growth of some individuals in spite of excess food, as demonstrated in fish, crustaceans and the aquatic stages of amphibians (West, 1960; Yu and Perlmutter, 1970; Malecha, 1980; Nelson *et al.*, 1980). Commercial food-producing enterprises aspire for populations of animals that exhibit uni-formly high rates of growth, not populations whereby just the more dominant individuals attain their true growth potential. Interestingly, Doyle and Talbot (1986) proposed that in most commercial fish enterprises, selection for growth rate does not favor more competitive or aggressive animals but rather fish that are generally nonreactive and ignore each other. Selection for growth rate favors aggressive fish only when resources are severely limited and aggression is independent of body size. Less reactive fish who ignore others and seldom initiate aggressive interactions are more likely to conserve energy by minimizing involve-ment in agonistic interactions and thus devote more energy to growth. Peebles (1979) found that both body size and growth rates were more variable in captive wild-genotype prawns (*M. rosenbergii*) than in populations of a semi-domestic stock 15 generations removed from the wild. The social relationships of individuals in wild-genotype populations were typically organized in a relatively strict linear hierarchy of dominance relationships, which led to a highly skewed exploitation of food resources among individuals and, thus, skewed growth rates. The semi-domestic prawns were less aggressive towards one another and, consequently, their hierarchies were less linear and rigid, and had many triangular relationships (e.g. A dominates B, who dominates C, who dominates A). The more 'relaxed' hierarchies of the semi-domestic stock allowed for a more uniform distribution of food among individuals and, consequently, more uniform growth rates. It would be interesting to determine if growth rate in newly captive populations of animals is greater when selection for growth rate is coupled with selection for reduced behavioral reactivity.

Experiential Effects on Behavioral Variability

The captive environment can either increase or decrease variability in the expression of behavioral traits. Henderson (1970) examined variability in the foraging behavior of domestic laboratory mice (*M. musculus*) representing six inbred strains and their respective hybrids after being reared in either standard or enriched laboratory cages (see p. 36 for an explanation of the rearing and test environments). Variability in the behavior of the different stocks was obscured by rearing in standard laboratory cages. Genotypic effects on the simple food-seeking task were considerably greater among mice reared in the enriched cage environment than the same stocks reared in the standard small-cage environment.

Conversely, behavior may become more variable due to deficits in early experience. For example, Price *et al.* (1994) demonstrated that the failure of many domestic yearling rams (*Ovis aries*) to show sexual interest in estrous ewes is because of lack of exposure to females in their first year of life. The sexual performance of inexperienced rams ranged from total abstinence to highly successful, whereas variability in the performance of experienced rams was clustered in the normal to high range with virtually all experienced rams exhibiting successful matings. Mason (1991a) and Lawrence and Rushen (1993) describe a number of behaviors performed by captive animals in an unnatural way in response to impoverished housing environments.

Conclusions

The assumption held by many that domestication and captive-rearing results in a reduction in genetic variation is only partly true. Recent investigations have shown that considerable genetic variation has been retained in many populations of captive and domestic animals. Studies revealing correlated effects of artificial selection have demonstrated the interconnectedness of many traits. The proposed 'resource allocation theory' may be particularly relevant to the domestication process since it hypothesizes a reallocation of available resources accompanying shifts in selective pressures. With a limited pool of resources, exploiting a relatively large proportion of the pool to support one trait results in fewer resources available to support other functions. In captivity, environmental resources normally used for survival in nature may become available for reproduction and artificially selected production traits. The animal's behavior can be an integral part of this process.

Chapter 11

Morphological and Physiological Traits

Morphological and physiological traits, as well as behavior, have been subjected to considerable selection under domestication and captive breeding. For example, domestication has brought about a reduction in the size of the skeleton, brain, spinal cord and eyes of the rabbit and increased the weight of the pelage, certain muscle parts and fatty tissue. A reduction in the body and brain size of animals during domestication has been a popular topic and is given special attention in this chapter. Possible explanations for these changes are proposed, including reduced functional demands on the organism in captivity. Changes in other structural and physiological traits in response to natural and artificial selection are also discussed, as well as correlated effects of such selection. Meyer *et al.* (2000) reported that the primary (guard) hairs of domestic mammals are more variable than for wild mammals not only with regard to length and diameter but also with respect to typical parameters of the hair cuticula. The correlation (r) of hair diameter with area of cuticula scales was 0.73 in the wild animals examined and 0.27 in the domesticated species sampled. Popovic (1988, 1992) noted that in wild pigs the nasal airways and the whole nasal cavity are relatively longer, narrower and more noticeably arched than in domestic pigs. As a result, air turbulence, and thus the efficiency of basic olfactory (i.e. nasal) functions, are believed to be greater in wild pigs than in their domestic counterparts. The chapter concludes with several examples of how the captive environment itself can influence variation under domestication.

Direct and Correlated Effects of Artificial Selection for Morphological Traits

As noted in Chapter 7, artificial selection has had a major impact on the genetic and phenotypic variation of captive animals as illustrated by the diversity of various breeds of domestic animals. For example, breeding for morphological characteristics in the domestic rabbit (*O. cuniculus*) accelerated changes in gene frequencies and fixation of different alleles at polymorphic enzyme loci (Hartl, 1987). The variety of rabbit coat colors has also increased during the last centuries by persons breeding them for show (Nachtsheim and Stengel, 1977).

Trut (1999) summarizes the correlated effects of 40 years of selection for tameness (towards humans) in silver foxes (*V. vulpes*) on the expression of various morphological traits. The first morphological characteristics affected (eighth to tenth generations) were changes in pelage color, chiefly a loss of pigmentation in certain areas of the body, leading in some cases to a star-shaped pattern on the face similar to that seen in some breeds of dogs. Next came traits such as floppy ears and rolled-back tails similar to those in some dog breeds. After 15–20 generations, foxes with shorter tails and legs and over- and under-bites appeared. The appearance of novel pelage colors may be attributable to changes in the timing of embryonic development, such as the migration rate of melanoblasts, the embryonic precursors of the pigment cells (melanocytes) that give color to an animal's fur. Changes in skull size and shape were noted in both sexes. The cranial height and width of domesticated foxes tended to be smaller and snouts tended to be shorter and wider than those of unselected control foxes. In addition, the shape and size of the skulls of domesticated male foxes gradually became more like those of females. Such reduction in sexual dimorphism between the sexes may be a result of reduced male–male competition for access to females (i.e. relaxed selection for male competitiveness).

Artificial selection for meat production in domestic chickens (*G. domesticus*) provides a classic example of correlated effects of selection (Jackson and Diamond, 1996). Broiler chickens have been consciously selected for large size, large pectoral (breast) muscles, rapid growth and high feed efficiency. At hatching, broiler chicks are three times larger than red jungle fowl chicks (*G. gallus*), wild ancestor of domestic chickens. At 9 weeks of age, broilers were 5.1 times larger than jungle fowl and at 41 weeks, when asymptotic body mass had been reached, broilers were still 4.3 times heavier than jungle fowl. Concomitant with selection for rapid growth rate and large size in broiler chickens are correlated effects on body organs associated with food processing. The broilers' gut (small intestine), gizzard (muscular grinding stomach) and proventriculus (acid-secreting stomach) are all larger than jungle fowl of similar body mass. For example, the wet mass of the small intestine of broilers exceeds that of jungle fowl of equal body mass by 2.8 times. The broiler intestine is not only longer but has a greater luminal diameter than the jungle fowl intestine. A large gut is required to efficiently process the broilers' relatively high rate of food intake. Interestingly, layer hens selected for egg production have lower growth rates, smaller pectoral muscles and lighter and shorter intestines than broilers. It is also of interest that broiler chickens have significantly shorter and smaller-diameter legs and smaller brains than jungle fowl. Both of these latter changes may be more related to relaxed natural selection in captivity or disuse during development than a correlated response to selection for growth rate. One could argue that jungle fowl need relatively strong legs and large brains to avoid predation and obtain needed resources in nature whereas in captivity, predation is nonexistent and metabolically expensive brains are relatively superfluous. The effects of selection for meat production in broiler chickens conform to predictions of the 'resource allocation theory' (Beilharz *et al.*, 1993; Rauw *et al.*, 1998) introduced in Chapter 10.

Changes in Body Size with Domestication

With few exceptions, domestication has resulted in a reduction in body size. One popular explanation is that smaller animals may have been selected by man to facilitate handling and control (Meadow, 1984). Smaller animals also mature at an earlier age (Tchernov and Horwitz, 1991). The practice of breeding very young animals using mating systems that have eliminated male–male competition and female choice minimize the importance of body size, vigor and aggressive potential in gaining access to mates. Tchernov and Horwitz (1991) proposed that the reduction in body size of domesticated animals may reflect a shift from selection for individual viability (i.e. competitive ability, longevity, etc.) to selection for a higher rate of reproduction as in a shift from K- to r-selection. The selection strategy of an organism may be related to the amount and nature of resources invested in reproduction vs. each offspring. In more stable environments with relatively high population densities, selection may favor directing resources into fewer offspring, thus ensuring their fitness and maximal exploitation of the environment (K-selection). In relatively unstable environments, fitness may be enhanced by directing high energy input into reproduction (i.e. many offspring) with correspondingly low input into each offspring (r-selection). K-selection is generally associated with larger size, slower development and lower fecundity, while r-selection is related to higher fecundity, rapid development and smaller body size (Fig. 11.1). Other explanations for reduced body size include inbreeding depression, pleiotropic effects of artificial selection, poorer nutrition associated with artificial diets, ability to survive on reduced quantities of feed and water, reduction in opportunities for activity and exercise, and reduced predation pressure.

There are some notable exceptions to the general pattern of size diminution accompanying domestication. The domestic horse (*E. caballus*) is generally larger in stature than its wild progenitors (Tchernov and Horwitz, 1991). Since the horse was domesticated somewhat later than our traditional livestock species (by 1000 years or more), its domestication may have followed a different path. The widespread use of the horse for transportation (Anthony *et al.*, 1991) and as a beast of burden may have encouraged selection for large body size (Bökönyi, 1984).

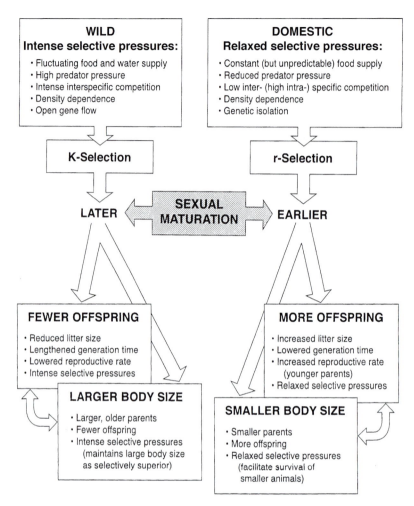

Fig. 11.1. Flow diagram of shift along the r–K-ecological strategy continuum in primeval domesticates, showing a more r-oriented selection strategy with the advent of domestication (Tchernov and Horwitz, 1991).

Hammond (1962) pointed out that domestic meat animals (e.g. pigs, sheep) have much thicker bones in proportion to their length than similar wild ungulates, a possible result of selection on a high plane of nutrition. Domestic pigs released in New Zealand by Captain Cook have reverted to a wild boar-like conformation and have relatively thin bones compared with modern-day domestic pigs of the same body weight. Bökönyi (1974) claimed that one of the first results of pig (*S. scrofa*) domestication was a reduction in skeletal size but that the smaller skeletal size of domesticated pigs may only have lasted until the beginning of breed formation when some of the early swine breeds soon approached the size of wild boars (Jones, 1998).

Changes in Brain Size

Brain size relative to body size has generally been reduced in domestic animals and is more variable than in their wild counterparts (Röhrs, 1986; Kruska, 1988; Hemmer, 1990; Ebinger, 1995). Coppinger and Coppinger (2001) summarize work done on head size and shape and brain size in domestic dogs (*C. familiaris*), wolves (*C. lupus*), coyotes (*Canis latrans*) and jackals (*Canis aureus*). All of these species have the same head and brain dimensions when they are born and they tend to have the same growth trajectories until they are 10 or 12 weeks old. After that time, wolves grow bigger heads and brains than dogs of equal size. If a 45-kg dog is compared with a 45-kg wolf, the dog's head is about 20% smaller. If skull sizes are matched (big dog and normal size wolf), the dog's brain is about 10% smaller. Stuermer *et al.* (1997) reported 18% smaller brain-to-carcass weight ratios in domesticated gerbils (*M. unguiculatus*) than in wild gerbils trapped in their native habitat. Kruska (1996) and Kruska and Schreiber (1999) compared the brain sizes of wild and domesticated (ranch) mink (*M. vison*) reared in either small cages or larger open-air enclosures. Mink have been in captivity since about 1866 (Shackelford, 1949, 1984; Enders, 1952) and, in Europe, interbreeding with wild mink has been relatively uncommon. Compared to wild mink, brain sizes of the ranch mink averaged 11% (open-air enclosures) or 20% (cages) less at any given carcass weight (fur, viscera and fat removed). In addition, brain size and parts of brains were more variable in size in ranch mink than in wild mink (Kruska, 1996). Kruska (1982) reported that llamas (*L. glama*) and alpacas (*Vicugna pacos*) have 17% smaller brains than their wild progenitor, the guanaco (*Lama guanicoe*) of South America, when animals of comparable body weight are compared. The telencephalon and neocortex are especially reduced in size. Röhrs and Ebinger (1998) compared the brain weights and brain case capacities of wild Przewalski horses (*Equus przewalski*), Przewalski horses from zoological gardens and domestic horses (*E. caballus*). Brain weights of domestic horses averaged about 16% less than wild Przewalski horses and the brain weights of Przewalski horses from zoological gardens did not differ significantly from the domestic stock sampled. Brain case capacity was about 14% smaller in both domestic and captive Przewalski horses than in wild Przewalski horses. Röhrs and Kruska (1969) determined that the brains of domestic pigs (*S. scrofa*) averaged 34% smaller than the brains of European wild boars. Kruska and Röhrs (1974) contrasted the brain sizes of domestic pigs, wild boars and feral pigs introduced to the Galapagos Islands some 70–140 years ago. Interestingly, the brains of the feral pigs were the same size as the domestic pigs, with both about one-third smaller than the wild boars. The authors concluded that the effect of domestication on brain size is not reversed when domestic animals are returned to the wild. Herre and Röhrs (1990; cited in Kruska and Röhrs, 1974) reached a similar conclusion when studying feral populations of other domesticated species (e.g. feral dogs, cats, goats and donkeys). The failure of brain size to increase in domestic animals returned to the wild (i.e. feralized) is an intriguing result and one that on the surface does not support the resource allocation theory (see Chapter 10 for an explanation) proposed by

Beilharz *et al.* (1993). It may be that in these cases, feralization has not been in progress long enough for an increase in brain size to become measurable (i.e. it will become noticeable in the future). Alternatively, the environment in which the animals have become feral may not be as demanding as the environment of their wild ancestors. For example, the feral pig study reported above (Kruska and Röhrs, 1974) took place on the Galapagos Islands, where there are no natural predators of pigs.

Kruska and Röhrs (1974) also discussed the reduced variability in size of almost all brain structures in wild and feral pigs compared to their domestic counterparts. An important exception was the size of the hippocampus, which was more variable among wild and feral pigs. The reduced variability of the limbic main center among individual domestic pigs may be due to selection for reduced aggression and emotional reactivity under conditions of domestication.

Of the five fundamental parts of the brain, the telencephalon consistently shows the greatest decrease in size in all species examined to date, except for the mink (*M. vison*; Kruska, 1988). In the mink, the mesencephalon and cerebellum show the greatest decrease in size. These structures are related to motor activity. Kruska (1988) speculates that domestication of the highly active mink may have favored individuals who are more likely to tolerate the spatial restrictions of cages and forced inactivity. Interestingly, the telencephalon and neocortex of the mink's brain go through a natural ontogenetic reduction in size in the late juvenile phase of development, a phenomenon that may occur in most mustelids (Steffen *et al.*, 2001). Could this process in mink offer clues to the reduction in brain size during domestication?

Auditory structures in the brain

Plogmann and Kruska (1990) compared the size of the main auditory structures in the central nervous system of six European wild boars (*S. scrofa*) and six domestic pigs. In general, all measured auditory-related structures were smaller in the brains of domestic pigs than in their wild counterparts, even when adjustments were made for the smaller brains of the domestic pigs. The cochlear nucleus, which is the principal sensory nucleus of audition, showed the smallest decrease in size (15%) of all the auditory structures examined. Functional changes in auditory capacity were not measured.

Burda *et al.* (1988) reported that young (2 months) laboratory rats (*R. norvegicus*; Wistar strain of the Norway rat) possess 6% fewer outer and inner hair cells in the organ of Corti than young wild-caught Norway rats (*R. norvegicus*) and about 13% fewer neurons in the spiral ganglion. In addition, Wistar rats had more densely packed hair cells in the cochlea.

Visual structures in the brain

A significant reduction in the size of visual structures of the brain has been noted in domesticated populations of several species including pigs (*S. scrofa*), sheep (*O. aries*; Ebinger, 1975) and rats (*R. norvegicus*; Kruska, 1980). Compared with the European wild boar (*S. scrofa*), the parts of the domestic pig brain associated with vision have decreased more than the parts associated with olfaction and audition (41 vs. 32 and 30%, respectively; Plogmann and Kruska, 1990). The greater reduction in the visual structures of the brain and the belief that olfaction and audition are more important than vision in the overall adaptation of this species suggest that domestication may have had a greater influence over structures underlying less critical functions.

Species-specific differences

There are species-specific differences in the degree to which domestication has affected brain size. Frick and Nord (1963; cited in Kruska, 1987) found no differences in the brain size of wild and domestic mice. This observation should be confirmed considering that domestication has reduced the brain size of other rodents maintained under similar laboratory conditions. Röhrs (1985) hypothesized that the brains of highly cerebralized species are affected more by domestication than are those of less cerebralized species, and phylogenetically 'newer' brain parts are affected more strongly than are 'older' structures. Consequently, Röhrs proposes that domestication is a special 'regressive evolutionary process'. Are phylogenetically advanced brain parts more subject to change during domestication because they are more plastic, or is it that man reduces the need for higher-level brain functioning by serving as a buffer between the animal and its environment in captivity? The limbic system, particularly the hippocampus, is typically reduced more than other brain components in highly domesticated mammals such as the dog (*C. familiaris*), sheep (*O. aries*) and pig (*S. scrofa*) (Kruska, 1980, 1988; Herre and Röhrs, 1990) and is believed to be associated with a reduction in aggression and emotional reactivity to facilitate living in close proximity to man, impoverished physical environments and crowded social groupings (Herre and Röhrs, 1977).

Mammals vs. birds

Reductions in brain size and functions have tended to follow the same trends in domesticated birds as in mammalian species. Ebinger (1995) found that the whole brain volume of domestic ducks (*Anas platyrhynchos domestica*) averages 14.3% less than their wild counterpart, the northern mallard (*A. platyrhynchos*). Likewise, the brain weight of domestic ducks is 15.8% lighter than for wild mallards of similar body size. Figure 11.2 illustrates the percentage reduction in various parts of the duck brain by volume. Ebinger (1995) noted that variability in the size of brain parts is generally greater in domestic ducks possibly because there are a number of domestic duck breeds, whereas the wild population is more uniform. A reduction in brain size with domestication has also been reported in turkeys (24–35%; Ebinger *et al.*, 1989), pigeons (7%; Ebinger and Löhmer, 1984) and geese (16%; Schudnagis, 1975; Ebinger and Löhmer, 1987). Most of our domesticated birds have come from three families, *Phasianidae, Anatidae* and *Columbidae*. Within these groups, it appears that birds with the larger brains (e.g. *Anatidae*) show the greatest reduction in brain size with domestication (Ebinger, 1995). A notable exception to this rule is the turkey, which has one of the smallest brains but has exhibited one of the greatest percentage reductions in brain size from its domestication.

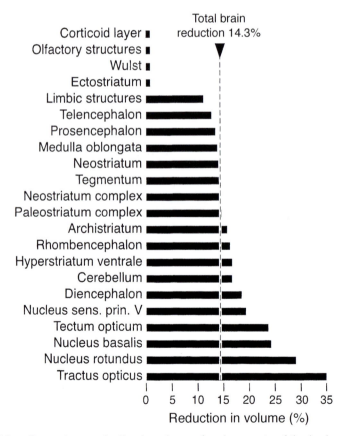

Fig. 11.2. Percentage reduction in volume of various parts of the brain of the domestic mallard duck. The vertical arrow line indicates the percentage decrease in total brain volume.

Functional associations

The reduction in brain size accompanying domestication may reflect adaptation to the captive environment and the plasticity of the central nervous system. A reduction in sensory input may explain why those brain parts of the duck that coordinate visual functions have been significantly reduced during domestication, whereas olfactory centers have remained unchanged (Ebinger, 1995). Reduced sensory input as well as a reduction in need for integrative capacities may explain why the size of the ventral hyperstriatum is so reduced in domesticated birds. The ventral hyperstriatum is known to be involved in learning (McCabe *et al.*, 1982). Learning may be less important for survival and successful reproduction in captive environments, thus allowing a kind of 'regressive evolution' to take place (Röhrs, 1985). Comparative learning studies would be useful to determine if domestic ducks have reduced learning capacity. Dabbling, a feeding behavior involving rapid, splashing movements of the bill, is less common in domestic ducks than in their wild counterparts. Correspondingly, the telencephalon and medulla oblongata of the brain, which control such behavior, are noticeably smaller in the domestic duck breeds.

It is relatively easy to postulate that changes in brain size and composition during domestication are associated with a decrease in functional demands associated with life in captivity. However, until these associations are verified, the proposed relationship between reduced size and reduced function will remain conjecture. Gille and Salomon (2000) proposed that the smaller relative brain sizes of domesticated (Pekin) mallard ducks (*A. p. domesticus*) compared to wild mallards (*A. platyrhynchos*) can be explained, at least in part, by the fact that brain growth is more conservative and less easily influenced by selection than body weight. Although brain sizes of wild and domestic mallards were similar at hatching, during development, body size of the domesticated ducks more than doubled while brain size increased by only 20% (Gille and Salomon, 1994). Artificial selection for body size is not necessarily accompanied by a proportional increase in brain growth.

It is unlikely that animal breeders directly selected for reduced brain size. What would be their incentive for doing so? The fact that so many species of domestic animals have smaller brains than their wild progenitors suggests that it is a consequence of some underlying process of domestication (Coppinger and Schneider, 1995).

Changes in Other Structural Characteristics

In addition to the brain, other body structures may be smaller in domestic stocks than in their wild progenitors. Stuermer *et al.* (1997) reported that eye weights of domesticated gerbils (*M. unguiculatus*) were 26% lighter than for wild-caught gerbils after adjustments were made for differences in body weight. Kruska and Schreiber (1999) compared wild and domesticated mink (*M. vison*) reared under identical conditions and found that the domesticated mink possessed 8.1% smaller hearts and 28.2% smaller spleens. They postulated that increased capacity of the circulatory system is more important for survival in nature than in captivity. In contrast, reproductive organs are typically larger in domestic stocks. In dogs (*C. familiaris*), both testis size (sperm production) and size of the epididymis (sperm storage) are year-round more than a third larger than in comparably sized wolves during their normal winter breeding season (Haase, 2000). This difference may be related to the rather promiscuous sexual behaviors of dogs relative to their monogamous wild ancestors. Relatively large gonads and accessory structures are characteristic of species that engage in frequent copulation and species in which females are frequently mated by multiple males (Kenagy and Trombulak, 1986).

Some Adverse Side-effects of Artificial Selection for Morphological Traits

Artificial selection for economically important traits can have adverse correlated effects. Selection for growth rate, meat percentage and feed efficiency in pigs (*S. scrofa*) and broiler chickens (*G. domesticus*) has led to leg problems under intensive management (Jørgensen and Vestergaard, 1990; Sørensen, 1992; also see review by Rauw *et al.*, 1998). Certain breeds of mink (*M. vison*) selected for fur characteristics have an increased incidence of physical defects (Nes *et al.*, 1988, cited in Malmkvist and Hansen, 2001). In such cases, veterinary care and management practices allow animals that would normally perish at an early age in nature to survive and reproduce in captivity, thus increasing or preserving the diversity of the gene pool.

Some Adverse Side-effects of Artificial Selection on Physiological Traits

Artificial selection for high production efficiency in poultry, swine and dairy cattle has resulted in a number of undesirable side effects affecting physiological traits (see review by Rauw *et al.*, 1998). For example, broiler chickens (*G. domesticus*) selected for rapid growth rate have a higher incidence of heart failure, weakened immune system and a fourfold increase in mortality at 42 days of age compared with nonselected birds. Pigs (*S. scrofa*) selected for increased growth rate and dairy cattle (*B. taurus*) selected for increased milk production generally show poorer reproductive performance than unselected controls. Harrison *et al.* (1990) found that high-producing dairy cows had longer post-partum intervals to first visible estrus than control cows (66 vs. 43 days, respectively). Gallo *et al.* (1996) reported that loss of body condition score of Holstein cattle during early lactation was nearly twice as great for cows yielding 12,000 kg of milk per lactation than for cows producing less than 6000 kg.

Effects of Artificial Selection on Gene Expression

Artificial selection may also influence the degree of phenotypic expression of mutant alleles. Belyaev (1969) reported that selection for depigmentation (silver color) in the pelage coloration of foxes (*V. vulpes*) resulted in the appearance of platinum and white-nosed foxes through an increase in the 'expressivity' of a mutant allele that produced only a white patch on the paws of animals in the unselected population.

Environmental Influences on Phenotypic Variation

Of course, not all variation under domestication can be attributed to genetic change. The captive environment can have considerable influence on the expression of phenotypic traits. Constable *et al.* (1998) found differences in serum biochemistry between 11 wild (free-ranging) and 20 captive gray wolves (*C. lupus*). The free-ranging wolves were sampled in Alaska and the captive wolves were housed in Minnesota for approximately 2 years prior to sampling. Relative to captive wolves, the free-ranging wolves had significantly lower sodium, chloride and creatinine concentrations, and significantly higher potassium and blood urea nitrogen (BUN) concentrations, BUN to creatinine ratios, and alanine aminotransferase, aspartate aminotransferase and creatine kinase activities. Different nutrition, activity levels and environmental stress were cited as likely explanations for the differences observed rather than inherent differences between the two populations. Hoefs and Nowlan (1997) compared horn growth between free-ranging Dall mountain sheep (*Ovis dalli*) and captive animals derived from the same population. It was noted that the captive males developed larger and more massive horns, presumably due to better nutrition. Araújo *et al.* (2000) noted that captive common marmosets (*Callithrix jacchus*) are heavier than wild marmosets and cite improved diets and lack of physical activity as being responsible for the difference. Gross (1998) has cited a number of studies showing differences between farmed and wild Atlantic salmon (*S. salar*) attributed to the rearing environment (see Table 3.1). Swain *et al.* (1991) have demonstrated that relatively consistent morphological differences between hatchery and wild populations of coho salmon (*Oncorhynchus kisutch*) are environmentally induced rather than a consequence of genetic changes accompanying domestication. Hard *et al.* (2000) also linked differences in the morphology of wild and captive-reared coho salmon to their rearing environment. Captive-reared fish were collected in the wild as juveniles from the same streams that provided wild fish, and were reared in circular tanks at low densities on a conventional hatchery diet. Captive-reared fish differed from wild fish by sharply reduced sexual dimorphism, as well as smaller heads, less hooked snouts, increased trunk depth, larger caudal peduncles, shorter dorsal fins, larger hindbodies and a reduction in body streamlining. The magnitude and pattern of differences suggested that at least some of them were environmentally induced. The reduced streamlining and more forward-positioning of fins in the captive-reared fish may have been due to elimination of the opportunity and need for sustained swimming (Fleming and Gross, 1994). The reduced head length, reduced snout curvature and dorsal hump height in captive-reared fish may have resulted from a combination of environmental factors, such as lack of exposure to sea water or to a migratory experience, absence of natural food or elevated rearing densities. To the extent that the differences between wild and captive-reared fish are due to environmental factors, it demonstrates the plasticity of morphometric development in this species. It is believed that the morphometric changes in male coho salmon due to captive rearing may reduce their reproductive success, especially when competing with wild males, a result also reported by Berejikian *et al.* (1997). Other examples of environmental effects on phenotypic traits are provided in the ensuing chapters of this book.

Conclusions

Aside from species artificially selected for large size (e.g. horse), animal domestication has rather consistently reduced body size. A reduction in competition for essential resources and the removal of predation pressure are among the most plausible explanations for this phenomenon. Brain size has also been reduced under domestication, particularly the phylogenetically 'newer' parts of the brain, which may be more plastic in terms of development. Humans often serve as a buffer between domesticated animals and their environment, perhaps reducing the need for higher brain function in captivity. Reduction in the size of the limbic system of the brain, a structure involved in aggression and emotional reactivity, may also reflect reduced demand for those behaviors in captivity. Artificial selection for economically important production traits in domestic animals can result in undesirable side-effects on morphological development and physiology, such as structural deformities and reduced fecundity, providing further support for the resource allocation theory described in the last chapter. The captive environment itself may be responsible for development and variation in structural traits, illustrating the plasticity of morphometric development as well as the role of environmentally induced developmental events recurring during each generation of the domestication process.

Chapter 12

Feeding and Drinking

The next four chapters discuss issues associated with adaptation to the biological environment of captive animals. Topics covered include feeding and drinking, predation and infectious diseases/parasites, interaction with humans and the social environment. This chapter begins with a general discussion of the transition from nature to captivity and variation in captive environments. However, the major focus of the chapter concerns issues related to feeding and drinking by captive animals. Diet selection, methods of food delivery and feeding schedules are discussed as well as experimental work developing more naturalistic feeding regimes and methods for reducing competition for food. The chapter concludes with a discussion of self-imposed food deprivation by animals newly introduced to captivity.

The Transition from Nature to Captivity

The process of domestication is set in motion during the transition from nature to captivity. Carlstead (1996) points out that 'wild' and 'captivity' are only extremes on a continuum. Free-living wild populations are found from wilderness areas to large semi-natural preserves. Captive environments range from large semi-natural enclosures to small laboratory cages and aquaria.

The transition from nature to captivity is accompanied by many changes in the biological and physical environments of the population, some of which bring about important changes in natural selective pressures and phenotypic traits (Price, 1976a; Boice, 1981; Mason et al., 1987). Providing animals with food and medical care, protecting them from predators and assisting in the care of offspring are just a few of the functions served by humans that may improve survival and reproductive success in captivity. Loss of control over their environment can also affect the adaptation process (Mineka et al., 1986; Carlstead, 1996) and result in phenotypic changes during domestication.

In nature, animals learn that their behavior can exact changes in the environment, often in a consistent and predictable manner. Consequently, they remain motivated to respond to environmental stimuli. In contrast, captive animals have fewer opportunities to learn that their behavior can modify their environment and that the stimuli around them are subject to control (Mason, 1978). Under such circumstances, captive animals often lack the experience to respond optimally to novel stimuli and changes in their environment when they occur. As a consequence, cognitive processes may develop differently in captivity than in the wild (Carlstead, 1996).

Variation in Captive Environments

It is difficult to generalize when comparing the requirements of adaptation to the biological environments in nature and captivity. Variation in species-typical biological characteristics and variation in natural and captive environments create a myriad of challenges for each animal population undergoing domestication. The degree of husbandry applied in different captive environments can be highly variable (Smits, 1990). Crabbe *et al.* (1999) have recently demonstrated that even subtle environmental differences among laboratories can significantly affect the outcome of experiments testing the same genotypes and behaviors. Eight strains of domestic mice (*M. musculus*) were simultaneously tested for six behaviors at three different laboratories on the same day and same time. Despite the investigators efforts to equate laboratory environments, significant and, in some cases, large effects of site were found for nearly all variables. In addition, there were considerable differences in the pattern of strain differences at the three sites for several variables. In spite of the seemingly endless array of captive environments, a number of behavior patterns and adaptations are typically observed during the domestication process. The following chapters consider some of the more important environmental impacts associated with the transition from wild to captive environments.

Dependency on Humans for Food

In nature, animals spend a large share of their time and energy searching for and consuming food. Choices made by animals with regard to feeding sites, diet selection and foods consumed are often quite varied. Most captive animals, on the other hand, are dependent on humans to provide an appropriate diet, and foods offered are often relatively uniform on a daily and seasonal basis. Since diet selection of many species is largely learned (Lynch and Bell, 1987; Forbes, 1995; Galef and Allen, 1995; Provenza, 1995), the uniformity of diets fed to captive animals may result in changes in food preferences and/or a reluctance to sample unfamiliar food items and familiar foods presented in novel ways (Galef and Clark, 1971; Heinrich, 1988). Schoonhoven (1967) found that laboratory-reared tobacco hornworms (*Manduca sexta*) accepted host food plants that their wild counterparts rejected, based on a change in the sensitivity of contact chemoreceptors.

When man provides food for captive animals, there is little incentive for animals to exhibit optimal foraging strategies and energy-saving feeding behaviors. Consequently, relaxed selection on such traits may follow. Gustafsson *et al.* (1999a) and Andersson *et al.* (2001) examined the foraging behaviors of wild and domestic (hybrid) pigs (*S. scrofa*) and chickens (*G. domesticus*), respectively, but the results were equivocal. To a great extent, both wild and domestic strains behaved in accordance with general predictions of optimal foraging theory.

Food Provisioning in Captivity

Knowledge of animal feeding patterns, food preferences and nutritional requirements can be critical for the successful rearing of animals in captivity (Hickman and Tait, 2001). Mass rearing of insects on artificial diets is often accompanied by high rates of mortality in the early generations as natural selection eliminates individuals with maladaptive behavioral and physiological predispositions (Boller, 1979). Selecting a suitable artificial diet for predatory insects (e.g. spiders) is particularly difficult since carnivorous species may require behavioral cues from the prey to initiate attack and feeding, and some must feed on a variety of insect prey species to obtain the optimal nutrition for survival and reproduction (Amalin *et al.*, 2001). Feeding nonpredatory insects in captivity is less problematic but a knowledge of food preferences and nutritional requirements is still very important. Medfly larvae (*C. capitata*) reared on peaches, persimmons and cactus fruits had greater reproductive success than those reared on figs, pears and artificial diets (Cirio, 1970, cited in Leppla *et al.*, 1983). A critical event for the successful colonization of medflies in captivity is the initial transition from ovipositing in fruit to use of an artificial substrate for ovipositing and larval development (Rössler, 1975a). Nielson (1971) showed that the apple maggot (*Rhagoletis pomonella*) developed equally well on apples and artificial diet, but the flies from larvae reared on the artificial diet exhibited aberrant behavior. Wong and Nakahara (1978) found that the mating percentage of wild medflies increased from 30 to 40% by adding fruiting guava plants to their environment. Acidic artificial medfly diets containing protein, sucrose and preservatives have been successfully used for larval development (Schroeder *et al.*, 1972).

Providing the appropriate diet at the right time can be critical in rearing larval fish in captivity. Dou *et al.* (2000) reported that tank-reared Japanese flounder larvae, *Paralichthys olivaceus*, consumed rotifers in preference to *Artemia* up to 10 days after hatching. At 15 days post-hatch they began to avoid rotifers and after 25 days they fed almost solely on *Artemia* (Fig. 12.1). Feeding rates of the larvae prior to 10 days post-hatch depended on prey density, but were independent of prey density in older larvae. Al-Abdul-Elah *et al.* (2001) describe recent work to develop a suitable diet for rearing captive silver pomfret (*Pampus argenteus*), a highly desirable food fish and new candidate for aquaculture. A

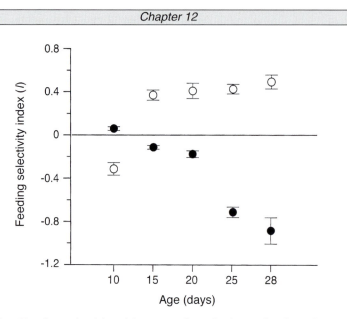

Fig. 12.1. Feeding selectivity of Japanese flounder larvae for *Artemia* prey (open circles) and rotifers (solid circles) throughout the larval stage at 09.00 h (Dou *et al.*, 2000).

mixture of different microalgae together with rotifers (*Brachionus plicatillis*) provided the best larval survival and growth. The microalgae did not improve growth of the developing fish even though they were eaten. Rather, the microalgae reduced light intensity on the water's surface resulting in greater dispersal of the larvae in the water column. In addition, the nutritional value of the rotifers was greater if they had been feeding on specific microalgae. A remaining hurdle in the rearing of captive silver pomfret concerns their behavior of consuming tiny air bubbles at the water surface of their tanks. Apparently, the young fish mistake the air bubbles for tiny jellyfishes and medusae, which are part of their diet at this stage in nature. The consumption of bubbles causes the fish to lose their balance in the water column and float. Experiments with Atlantic wolffish larvae (*Anarhichas lupus*) demonstrated that their dietary preferences shifted from mostly live prey to pelleted feed over an 8-week period (Brown *et al.*, 1997).

    Artificial diets may be more cost effective than natural diets, but growth and survival may be compromised. Cohen and Staten (1993) reported that predatory big-eyed bugs (*G. punctipes*) fed an artificial diet of ground beef and beef liver exhibited reduced metabolic efficiency of food utilization and reduced diet digestibility compared to big-eyed bugs fed a natural diet of insect eggs and larvae, even though the overall composition of protein, carbohydrates and lipids in the two diets were nearly identical. Sharma and Chakrabarti (1999) compared the growth of common carp (*Cyprinus carpio*) larvae fed either live food (plankton) or an artificial diet consisting of fish meal, wheat flour, rice bran, yeast, vitamins and calcium. Average weight of larvae after 40 days on the live food system was three- to fivefold greater than larvae fed the artificial diet. Enz *et al.* (2001) found that whitefish larvae (*Coregonus suidteri*) fed live zooplankton grew significantly larger than those fed frozen zooplankton. Larvae fed a commercial dry diet grew as well as those fed frozen zooplankton, but mortality rates were considerably higher (34% vs. 3%). Mischke and Morris (1998) showed that bluegill sunfish larvae (*Lepomis macrochirus*), like many fish larvae, had to initially be fed a diet of live brine shrimp (*Artemia franciscana*) to later grow and survive on a commercial diet. The initial feeding of live prey may facilitate the ingestion, digestion and assimilation of dry diets when subsequently offered and may be important in activating proteolytic enzymes (see references cited in Mischke *et al.*, 2001).

Hickman and Tait (2001) found that weaning New Zealand turbot larvae (*Colistium nudipinnis*) from a natural, live food diet (*Artemia*) onto inert, manufactured feeds was a major hurdle in attempts to domesticate turbot for large-scale commercial production in captivity. Hamlin and Kling (2001) describe the sensitivity of larval haddock (*Melanogrammus aeglefinus*) to the timing of weaning to a formulated diet for commercial production. Survival was only 2.5–6.3% when weaning was initiated between 14 and 35 days post-hatch, compared to 38% survival for larvae maintained on live food throughout the study. When weaning was initiated at 42 days, survival matched that of the live-food controls. Crawford (1984, cited in Hart and Purser, 1996) reported 100% mortality of hatchery-reared greenback flounder larvae, *Rhombosolea tapirina*, when abruptly weaned from live food (*Artemia*) to an artificial diet. In contrast, Hart and Purser (1996) found only 18% mortality when the larvae were gradually weaned from the live *Artemia* diet to the artificial diet (provision of live feed is reduced and the supply of artificial diet is increased) over a 10-day period starting at 23 days post-hatch. Mortality increased to 62% by waiting until 58 days to initiate the 10-day weaning process. The lack of a functional stomach or the lack of digestive enzymes prior to 20 days post-hatch prohibits feeding the artificial diet as the first feed. Waiting too long to initiate the weaning process allows the larvae to physiologically adapt to the uniform live diet, making the transition to the artificial diet more problematic. Duration of the gradual weaning process (5, 10 or 20 days) had no effect on survival, but growth rates improved with the longer weaning periods. Since feeding a live diet is expensive, the 10-day weaning period was believed to be most cost effective.

Brown *et al.* (1997) demonstrated that larval lumpfish (*Cyclopterus lumpus*) ingested more food and grew faster when they were fed live prey intermittently (two or three times per day) rather than the same total amount of prey fed continuously. Savory (1974) demonstrated that feeding chicks (*G. domesticus*) their food in pellets rather than as a mash reduced feeding time. Growth rates were also improved, perhaps due to the reduced energy expended in feeding, or in the case of Leghorns, improved digestibility. Deeming and Bubier (1999) noted that the failure of captive ostrich chicks (*Struthio camelus*) to recognize food items (e.g. pellets) is a frequent cause of starvation. The color of potential food items may influence acceptability. Bubier *et al.* (1996) found that 9-day-old captive ostrich chicks pecked at strips of green tape ten times more frequently than white tape, while red, blue, black and yellow were largely ignored. An interest in white objects may be related to the white urates (salts of uric acid) accompanying the feces of adult birds (Deeming and Bubier, 1999). In the wild, young ostriches may be inoculated with beneficial microflora through coprophagy.

Stereotypies exhibited by captive herbivores fed diets largely consisting of concentrated feeds, which are rapidly consumed, appear to result from the lack of opportunity to perform foraging, consumption and digestive behavior patterns (Appleby and Lawrence, 1987; Terlouw *et al.*, 1991). Baxter and Plowman (2001) reported that increasing the fiber content of the diet of giraffes (*Giraffa camelopardalis*) by feeding them hay resulted in increased time spent ruminating and decreased time spent performing oral stereotypies (tongue-playing and repetitive licking of walls, bars, trees, etc.) while feeding time did not significantly change, suggesting that oral stereotypies may be connected to rumination rather than feeding in this species.

Hand-feeding Young Animals

Techniques have been worked out for artificially (i.e. hand) feeding the young of many species. Getting the neonatal mammal to accept the nipple may be problematic. Lagomorphs (e.g. rabbits) resist even the most determined efforts to get them sucking (Evans, 1987). If repeated attempts are made to feed them against their will, they can be stressed to the point of death. Many newborn mammals use the odor of their own or mother's saliva to guide them to a nipple and to stimulate sucking (Blass and Teicher, 1980). The absence of such cues can make hand-rearing extremely difficult.

Kepler (1978) described the process of learning how to feed sandhill crane (*Grus canadensis*) and whooping crane (*Grus americana*) chicks in captivity. An appropriate starter diet was formulated but many chicks initially failed to recognize it as food. These recalcitrant birds were hand-fed moistened canned dog food with increasing amounts of starter mash mixed in until the birds learned to feed on the mash alone. In this process, the dog food stuck to the birds' bills, making ingestion difficult and stimulating hungry pen-mates to peck at the food adhering to the bills of other birds. Some birds even waited for the keepers to feed them and some became malnourished from eating too much dog food and not enough mash. In seeking a better way to start chicks on starter mash, it was noted that adult cranes in the wild would present to their chicks small food items held in their bills. It was theorized that perhaps the chicks would peck at bill-like objects, which could then be paired with food. Wooden dowels (12 mm diameter and 20 cm long) were found to stimulate the chicks' pecking response and the color red was preferred. Dowels were waved in front of the birds and moved toward food dishes when the chicks began following them. Over the food dishes the crane chicks would peck at the dowel and the force of their pecks would carry their bills into the starter mash, which they ingested. All chicks learned to feed on mash within an hour. It was subsequently found that the dowels did not need to be hand-manipulated. Dowels were fixed upright in the food dishes or suspended slightly above the food and the chicks learned to feed by pecking at their base. Most crane chicks learned to feed from the dowel within one or two training sessions as soon as they were ready to feed (1–3 days after hatching).

Feeding Schedules

The timing of feeding events may influence the behavior of captive animals. Bloomsmith and Lambeth (1995) showed that feeding fresh produce to captive chimpanzees (*Pan troglodytes*) on a predictable schedule (within a 30-min period) four times a day resulted in more time inactive and a tendency towards greater atypical behavior than feeding chimps on a more unpredictable schedule (within a 150-min period). Social interactions were not influenced by feeding schedule. The predictable schedule appeared to encourage inactivity during the pre-feeding period as if the animals were waiting for food to be delivered. Scheduling of feeding times in zoos to facilitate public viewing may actually create an undesirable situation in which visitors are more likely to see animals in an inactive state or possibly engaged in atypical behaviors. Other studies have shown that feeding captive animals on a predictable schedule may reduce social disturbances. Carlstead (1986) demonstrated that domestic pigs (*S. scrofa*) fed on a regular schedule each day were less aggressive towards one another than pigs fed on an unpredictable schedule.

Methods of Food Delivery

In captivity, food and water are often provided at predictable locations and in sufficient quantities so that the time and energy spent by captive animals in finding food and water are greatly reduced relative to their wild counterparts (Newberry, 1995). As a result, captive animals often have a relatively large amount of discretionary time at their disposal, which may lead to 'boredom' and atypical behaviors. Interest has been shown in delivering food to captive animals in a way that consumes more of their time and provides a form of stimulus enrichment. The 'Edinburgh foodball', which dispenses food rewards (food pellets or cubes) through small holes as it is rolled about the floor by the animal, is one such device that has been successfully used with pigs (*S. scrofa*; Young *et al.*, 1994). A similar device called the 'Equiball' has been successfully used with horses (*E. caballus*; Winskill *et al.*, 1996; Fig. 12.2). Henderson and Waran (2001) stated that feeding time for their horses increased by about 260% using the Equiball. Stoinski *et al.* (2000) describe the use of browse as a form of stimulus enrichment for elephants (*Loxodonta africana*) normally fed hay. When browse was present, the elephants showed a significant increase in feeding and significant decreases in drinking and inactivity. Hubrecht *et al.* (1992) found that dogs (*C. familiaris*) fed food pellets from a hopper that dispensed only small amounts of food at a time spent considerably

Fig. 12.2. 'Equiball' feeding device for horses. (Photo courtesy of Natalie Waran, University of Edinburgh, UK.)

more time engaged in feeding than dogs fed from bowls. Interestingly, repetitive behaviors (including stereotypic behaviors such as jumping, circling and pacing) were considerably less in the pens with hoppers. Anderson and Chamove (1984) reported that providing macaque monkeys with litter in which to forage for grain increased foraging behavior fivefold and reduced aggression within the group. Shepherdson *et al.* (1990) noted that a captive kinkajou's (*Potos flavus*) stereotypic behavior was reduced tenfold by requiring it to search and climb for food hidden in its enclosure.

Keepers of captive animals do not have to replicate the natural environment of the species to provide a behaviorally relevant captive habitat (Markowitz, 1982). The principles of operant conditioning and behavioral engineering can be used to create an environment that, although unnatural, provides the animal with opportunity to exercise species-typical behavior patterns (Markowitz and LaForse, 1987). The use of various operant conditioning devices (e.g. Skinner box) to force animals to work to obtain food represents an acceptable and behaviorally relevant management tool to assist many animal species in adapting to confinement (Moore *et al.*, 1975; Markowitz and Woodworth, 1978; Stevens, 1978). Such feeding techniques provide captive animals with a biologically acceptable means of adapting to relatively sterile environments while giving them greater control over their living environment, a condition that has a natural intuitive appeal to animal keepers. A number of researchers have demonstrated that captive animals will work to obtain food that is otherwise freely available (Carder and Berkowitz, 1970; Duncan and Hughes, 1972; Inglis and Ferguson, 1986; Shepherdson *et al.*, 1993; Reinhardt, 1994). Carlstead *et al.* (1991) found that providing food snacks to a captive American black bear (*Ursus americanus*) by a machine that automatically dispensed food at variable intervals during the day was less successful in reducing stereotypic behavior than when the animal had to search for snacks hidden in manipulable exhibit furnishings (e.g. hollow logs, under rocks, etc.).

Naturalistic Feeding Regimes

Shepherdson *et al.* (1993) proposed that captive animals should be provided with an environment in which the animal can find food as a consequence of its natural exploration and foraging behaviors. They tested this proposal by offering live food to a captive fishing cat (*Felis viverrina*) and by offering multiple feeding opportunities, sometimes with food hidden, to several leopard cats (*Felis bengalensis*). Feeding live prey (fish in a pool) to the fishing cat increased food-finding/capture behaviors from 0 to nearly 40% of the animal's daily activity budget. At the same time, percentage time spent sleeping decreased from 67 to 21%. The diversity of enclosure use also increased when live fish were provided. These changes were seen up to 8 days following the last presentation of live prey.

In nature, small felids like leopard cats typically consume small prey and engage in several feeding bouts each day. In captivity, food is typically provided once per day in an exposed location. In the Shepherdson *et al.* (1993) study, locomotory/exploratory behaviors of four leopard cats increased nearly threefold (from 5.5% to over 14%) when their daily ration was provided in four smaller meals hidden in piles of branches. This increased exploratory behavior was not simply due to an increase in time spent searching for and locating hidden food, but rather it occurred at all times of the day. Interestingly, this multiple-feeding regime decreased their repetitive stereotypic pacing behavior by about 50%, from 18% to less than 9% of the animals' activity budget.

Providing live prey and/or multiple hidden meals combined two important features. Firstly, the animals were required to perform some degree of functionally naturalistic foraging behavior to acquire food and, secondly, food presentation was unpredictable. The animals could not be sure if, when and where any food remained. As a consequence, the animals displayed more naturalistic behavior patterns and atypical behaviors, such as stereotypic pacing, became less common. Such enriched feeding regimes can improve animal welfare.

Reducing Competition for Food

In general, food is delivered to intensively reared captive animals in relatively small spaces (i.e. clumped). For animals housed in groups, this frequently leads to competition and aggression, whereby dominant animals have prior access to food

resources and subordinate animals get what is left over. Boussiou (1970) showed
that differences in the feeding times of dominant and subordinate pairs of cattle
(*B. taurus*) are reduced when feeding troughs are partitioned to protect the head of
the subordinate animal (Fig. 12.3). Holmes *et al.* (1987) obtained a similar result
with horses (*E. caballus*) and demonstrated that a wire-mesh feed-trough partition,
which permits the subordinate to see the dominant animal at all times, is more
effective than a solid partition. Hamill *et al.* (2001, unpublished results) found that
impeding the vision of dominant horses using fine-mesh fly masks doubled the
feeding time of subordinate mares in close proximity. Huon *et al.* (1986) reported
that partitioned feed troughs did not reduce the aggressive behavior of feeding
chickens (*G. domesticus*) but feeding time was reduced by 36% and feed consumption was lower (21%) relative to chickens feeding in unpartitioned feeders.

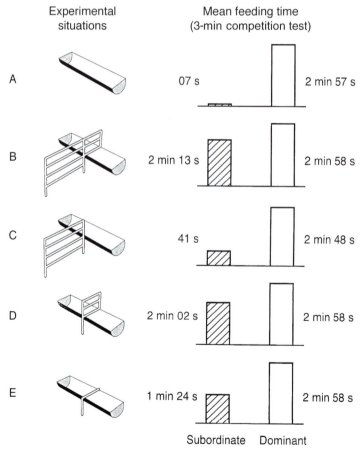

Fig. 12.3.   Effect of different types of physical barriers on side-by-side feeding
time of hungry cows classified by dominance status. Barriers providing complete
protection of the head resulted in the highest feeding times for subordinates
(Bouissou, 1970).

Self-imposed Food Deprivation

Field-trapped wild animals may exhibit self-imposed food deprivation when first brought into captivity. Hickman and Tait (2001) reported that attempts to acclimate adult wild New Zealand turbot (*Colistium nudipinnis*) to captivity were unsuccessful because the fish did not adapt to any of the wide variety of natural and manufactured food materials offered to them over periods of up to 6 months. Price and Stehn (1977) determined how time in captivity and age (juvenile vs. adult) at the time of capture influenced the tendency of wild-caught Norway rats (*R. norvegicus*) to exhibit self-induced food deprivation in response to forced occupation of a novel environment (activity wheels). A marked drop in food consumption was noted when rats in captivity for 4 and 25 weeks were taken from their home cages and placed in the activity wheels (Fig. 12.4). Food consumption increased over the first 3–4 days in the wheels and then gradually reached levels comparable to the amount consumed in their home cages before and after being placed in the wheels. The greatest change in food consumption was exhibited by animals placed in the wheels immediately following capture. These animals also exhibited a drop in food consumption when placed in home cages ($20 \times 28 \times 20$ cm) for the first time after removal from the wheels (Fig. 12.4). Rats in captivity for 25 weeks showed an increase in consumption when returned to their home cages. Changes

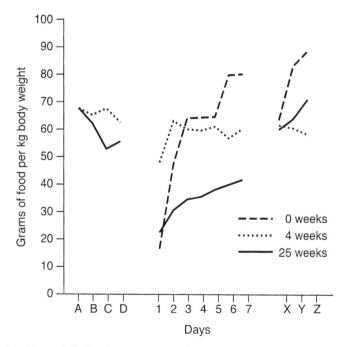

Fig. 12.4. Mean daily food consumption of male Norway rats trapped as adults and forced to occupy activity wheels starting at 0, 4 and 25 weeks of age (Price and Stehn, 1977). (A–D = pre-test; 1–7 = days in wheels; X–Z = post-test days.)

in food consumption of both juvenile and adult rats placed in home cages rather than wheels immediately after capture were similar to the animals immediately placed in wheels. Rats trapped as juveniles consumed significantly more food per unit body weight than rats trapped as adults, and gained more weight while in the wheels. Bringing a wild Norway rat into captivity as a juvenile rather than an adult had little influence on its response to being placed in the novel environment after 25 weeks in the laboratory. In a similar study, Price (1972) found that first-generation captive-reared prairie deermice (*P. m. bairdii*) exhibited a reduction in food consumption during the first 2 days of forced occupation of activity wheels, whereas food consumption did not change for a comparable stock of deermice 20–25 generations removed from the wild. Reduced reactivity resulting from relaxed and/or directed natural selection in captivity provided the best explana-tion for the more rapid adaptation of the semi-domestic stock to their new living quarters.

Conclusions

Selection and effective delivery of appropriate diets at each stage of development is critical for the adaptation and well-being of animals in captivity. Accomplishing this objective in new domesticates may require a thorough study of the species' natural history and a degree of experimentation. Effective delivery of food includes proper timing and predictable scheduling of meals to facilitate intake and minimize social competition. Experimental feed delivery systems designed to mimic naturalistic foraging and hunting strategies and increase feeding time promise to improve the well-being of captive animals. Animal welfare may be jeopardized by self-imposed food deprivation frequently displayed by wild-caught animals when first brought into captivity.

Chapter 13

Predation, Infectious Diseases and Parasites

The absence of predation in captivity can have important consequences for both prey and predator. This chapter discusses the effect of captive rearing on the anti-predator behaviors of prey species and the ability of predatory species to capture prey. The second part of the chapter addresses problems associated with infectious agents, disease and parasitism in captivity compared with the natural environment.

Lack of Predators in Captivity

Predation on free-living populations of wild animals can be intense (Wilson *et al.*, 1992). Unconfined populations of domestic animals (e.g. sheep, goats) in range environments can also experience severe predation (Shelton and Wade, 1979). For most captive animals, however, contact with predators has been eliminated or greatly reduced by the use of enclosures of various construction and the close proximity of captive animals to humans.

Effect of Captive-rearing on Anti-predator Behaviors

It seems reasonable to hypothesize that domestic animals are inherently more susceptible to predation than their wild counterparts due to a lack of experience with predators and relaxed selection on anti-predator behaviors. Fleming and Einum (1997) found that wild and farmed Atlantic salmon (*S. salar*) did not differ in their latencies to swim to shelter when exposed to a trout model (predator) but farmed fish emerged from the hide more quickly afterwards. Johnsson and Abrahams (1991) showed that juvenile steelhead × domesticated rainbow trout hybrids (*Oncorhynchus mykiss*) exposed themselves to predators more frequently than juvenile steelhead (pure strain). Johnsson *et al.* (1996) demonstrated that cultured brown trout (*S. trutta*) had a less-pronounced anti-predator response than wild conspecifics. Berejikian (1995) reported that hatchery-reared wild steelhead fry (*O. mykiss*) were more successful in avoiding predation by sculpin (*Cottus asper*) than the fry of hatchery stock originally obtained from the same river system as the wild population sampled. Kellison *et al.* (2000) compared the anti-predator responses of hatchery-reared and wild summer flounder (*Paralichthys dentatus*) and their susceptibility to predation by blue crabs (*Callinectes sapidus*). Hatchery-reared fish buried less often, tended to spend less time buried and took longer to become cryptic (burial or pigmentation response) than wild fish in either sand or mud. Predator-naive hatchery-reared fish were more susceptible to predation by blue crabs than anti-predator-conditioned hatchery fish and the latter were more susceptible to predation than wild fish (Fig. 13.1). The longer time periods used by hatchery-reared fish to become cryptic on the benthos may have been due to lack of experience with natural substrates. High-density culture techniques such as those used in the hatchery-rearing of summer flounder necessitate high feeding rates and frequent cleaning of holding tanks. Hence, the use of natural substrates is problematic. It is also possible that dietary deficiencies in captive-reared flounder could impair the fish's ability to use its pigmentation in becoming cryptic with its environment.

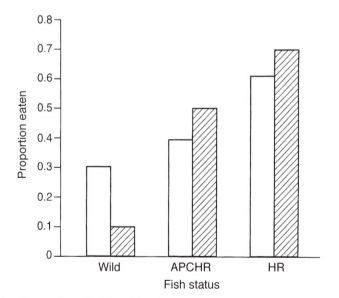

Fig. 13.1.   Proportion of wild and hatchery-reared summer flounder eaten by blue crabs in laboratory predation trials on sand (open bars) and mud (hatched bars) substrates. Hatchery-reared fish were either anti-predator conditioned (APCHR) or predator-naive (HR) (Kellison *et al.*, 2000).

In contrast, Fix (1975) and Waltz (1976) found laboratory-reared wild and domestic Norway rats (*R. norvegicus*) equally susceptible to predation by ferrets (*M. furo*) in both indoor and outdoor enclosures. Spiegel *et al.* (1974) reported that captive-reared wild Norway rats were more susceptible than domestic Norway rats to predation by great-horned owls (*Bubo virginianus*) in a large semi-natural enclosure. Neither stock actively avoided the aerial predator and differences in availability of the two stocks to the predator (time away from protective cover) could not explain the differential mortality. Johnsson *et al.* (2001) compared the anti-predator behaviors and heart rates of captive-reared wild and farmed Atlantic salmon (*S. salar*) in response to a simulated attack by a heron model. Age 1+ wild stock exhibited more pronounced flight and heart-rate responses than the domesticated stock to the simulated predator attack, but differences were negligible in age 2+ fish. Responses to predators can be readily shaped by experience (e.g. learning). The basic reactivity of animals may determine their initial responses to novel stimuli, after which experience may dictate which stimuli are avoided and which are ignored. More studies are needed to determine whether wild and domestic stocks differ in their ability to learn to avoid predators. In the Johnsson *et al.* (2001) study described above, wild and domestic salmon habituated to artificial heron attacks in a similar manner.

Domestic Animals Experience Higher Losses to Predators

When actual predation on wild-caught animals (born and reared in nature) is compared with predation on their captive-reared counterparts, the latter almost always experience the heavier losses (Schroth, 1991). Hill and Robertson (1988) reported that captive-reared ring-necked pheasants (*Phasinanus colchicus*) introduced to the wild were three times more susceptable to predation than their wild-born counterparts. Waltz (1976) found wild-caught Norway rats more effective in avoiding predation by ferrets than first-generation laboratory-reared wild rats. Kardong (1993) found wild-caught deermice (*P. maniculatus*) less susceptible than laboratory mice (*M. musculus*) to predation by blindfolded rattlesnakes (*Crotalus viridis oreganus*). Stunz and Minello (2001) and Stunz *et al.* (2001) compared habitat preferences and predation by pinfish (*Lagodon rhomboides*) on wild-caught and first-generation hatchery-reared red drum (*Sciaenops ocellatus*) in laboratory tanks simulating estuarine habitat types of different structural complexities. In general, wild-caught red drum showed a preference for structured habitats, while hatchery-reared fish did not show strong selection for any habitat type. Predation rates were higher on the hatchery-reared stock regardless of habitat types largely because they schooled near the surface and did not use cover as refuge. When wild-caught red drum were introduced to the tanks, they immediately swam to the bottom and used habitat structure to remain cryptic. Mortality of the wild fish was greatest in tanks with the least amount of structure (nonvegetated sand bottom). The hatchery-reared fish had been reared in ponds with little habitat structure and few predators, while the wild fish were captured in a structured benthic habitat. Olla and Davis (1989) demonstrated that captive-reared coho salmon (*O. kisutch*) were better at avoiding predation by lingcod (*Ophiodon elongatus*) after being conditioned to live predators and predation-associated stimuli. It appears that the failure of captive-reared animals to exhibit appropriate predator avoidance behaviors is largely mediated by rearing environments that fail to provide certain key experiences during development.

Effects of Captive-rearing on Prey-catching Behaviors of Predators

While it is clear that captive-rearing can affect the propensity of prey to successfully avoid predators, what can be said about the effects of domestication on the predatory behaviors of predatory species? As noted in Chapter 8, Cohen (2000) compared the predatory capabilities of a population of predatory big-eyed bugs, *G. punctipes*, that had been reared in the laboratory for 90 generations and fed an artificial beef-product diet for more than 60 generations, with the F1 offspring of wild-caught bugs fed a natural diet of insect eggs, larvae and green beans. For testing, individuals of both populations were exposed to either tobacco budworm larvae (*H. virescens*) or pea aphids (*Acyrthosiphon pisum*). Domestication did not significantly influence predatory feeding efficiency as measured by prey handling times (total period of contact with prey including attack, prey preparation by extraoral digestion and ingestion of prey biomass), amounts of prey extracted and extraction rates.

Infectious Agents and Disease

Captive-bred animals may lack immunity to pathogens found in the wild (McVicar, 1997; Frölich and Flack, 1998). Conversely, captive animals may carry pathogenic organisms that can endanger wild populations and humans (Tanabe, 2001). Management practices that compromise the welfare of captive animals may increase their susceptibility to disease (Broom, 1988; Gross and Siegel, 1988). Even routine management procedures (e.g. handling, mixing of strange conspecifics) can put animals at risk. Hanlon *et al.* (1994) found consistent differences between the immune responsiveness of red deer calves (*C. elaphus*) born in captivity to wild-caught and 'farmed' mothers. After an abrupt weaning at 3 months of age, wild calves were less able than farmed calves to establish and maintain an immune defense to a foreign antigen. Wild calves had higher mean cortisol concentrations after weaning than farmed calves, suggesting that the events associated with weaning (i.e. separation from the dam, transportation, novel environment and diet) were more stressful to the wild calves. Wild calves also spent proportionately less time feeding and more time lying inactive than farmed calves. All of these factors point to higher stress levels in the wild calves, which probably explains their compromised immune systems. Hameed (1997) reported that bacterial numbers associated with eggs of pond-reared shrimp (*Penaeus indicus*) were greater than for wild shrimp. Bacterial numbers were significantly correlated with percentage of abnormal eggs, and the latter was significantly correlated with hatching rates. The percentage of abnormal eggs in the spawnings of wild- and pond-reared shrimp was 28 and 64%, respectively, and mean hatching rates were 63 and 23%, respectively.

Parasites

Parasite loads in wild and captive animal populations can be considerable and highly variable (Pakandl, 1994). In fish culture conditions, losses due to disease have been estimated at 10–20%, with parasitic infections accounting for about 25% of the total losses (Bauer *et al.*, 1981). In general, intensive fish culture results in an increase in parasite populations and serious epizootic outbreaks can occur. This is especially true for certain species of Protozoa, Monogenea and Copepoda, which lack intermediate hosts and have a direct life cycle which can be easily completed in a closed system (Bauer *et al.*,1981). Özer and Erdem (1999) reported that of nine species of ectoparasites found on farmed and wild carp (*C. cario*), the farmed carp had higher infestations of six species.

Cassinello *et al.* (2001) studied the relationship between inbreeding and parasite loads in three species of endangered gazelles, *Gazella cuvieri*, *Gazella dama* and *Gazella dorcas*, maintained under identical environmental conditions. The availability of complete genealogies permitted calculation of coefficients of inbreeding for each individual and population. The species with the highest mean level of inbreeding (Cuvier's gazelle) showed the greatest susceptibility to nematode infection. At the individual level, coefficients of inbreeding were positively related to levels of parasitism in the Cuvier's gazelles but not in the other two species with lower levels of inbreeding. Thus, animal breeders can put captive animals at increased risk of parasitism if levels of inbreeding are not carefully controlled. This may occur through genome-wide effects (of reduced heterozygosity) or through an effect on specific loci related to pathogen resistance.

Under some conditions, parasite loads are less significant in captive populations. Hemmingsen *et al.* (1993) conducted a field experiment on Atlantic cod (*Gadus morhua*) to compare the ascaridoid worm burden of caged and wild-caught fish over a 2-year period. Wild-caught fish showed an increase in nematode worms in their liver and muscle over the study period, whereas the worm burden of the caged fish did not change. Transmission of the ascaridoid nematodes among the caged fish was prevented by feeding an artificial uninfected food.

Conclusions

Prey species typically learn to identify predators by sight, sound or smell and learn to recognize when predators are actively hunting. Learning anti-predatory behaviors is also facilitated by a generalized emotional reactivity and sensitivity to novel stimuli as well as observing the behavior of conspecifics toward predators. Learning the cues associated with predators does not occur in a predator-free environment. Furthermore, reduced levels of emotional reactivity are probably selected for by caretakers to improve ease of handling. Reduced emotional reactivity is also believed to improve reproductive performance in captivity by reducing stress and thus may be favored by natural selection in captivity. Consequently, domestic animals tend to be grossly deficient in anti-predator behaviors.

Learning appears to play a lesser role in the development of predatory behaviors *per se*; they are more 'hard wired'. When domesticated predators fail to respond to prey, it is usually due to a lack of motivation rather than a lack of motor skills. This topic is visited again in Chapter 19.

Few conclusions can be reached about the effects of domestication on the incidence of infectious agents, disease and parasitism in captive populations. The close quarters in which most captive animals live should increase their susceptibility to health-related problems. On the other hand, routine health care afforded captive animals and the ability to treat infected individuals may offset any increased vulnerability associated with the captive environment. The loss of vigor associated with close inbreeding appears to increase susceptibility to parasitism.

Chapter 14

Interactions with Humans

The interaction of captive animals with humans is a major element in the process of domestication. The degree of tameness of individual animals to a great extent determines the nature of that interaction. The purpose of this chapter is to review what is known about the concept of tameness and its development in captive animals. Genetic and experiential factors contributing to tameness are discussed. The basic premise is that individuals inherit a capacity to be tamed (tameability) and that the experiences each animal has with humans and other animals determines the extent to which its potential tameness is reached. One of the more interesting aspects of tameness is the close linkage of tameness to brain biochemistry. This chapter reviews information demonstrating the relationship of tameness to the serotonergic and catecholamine systems of the brain and the ability to modify the degree of tameness pharmacologically and with brain lesions.

Concepts of Flight Distance and Taming

Most wild animals naturally avoid close contact with man unless they have been habituated to the presence of humans (Van Lawick-Goodall, 1968; Geist, 1971). In captivity, the capacity to adapt to the presence of people and frequent handling is an important fitness-determining factor. The ease with which wild golden hamsters (*M. auratus*) can be handled was a key feature for their domestication and adoption as a laboratory animal and pet species (Murphy, 1985). When interacting with humans, domestic animals typically exhibit shorter 'flight distances' (how close a person can approach an animal before it moves away or takes flight) than their wild counterparts. This difference can be greatly influenced by taming. The process of taming is an experiential (learning) phenomenon occurring during the lifetime of an individual, in which an animal's avoidance of people is reduced and willingness to approach people is increased. Tameness is a measure of the extent to which an individual is reluctant to avoid or motivated to approach humans. It is an important behavioral trait of captive animals since it facilitates animal handling and management (Grandin, 1993), thus minimizing the negative effects of these processes on animal welfare (Gross and Siegel, 1979; Hemsworth and Barnett, 1987; Barnett *et al.*, 1994; Hemsworth *et al.*, 1994a).

Heritability of Fear

Some animals (species, breeds or individuals) are naturally more tame than others even when reared under identical conditions, suggesting an important genetic component to tameness (Grandin, 1998). For example, breed differences in tameness have been reported for sheep (*O. aries*; Roumeyer and Bouissou, 1992; Le Neindre *et al.*, 1993) and chickens (*G. domesticus*; Craig *et al.*, 1983; Craig and Muir, 1989). Le Neindre *et al.* (1995) found a significant sire effect (34 sires) and a heritability estimate of 0.22 for docility scores obtained in handling tests administered to 906 Limousin heifers (*B. taurus*). The test consisted of leading individual animals to a corner during a 2-min period, keeping it there for 30 s and then stroking it. Hemsworth *et al.* (1990) reported that the behavioral variable, time to physically interact with humans, in domestic pigs (*S. scrofa*) is a moderately heritable characteristic (0.38 ± 0.19 SE).

Selection for Tameness

Selection studies have also highlighted the genetic contribution to tameness. One of the most celebrated is a 40-year study selecting silver foxes (*V. vulpes*) for tameability initiated in Siberia in 1959 by Dmitry Belyaev (see summary by Trut, 1999). This research has created a population of tame foxes fundamentally different in temperament and behavior from unselected farmed foxes and which exhibit a number of physiological and morphological traits similar to other domesticated animals. In each generation, fox pups were monthly offered food from the experimenter's hand while the experimenter attempted to stroke and handle the pup (Fig. 14.1). At 7 or 8 months, they were assigned to one of four tameness categories. Foxes in the least-tame category would flee from the handler or bite when stroked or handled. The most-tame foxes were eager to establish human contact, whimpering to attract attention and sniffing and licking experimenters like domestic dogs (*C. familiaris*; Fig. 14.2). Eighteen per cent of the foxes in the tenth generation of selection qualified for this most-tame category. By the 20th generation, 35% had qualified, and after 40 years of selection the figure had reached 70–80%. Interestingly, degree of tameness is first noticed before the pups are 1 month of age. It was determined that about 35% of the variation in the foxes' defense response to the experimenter could be accounted for by genetics. In another noteworthy study, successful selection of mink (*M. vison*) for 'fearful' and 'confident' behaviors toward humans has been practiced at the Danish Institute of Agricultural Sciences since 1988 (Malmkvist and Hansen, 2001). In each

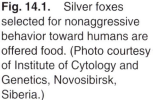

Fig. 14.1. Silver foxes selected for nonaggressive behavior toward humans are offered food. (Photo courtesy of Institute of Cytology and Genetics, Novosibirsk, Siberia.)

generation, mink are subjected to a handling test (Fig. 14.3) and a 'stick' test (Fig. 14.4), in which a wooden tongue spatula is inserted through the wire-mesh wall of the animal's cage. Figure 14.5 illustrates the response to selection of the fearful, 'confident' and nonselected control lines over the first 11 generations. Fearful mink exhibited higher plasma cortisol levels after handling than mink from the 'confident' line, even though the capacity of the adrenals to secrete cortisol (as evaluated in an ACTH challenge test) was not affected by selection (Hansen, 1997, cited in Malmkvist and Hansen, 2001). Furthermore, cross-fostering of offspring of 'confident' mothers on fearful mothers and vice versa had no effect on the subsequent timidity scores of the mink, essentially ruling out maternal effects as a significant contributor to the strain differences obtained. Jones *et al.* (1994) reported that a line of Japanese quail (*C. japonica*) artificially selected for an exaggerated (high stress) plasma corticosterone response to brief mechanical restraint exhibited greater fear of humans than a companion selection line selected for low (low stress) corticosterone response.

Twin Study on Tameness

Twin studies also point to important genetic effects on tameness. Lyons *et al.* (1988a) examined the tameness of dairy goats (*Capra hircus*) toward humans both

Fig. 14.2.   Silver fox selected for nonaggressive behavior toward humans sniffs an experimenter's hand. (Photo courtesy of Institute of Cytology and Genetics, Novosibirsk, Siberia.)

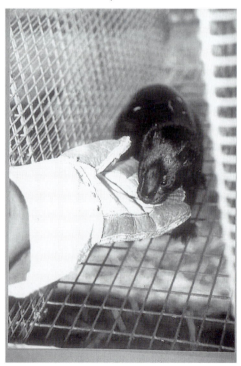

Fig. 14.3.   Handling test for mink used in selecting mink for 'fearful' and 'confident' behavior toward humans. (Photo courtesy of the Danish Institute of Agricultural Sciences.)

Fig. 14.4. 'Stick' (spatula) test for mink used in selecting mink for 'fearful' and 'confident' behavior toward humans. (Photo courtesy of the Danish Institute of Agricultural Sciences.)

within and between twin sets. One individual of each twin set was reared by its natural mother and the other individual was hand-reared apart from its mother. The hand-reared and mother-reared goats were separately ranked for tameness based on their scores in a standard human interaction test. In all cases, the absolute tameness scores of the hand-reared goats were lower than for their dam-reared twins. Moreover, the most-tame hand-reared goats had co-twins that ranked most tame within their respective dam-reared group (Fig. 14.6). Likewise, the least-tame hand-reared goats had co-twins that ranked least tame among the mother-reared animals. Although there was no control for pre-natal maternal effects in this study, the results suggested a significant genetic contribution to the development of behaviors characterizing tameness in this species.

Correlations Between Emotional Reactivity and Tameness

Animals that are less reactive (emotionally responsive to changes in their environment) generally exhibit greater tameness toward people. Lankin (1997) studied the behavior of 11 breeds of sheep (*O. aries*) and claimed that breeds subjected to intensive selection for commercial purposes tend to be tamer toward humans than breeds that have not been subjected to intense artificial selection (Fig. 14.7). The East Friesian breed has been intensively selected for meat and milk production and is particularly tame toward humans. More recently, Lankin (1999) compared

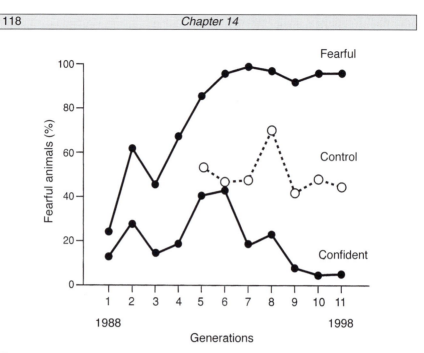

Fig. 14.5. Percentage of mink reacting fearfully (i.e. showing avoidance in the 'stick' test) over 11 generations of selection for 'confident' and 'fearful' behavior (Malmkvist and Hansen, 2001).

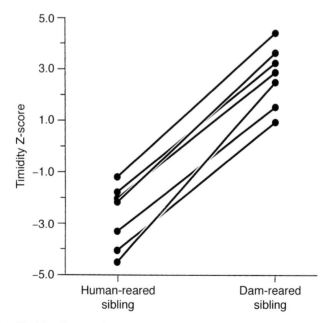

Fig. 14.6. Timidity Z-score for seven sets of twin dairy goats. In each twin set, one was human-reared and the other dam-reared (Lyons *et al.*, 1988a).

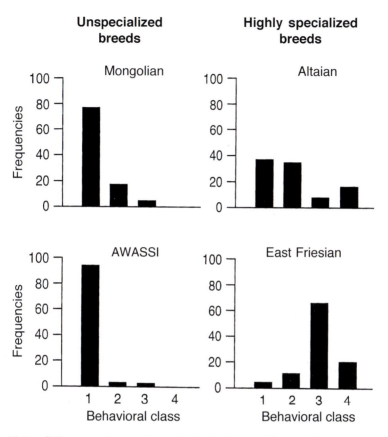

Fig. 14.7. Differences (percentage distribution) in avoidance of humans between two relatively unspecialized breeds of sheep not selected for productivity and two commercial breeds of sheep selected for productivity. Animals in behavioral class 1 show the greatest avoidance of humans and those in behavioral class 4 exhibit the least avoidance of humans (Lankin, 1997).

the reactivity rating of ewes to their respective corticosterone responses to various stressors. Highly reactive 'wild' ewes exhibited the greatest corticosteroid responses to social isolation, transportation and competition for feed, whereas 'domestic' ewes exhibiting low reactivity showed the greatest corticosteroid response to being paired with another sheep in a cage. Interestingly, there were no group differences in their corticosteroid response to 24-h feed deprivation and being separated from their offspring at weaning.

The Concept of Tameability

A conceptual distinction should be made between tameness and tameability, the capacity to be tamed. Tameability, like tameness, is obviously a desirable trait of animals undergoing domestication (Hediger, 1938; Hale, 1969). It is a reasonable expectation that some species, breeds and individuals are more tameable than others and that tameability is heritable. However, comparative studies on tameability have not received much attention. Differences in tameability could be studied by monitoring species, breed and individual differences in response to various taming procedures. Using base populations exhibiting similar degrees of tameness at the onset of treatment would yield the clearest results.

Social Influences on Tameness

The component of tameness acquired by contact with humans is not transmitted from dam to offspring. The *offspring* of hand-reared (very tame) and mother-reared (relatively untamed) ungulates exhibit similar flight distances from humans when exposed to people in the absence of conspecifics (Blaxter, 1974; Lyons *et al.*, 1988a). However, relatively untamed animals often show less fear of people in the presence of tame conspecifics. Lyons *et al.* (1988b) reported that relatively untamed mother-reared dairy goats (*C. hircus*) exhibited shorter flight distances from humans when exposed to people in the presence of tame herd-mates.

The Roles of Habituation and Positive Associative Conditioning in the Taming Process

Acquired tameness can be achieved by habituation or positive associative conditioning and it may be attained without any deliberate effort on the part of the animal caretaker. In habituation, the animal's fear of humans is gradually reduced by repeated exposures in a neutral context; that is, man's presence has no obvious reinforcement properties. Jones (1993) found that domestic chicks (*G. domesticus*) showed decreased avoidance of humans following twice daily exposures to either: (i) placing the experimenter's hand in the chick's cage with no attempt to initiate physical contact; (ii) allowing chicks to regularly observe their neighbors being picked up and handled; or (iii) allowing visual contact with a person who simply stood in front of and touched the wire-mesh wall of the chick's home cage. Wolves (*C. lupus*) maintained in enclosures with minimal exposure to people were more reactive toward unfamiliar humans than wolves reared in enclosures permitting more exposure to humans (Ginsburg and Hiestand, 1992). Malmkvist and Hansen (2001) noted that mink (*M. vison*) in a line selected for 'fearfulness' toward people exhibited increasingly more 'confident' behaviors over repeated exposures to people (Fig. 14.8). Tsutsumi *et al.* (2001) reported that minipigs (*S. scrofa*) subjected to routine care (feeding and cage cleaning) gradually became tamer with age and met established criteria for tameness at about 10 months of age. Comparable animals given 'positive contact' (brushing, patting and talking to the animals) upon opening their cage door reached the same degree of tameness in only 4 weeks after the commencement of contact.

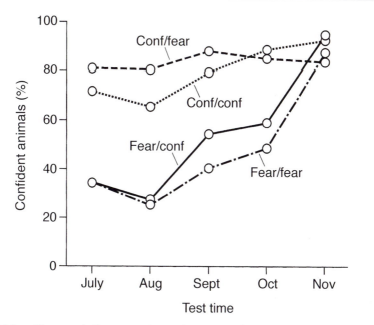

Fig. 14.8. Changes in the percentage of post-weaning mink reacting confidently towards humans over time (i.e. approach experimenter in the 'stick' test). Conf/conf, kits ($n = 106$) stayed with mothers from confident line; conf/fear, kits ($n = 82$) from confident line crossfostered on mothers from fearful line on day 1–2 after birth; fear/fear, kits ($n = 164$) from fearful line stayed with mothers from fearful line; fear/conf, kits ($n = 78$) from fearful line crossfostered on mothers from confident line (Malmkvist and Hansen, 2001).

Taming is Facilitated when People Become Associated with Positive Reinforcers

Taming may also be achieved by positive reinforcement in which the animal's fear of humans is reduced by the latter's role as a secondary reinforcer. Lankin (1997) noted that the flight distance of sheep (*O. aries*) from humans decreased with increasing hunger. As providers of such necessities as food, water and shelter, people readily become associated with such positive stimuli (primary reinforcers). Consequently animals' fear of people and avoidance behaviors are reduced. For some species such as the domestic dog (*C. familiaris*), humans can also take on the role of a social object (i.e. 'companion') whose presence itself is rewarding. Wolves (*C. lupus*) have evolved behaviors predisposing them for human companionship (Gould, 1986). Contact with humans during a 'sensitive period' for socialization early in life can enhance and accelerate the process of taming (Hediger, 1938; Mateo *et al.*, 1991). Artificial (hand) rearing of captive animals during the sensitive period of socialization is effective in taming animals (Lyons, 1989) but is labor intensive.

Positive human contact and gentle handling in the days immediately following birth is an effective tool for taming animals. Olson-Rutz *et al.* (1986) reported that mule deer fawns (*Odocoileus hemionus*) reared by their mothers but exposed to a minimum of 3 h of human contact and gentle handling each day for the first week after birth were as tame toward people as hand-reared fawns and were larger and generally healthier. Markowitz *et al.* (1998) demonstrated that mother-reared lambs (*O. aries*) handled during the first 3 days following birth subsequently exhibited shorter flight distances toward people than lambs handled on days 3–5, 5–7 or 7–9. In a similar study, Krohn *et al.* (2001) found that dairy calves (*B. taurus*) removed from their dams at birth and given positive experiences with humans (hand feeding and tactile contact) on days 1–4 exhibited shorter latencies to approach an unfamiliar person on days 20 and 40 than calves given the same treatment on days 6–9 or 11–14. Pedersen and Jeppesen (1990) found that handling young silver foxes (*V. vulpes*) from 2 to 8 weeks of age reduced their subsequent fear responses toward people (see also review by Nimon and Broom, 2001). Aengus and Millam (1999) showed that parent-reared orange-winged Amazon parrot chicks (*Amazona amazonica*) could be tamed by occasional handling without risk of imprinting on humans. Chicks incubated and hatched by wild-caught parents were handled periodically during the first 2 months after hatching and then evaluated for tameness as fledglings by their willingness to approach the handler, perch on a finger, be touched on the head and by their respiratory rate in the presence of the handler. Handled birds were more tame in all measures, thus supporting the position that neonatal handling of parent-reared parrots provides a low-labor and low-technology alternative to artificial rearing for improving their adaptation to the presence of humans.

Early Handling Can Improve the Reproductive Success of Captive Wild Animals

Early handling of wild animals not only increases tameness but can also improve reproductive success. Dalsgaard and Pedersen (1999, cited in Mononen *et al.*, 2001) reported that blue foxes (*Alopex lagopus*) handled repeatedly by humans at 7–10 weeks of age were less fearful of people and reproduced more reliably than foxes that had only been subjected to normal farm activities at a young age. Clark and Price (1981) studied the effects of handling on the reproductive success of first-generation captive-reared male wild Norway rats (*R. norvegicus*). Handling was for 1–2 min daily during the post-weaning period from day 26 to day 32. Twenty-three of 24 (96%) handled rats successfully copulated when left overnight with a hormone-induced estrous female domestic rat (*R. norvegicus*), a result that compared favorably with the 96% copulatory success attained when field-caught male Norway rats were exposed to females in captivity (Price, 1980). In contrast, only 27 of 61 (44%) nonhandled wild rats mated with an estrous domestic female. In addition, 24 of 25 (96%) handled wild females conceived and 21 of those gave birth. This is an extraordinarily high reproductive success rate for first-generation captive-reared wild female rats. Although not formally tested, these results suggest that fear of humans represents one of the principal factors (i.e. stressors) constraining reproduction in the early stages of domestication in this species.

Other studies have shown that early handling of domestic rat pups (*R. norvegicus*) can affect the development of the hypothalamic–pituitary–adrenal (HPA) responses to stress (see review by Liu *et al.*, 1997). Liu *et al.* (1997) demonstrated an inverse relationship between the frequency with which rat mothers groomed and licked their offspring and their offsprings' HPA-mediated behavioral and endocrine responses to stress during adulthood. Rat pups handled by humans receive copious amounts of grooming and licking by their mothers, and demonstrate the same reduced HPA responsiveness to stress as adults. To what extent could the improved reproductive success of captive rodents handled early in life be due to reduced responsiveness to the array of stressors typically associated with the captive environment?

Taming is Situation-specific

The taming process appears to be relatively context-specific. Galef (1970) tested the effects of several rearing experiences on the ease with which wild Norway rats (*R. norvegicus*) could be handled by humans. Second- and third-generation laboratory-reared wild rats were reared by either wild or domestic mothers, reared with either wild or domestic litter mates, given either minimal or maximal exposure to humans in a laboratory environment, and either not handled at all or handled for 2 min daily from days 10 to 23. At weaning (about 23 days of age), each rat was subjected to a handling test which scored such behaviors as difficulty of capture, escape behaviors, vocalizations, and bites directed toward the hand that restrained the animal. Only direct handling experience improved the ease of capture and handling.

Human Behavior and Posture Affect Tameness

The way in which humans approach animals and interact with them may determine the degree of tameness exhibited. Hemsworth *et al.* (1986) compared the behavior of pigs (*S. scrofa*) in response to a passive, stationary person compared to a person who actively approached, as well as varying the person's body posture. Pigs approached the experimenter significantly more when he stood passively in the test arena rather than when the pigs were actively approached. Squatting elicited more approach behavior by the pigs than when the experimenter stood erect (Fig. 14.9). It is not surprising that large, rapidly approaching objects are more threatening to animals than smaller, passive objects and that they elicit more intense reactions. Animal handlers can minimize flight distance by exhibiting behaviors that are perceived as nonthreatening.

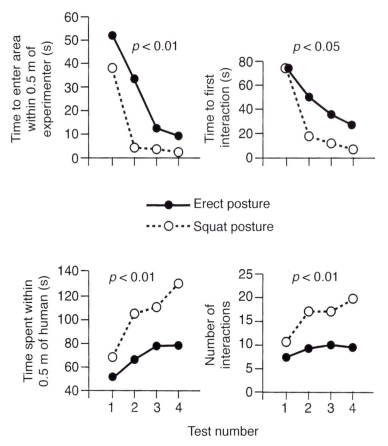

Fig. 14.9. Effect of human body posture (erect vs. squat) on the approach behavior of 48 pigs over four test trials (Hemsworth *et al.*, 1986).

Handlers' Attitudes and Behavior Affect Animals' Fear of People and Productivity

The animal handler's basic attitude toward animals can affect how he/she responds to animals and the body language he/she displays in their presence. These factors, in turn, influence how animals respond to their handler and their general fearfulness (i.e. tameness) of people. Fearfulness can affect productivity (Hemsworth *et al.*, 1989; Coleman *et al.*, 1998; Breuer *et al.*, 2000; Fig. 14.10). With the exception of Paterson and Pearce (1992), researchers have found that rough or unpleasant handling of animals consistently increases their fear of people, and fear is negatively correlated with productivity indices such as meat, milk and egg production, conception rates and litter size (see reviews by Rushen *et al.*, 1999; Hemsworth and Barnett, 2000). This result has been demonstrated with swine (*S. scrofa*; Hemsworth *et al.*, 1981), poultry (*G. domesticus*; Barnett *et al.*, 1992;

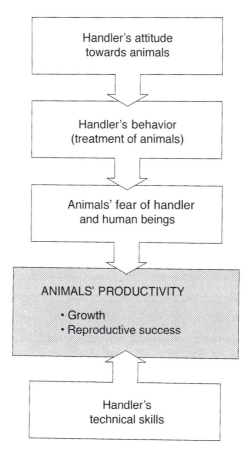

Fig. 14.10. Effect of the handler's attitude and behavior toward animals on the animals' fear of humans and productivity (Hemsworth *et al.*, 1989).

Table 14.1. Percentage of the variation in animal productivity between farms accounted for by animals' fear of human caretakers in standardized tests. (From Breuer *et al.*, 2000.)

Species	Product	% Variation	Reference
Cattle	Milk yield	19	Breuer *et al.* (2000)
Swine	Reproductive success	23	Hemsworth *et al.* (1989a)
Broiler chickens	Feed efficiency	28	Hemsworth *et al.* (1994b)
Layer hens	Egg production	28	Barnett *et al.* (1992)

Hemsworth *et al.*, 1994b) and dairy cattle (*B. taurus*; Seabrook, 1972; Breuer *et al.*, 2000; Hemsworth *et al.*, 2000; see Table 14.1). Even quality of animal products can be related to treatment received from the farmer. Lensink *et al.* (2001) found that the quality of veal calf (*B. taurus*) meat (color) was higher on farms where the animals were cared for with 'positive' behaviors during rearing. Table 14.1 shows the percentage of variation in productivity between farms accounted for by fear of human caretakers in standardized tests.

Recognition of Individual Humans

Tameness may vary depending on the specific person encountered and the location of the interaction. Although individual animals do not always behave differently toward different people, it has been demonstrated that many species of animals (cattle, sheep, pigs, rabbits, seals, rats and chickens) can recognize individual humans (see reviews by Rushen *et al.*, 1999; Davis and Taylor, 2001). It appears that visual cues (e.g. color of clothing, human behavior) are particularly important in identifying people but other sensory modalities (e.g. auditory or olfactory cues) may also be involved, depending on the circumstances (Tanida and Nagano, 1996; Tanida and Koba, 1997, both cited in Rushen *et al.*, 1999). Cattle (*B. taurus*) can use the color of clothing alone to distinguish between different humans; their response to individuals may change as clothing color is changed. Cattle can also use relative body height or facial features in recognizing individual humans (Rybarczyk *et al.*, 2001). The context of the interaction (e.g. place) may influence the animal's response to individual people. Cows readily learn to approach or avoid the same person in different places if they were treated differently (pleasantly or harshly) by that person in each place (Rushen *et al.*, 1999).

Development of Tame Behaviors

Hemmer (1988) distinguishes three basic types of taming: (i) tameness by acquiring confidence in the presence of people, as in hand-rearing; (ii) tameness by reduction of distrust, as attained by taming adult animals; and (iii) a genetically conferred natural tameness. Hemmer states that 'real' domestic animals will possess genetically conferred tameness. There is obviously a fine line between enhancing an animal's confidence with people and reducing its distrust of people (points i and ii). Both are acquired and can be achieved by habituation, primary associative conditioning or some combination of the two at any stage of development. Hemmer (1988) developed the first domesticated fallow deer herd in just a few generations by using strict selection of pre-adapted individuals for the foundation stock and by selecting for desirable traits including tameness in each generation.

Undesirable Aspects of Tameness

A high degree of tameness can be undesirable in certain circumstances. Highly tamed animals that are potentially dangerous to people because of their size and/or aggressiveness may direct behaviors toward humans that they would be reluctant to exhibit if they were more fearful of people. Wolf–dog hybrids (*C. lupus* × *C. familiaris*) inherit the capacity to become bonded to people but retain a stronger predatory drive than most dog breeds (Polsky, 1995). The combination of tameness toward humans and relatively strong predatory inclinations sets the stage for potentially dangerous aggressive interactions directed toward people, particularly children and strangers, who may inadvertently trigger an attack. Jenkins (cited in Gloyd, 1992), a former breeder of wolf–dog hybrids, estimated that only 5% make truly satisfactory pets.

Hand-reared male ungulates, particularly those reared in physical isolation from other animals, can be particularly dangerous to humans and may inflict serious or fatal injuries (Hemmer, 1988). Price and Wallach (1990) reported that Hereford bulls (*B. taurus*) hand-reared by humans in physical isolation from other cattle were more aggressive toward people than hand-reared bulls reared in small groups of contemporary peers. They hypothesized that the physically isolated bulls were socially uninhibited by never experiencing the negative consequences of their aggressive behaviors. Bulls reared in groups learned at a young age that aggressive acts were frequently followed by retaliation (i.e. when they exhibited aggressive behavior toward a penmate, the latter was likely to attack them in return). Consequently, group-reared animals may be more reluctant to initiate aggressive behaviors toward any social stimulus that they perceive as potentially dominant. In light of this discussion, it is noteworthy that dairy bulls have a greater reputation for being dangerous to humans than beef bulls. Dairy bulls are very often hand-reared in physical isolation from peers, whereas beef bulls are typically reared by their mothers in social groups.

Tameness is also undesirable when rearing animals in captivity for release in the wild. Released animals which lack fear of humans may be more approachable in the field and thus more vulnerable to a 'take' by humans. One can speculate that tameness acquired in captivity could generalize to other species encountered in the field, such as natural predators.

Selection for Tameness and Links to Brain Biochemistry

Strains of animals artificially selected for and against aggressiveness toward humans have been used to investigate the physiological basis of tameness. Belyaev (see Trut, 1999) believed that the key factor selected for in domestication was tameability and that selection for tameness and against aggressiveness would result in important changes in neurophysiology. Popova *et al.* (1991a) supported this hypothesis by demonstrating that the tameness of silver foxes (*V. vulpes*) can be directly linked to the brain serotonergic system. Foxes selected for tameness had higher levels of the neurotransmitter serotonin and its major metabolite, 5-oxyindolacetic acid, and greater tryptophan hydroxylase (key enzyme of serotonin synthesis) activity in the midbrain and hypothalamus than their unselected wild counterparts bred in captivity. Interestingly, similar changes in brain chemistry have been found in mink (*M. vison*; Nikulina *et al.*, 1985a) and Norway rats (*R. norvegicus*; Naumenko *et al.*, 1989; Popova *et al.*,1991b) selected for reduced aggressiveness toward humans. These findings are in agreement with other studies implicating serotonin as an inhibitory factor in the mechanism of fear-induced defensive aggression. Blanchard *et al.* (1988) reported that serotonin-1O agonist treatment reduced defensive threat and attack behavior in wild rats (*Rattus rattus*). Hammer *et al.* (1992) found greater serotonin-1A receptor density in the median raphe and hippocampal circuits of domestic Norway rats (*R. norvegicus*) than in their wild counterparts. Enhanced action of serotonin in the hippocampal circuits may be related to the attenuated defensive behaviors exhibited by domesticated animals. Hypothalamic serotonin (receptors) may not only be associated with tameness in animals but also with changes in gonadal hormone regulation and stress responses associated with domesticated silver foxes (*V. vulpes*; Popova *et al.*, 1991a).

Selection for tameness in foxes (*Vulpes*) and rats (*Rattus*) have also influenced the catecholamine system of the brain (Cuomo-Benzo *et al.*, 1977; Nikulina *et al.*, 1985b; Nikulina, 1990). In both species, selected (tame) animals possessed higher levels of noradrenaline in the anterior hypothalamus than was found in control animals, which exhibited aggression toward humans. No differences were found between domesticated and captive-bred wild foxes in the content of noradrenaline in the frontal cortex, hippocampus, posterior hypothalamus and the midbrain. Twelve generations of selection for tameness in silver foxes (*V. vulpes*) halved basal levels of plasma corticosteroids relative to their unselected controls (see summary by Trut, 1999). Levels were halved again after 28–30 generations of selection. Levels of dopamine, a protein that stimulates brain cells that influence body motion, motivation and the sensation of pleasure, were lower in the striatum and n. accumbens of tame foxes but higher in the tuberculum olfactorium. Tame and untamed foxes did not differ in dopamine levels in the anterior or posterior hypothalamus. Likewise, the level of dopamine metabolites, DOPAA and HVA, remained unchanged in the structures of the brain studied.

Pharmacological Induction of Tameness

Certain pharmacological treatments can have a 'taming' effect on wild animals. Blanchard *et al.* (1988) showed that the serotonergic $5HT_{1A}$ agonists buspirone and gepirone reduced many aspects of defensive behaviors in wild roof rats (*R. rattus*) without producing symptoms of motor impairment. Reactions were assessed in four situations: approach by the experimenter in an inescapable runway, stimulation of the vibrissae, contact with an anesthetized conspecific and attempted handling/pickup by the experimenter. While flight responses remained largely unaffected by both drugs, freezing and, particularly, defensive threat and attacks were inhibited. Both compounds significantly increased the number of subjects that could be approached by the experimenter to the point of physical contact and gepirone additionally reduced freezing at short (0.5 m) experimenter–subject distances. Malmkvist and Hansen (1999, cited in Malmkvist and Hansen, 2001) found that lower doses of an anxiolytic drug were needed to reduce fearfulness towards people in a strain of mink (*M. vison*) selected for 'confident' behaviors toward people than in a line selected for fearfulness.

Brain Lesions and Taming

Other work has shown that impairment of certain brain structures can have rather dramatic 'taming' effects on animals. 'Taming' of wild rats (*R. norvegicus*) has been demonstrated by lesioning the basal ganglia (Blanchard *et al.*, 1981a), mesencephalic central gray (Blanchard *et al.*, 1981b) or amygdaloid complex (Kemble *et al.*, 1984). Phillips (1964) demonstrated that wild mallard ducks (*A. platyrhynchos*) can be significantly 'tamed' by electrolytic lesions of the medial archistriatum and tractus occipitomesencephalicus (TOM) of the brain. Conversely, electrical stimulation of the medial archistriatum evoked fear responses. Wright and Spence (1976) obtained a similar result from the destruction of the TOM in Barbary doves (*Streptopelia risoria*).

Conclusions

Captive animals typically experience frequent interactions with humans. Reduced flight distance in the presence of people is one of the most obvious behavioral changes accompanying the domestication process. Tameness is inversely related to flight distance. The capacity to be tamed (tameability) is moderately heritable and responds well to artificial selection. The degree of tameness attained is heavily influenced by the animal's experiences with people. Taming is facilitated when people become associated with positive reinforcers such as food or pleasurable tactile contact. Tame animals are normally easier to handle and manage than flighty, untamed animals. As a result they are stressed less by interactions with people and may experience greater reproductive success and productivity. Researchers selecting for tameness in wild silver foxes (*V. vulpes*), mink (*M. vison*) and Norway rats (*R. norvegicus*) have found predictable changes in the activity of the serotonergic and catecholamine systems of the brain. Tameness can be induced pharmacologically and by appropriate brain lesions. Overall, tameness is becoming one of the better understood behaviors associated with the domestic phenotype.

Chapter 15

Social Environment

Most domesticated species are gregarious by nature. Thus, it is important to consider how the captive environment has influenced the social dynamics of animal populations. The purpose of this chapter is to discuss how the captive environment, especially animal management systems, limits the social choices of animals and how this affects their mating systems, reproductive success, agonistic behaviors, social organization and population dynamics.

Captivity Limits Social Options

One of the greatest differences between the social environments of wild and captive animals is the lack of social choices available to captive animals. In captivity, management decisions affecting the density and composition of populations are often made for the sake of human convenience or to expedite economic considerations. Humans often force captive animals to live under higher population densities than they would normally tolerate in nature (Hediger, 1964; Dawkins, 1980). Although increased population density may result in varying degrees of social stress, management practices and facility designs that minimize competition for food, water, shelter, mates and personal space should theoretically permit captive animal populations to exist at relatively high densities with a minimum of social strife. The fact that humans are in control of the social environment of captive populations places the responsibility for management errors on the animal caretaker. A thorough understanding of the behavioral biology of the species will help to minimize such tactical errors (Baker, 1994). As an example, the capybara (*H. hydrochaeris*) is a rodent species bred and raised in captivity for its meat, particularly in Venezuela and Brazil (see citations in Nogueira *et al.*, 1999). Until the behavioral biology of the species was better known, a sizable proportion (28%) of the offspring born in group rearing pens were killed by adult females other than the mother. Isolation of the mother at birth was not a solution since she would be attacked upon her reinstatement in the group following weaning of her young. Permanent individual housing of breeding females was impractical. Nogueira *et al.* (1999) solved the problem when they discovered that females living in groups would not practice infanticide if they had been reared with the mother starting no later than weaning. Females placed together later in life would appear socially compatible but would kill one another's young. No infanticides were recorded in 15 litters born to familiar female groups, whereas infanticides of one or more young occurred in 24 of 34 (71%) litters born to 'unfamiliar' female groups.

Changes in Social Systems and Dynamics in Captivity

Many species have complex social systems, which may change when populations are brought into captivity. Chimpanzees (*P. troglodytes*) in the wild exhibit temporary fission–fusion societies (de Waal, 1994). Although part of larger groups called communities, chimpanzees are normally found in small groups which are constantly changing in composition. Except for the mother–offspring relationship, social associations are temporary. Members of different communities do not associate and males of different communities may engage in lethal combat. In captivity, individuals are in constant contact and do not have the choice of companionship found in the wild. This does not pose a serious problem for captive males since they have evolved social behaviors that constantly keep social tensions in check. The formalized status rituals and reconciliation behaviors of males keep them strongly bonded in spite of relatively high levels of aggressive behavior. The male hierarchy regulates conflict in a way that unifies the competitors. In nature, it may be important for males to maintain a unified front when interacting with males of neighboring communities. In contrast, the social lives of female chimpanzees are greatly changed by captivity. In the wild, female chimpanzees live rather dispersed and solitary lives. In captivity, limited space forces females to live in closer proximity. As a consequence, female–female social bonds are stronger, with politically powerful alliances sometimes formed to curb male domination. Male–female relationships may secondarily change in captivity as a result of the closer female–female relationships. At the same time, when competition and aggression between females occur, conflict resolution through ritualization of dominance relationships (characteristic of male chimpanzees) is not observed. Consequently, reconciliations are less common among females than males, and the female hierarchy is less consistent and clear cut than the male hierarchy.

Interspecies Competition in Captivity

Researchers have noted the differential success of competing populations of closely related species in captivity. Pascual *et al.* (2000) studied competition between *D. subobscura* and *Drosophila azteca* in the laboratory at different initial proportions of each species (20%, 50%, 80% and 100%) and at two densities

(20 and 100 individuals). Compared to *D. subobscura*, *D. azteca* showed lower carrying capacity, lower fecundity, longer developmental time and longer time to initiate oviposition in all initial proportions and densities. In general, *D. azteca* exhibited a poorer adaptation to the laboratory environment.

Competition Within Captive Populations

As in nature, social competition within populations of captive animals can increase interindividual variability in growth rates and reduce the growth rate of the group as a whole by fostering individual differences in food consumption (Koebele, 1985) and differential growth efficiency resulting from the elevated activity and stress accompanying social interactions (Christiansen *et al.*, 1991). Wang *et al.* (2000) studied the effects of competitive social interactions on food consumption, growth rate, growth efficiency and interindividual body weight variation in juvenile hybrid sunfish (*Lepomis cyanellus* × *L. macrochirus*) maintained either individually or at group densities of five and 20 fish (per 25 l). Relative to the individually housed controls, coefficients of variation for body weight for the fish housed in groups increased by 81 and 74%, respectively, during the 50-day trial period. Individual food consumption was lower in group-housed fish, especially in the 20-fish groups, where the difference was statistically significant.

Social Organization of Captive Animals

There are two basic forms of social organization exhibited in captive animal populations. In the *dominance hierarchy*, often referred to as the *pecking order*, individuals within the population establish a dominant–subordinate relationship with other individuals in the group. The dominance hierarchy is the most common form of social organization exhibited in domesticated animals since it constitutes a pre-adaptation for domestication (i.e. it facilitates adaptation to living in social groups). *Despotism* is a variation of the dominance hierarchy in which one individual is clearly dominant over all other individuals, and individuals subordinate to it are sufficiently inhibited by the dominant animal that they do not establish dominant–subordinate relationships among themselves. Despotism is most frequently exhibited in small groups of animals confined to relatively small spaces (e.g. wild male house mice in small laboratory cages). *Territoriality* can be defined as the process by which a geographical area is delineated and maintained, by the behavior of animals acting individually or in groups, as an area for exclusive use of one or more resources. It is not very common in captive populations because territorial species are normally not very well pre-adapted to living in relatively small spaces. Domesticated llamas (*L. glama*) exhibit territorial behaviors similar to their wild relatives, vicunas (*Vicugna vicugna*) and guanacos (*L. guanicoe*), under range (extensive) conditions. This is considered an advantage to their human caretakers because the predictable use of space by extensively reared llama groups reduces the effort required to locate, control and exploit the herds (Tomka, 1992). Because territorial males exclude other adult males from their territories and female herds, herd owners often utilize surplus males (usually castrated) for meat and as beasts of burden.

Shifts in Social Organization in Captivity

Interestingly, individuals of some species/populations that are territorial in nature may find resources in small enclosures indefensible and shift to the dominance hierarchy form of social organization (Lott, 1991). Butler (1980) demonstrated such a shift in social organization in wild house mice (*M. musculus*) by systematically reducing the size of their living environment. Although territorial by nature, male vicunas (*V. vicugna*) and guanacos (*L. guanicoe*) may not express their territorial behavior under conditions of high density.

Effects of Social Isolation

Captive animals may be socially isolated for a variety of reasons, such as to prevent the spread of infectious diseases or to preclude fighting or breeding. Prolonged social isolation can cause stress, impaired health and reproduction, and the development of atypical behaviors. Certain species may be affected more than others by social isolation. Rhesus monkeys (*Macaca mulatta*) reared from an early age in social isolation are described as inactive, fearful and withdrawn, and nonsocial, and have difficulty in avoiding aggression from other animals (Harlow, 1965). Some rehabilitation can be achieved with behavioral 'therapy' but social deficits still persist (Cummins and Suomi, 1976). Veissier *et al.* (1994) have shown that veal calves (*B. taurus*) individually penned from birth are initially dominated by group-reared counterparts when placed in a social group.

The negative effects of a deprived environment are not necessarily permanent. The development of social behavior in domestic dogs (*C. familiaris*) and wolves (*C. lupus*) is well buffered against the effects of social isolation in the sense that social deficits are quickly corrected (MacDonald and Ginsburg, 1981). Animals placed in social groups after months of prepubertal social isolation exhibit normal species-typical social behaviors in just a few days of social interaction, and atypical behaviors exhibited during isolation (e.g. extreme withdrawal, extreme or violent behaviors, social indifference to other animals, stereotyped tail chasing and self-biting) generally disappear. Deprived canids are not totally immune from the effects of social isolation; dominance–submission within the group may be affected (Fuller and Clark, 1968). Noon (1991) reported on the socialization of a group of eight wild-born chimpanzees (*P. troglodytes*) that had been housed in individual cages for a prolonged period of time. The chimps were gradually introduced to one another over an 8-month period through a system of six interconnected cages and then placed together on an island in a South Florida tourist park. Twelve years after their release on the island, group social behavior was described as relatively normal. Two females produced a total of seven infants and exhibited normal maternal care. Two male chimpanzees, who had exhibited stereotyped rocking behavior in their cages, showed a reduction in rocking from ten episodes an hour to three episodes an hour after 6 months on the island. Kessel and Brent (2001) reported similar results when 11 baboons (*Papio hamadryas anubis*, *P. h. papio* and *P. h. hamadryas*) which had been singly reared indoors for an average of 5 years were moved to outdoor pens (3.85 m diameter corn cribs) in social groups. Stable social groups were formed without injury. During the first month in social groups, the male baboons displayed inappropriate sexual behaviors (masturbation and disoriented mounts). With time, several of the males exhibited normal sexual behavior and three of the females became pregnant within a year. Two of these females successfully reared their infants. Abnormal behaviors declined from 14% of observation time in the single cages to 3% in the sixth month of social housing. Cage manipulation and self-directed behaviors also significantly decreased, while social behavior, enrichment-directed behavior (to toys placed in their cages) and locomotion increased.

Effectiveness of humans as surrogate parents

Humans can sometimes serve as surrogate social partners for animals isolated from conspecifics (Davis and Balfour, 1992). Artificial (hand) rearing of young animals is effective in socializing captive animals to humans, but when combined with physical isolation from conspecifics can result in heightened aggressiveness toward people (Price and Wallach, 1990), inappropriate social interactions with conspecifics and impaired reproductive success (Beck and Power, 1988). Female rhesus monkeys (*M. mulatta*) separated from their mothers after infancy are more likely to reject or neglect their own offspring (Berman, 1990; Champoux *et al.*, 1992). Hand-rearing can also limit opportunities for social learning by preventing young from experiencing critical early experiences with parents such as in the development of food preferences (Altbacker *et al.*, 1995).

Use of Social Enrichment Devices

Various enrichment devices have been used to relieve the boredom and stress of social isolation (Shepherdson *et al.*, 1997). Bloomsmith and Lambeth (2000) exposed individually housed and socially housed chimpanzees (*P. troglodytes*) to videotapes of chimpanzees engaging in a variety of behaviors, other animals and humans and standard television programs as a form of enrichment. Individually housed chimps spent a significantly greater portion of available viewing time watching the monitor than socially housed chimps (67 vs. 19%, respectively). Washburn and Hopkins (1994) and Andrews and Rosenblum (1993) have demonstrated the enrichment properties of videotape/television as a reward for rhesus macaques (*M. mulatta*) and bonnet macaques (*Macaca radiata*), respectively.

Effect of Age and Sex Structure of Populations on Behavior

In captivity, the age and sex structure of animal populations can be very different from what is typically observed in nature. The average age of captive wild animals is generally much older than their free-living counterparts (Conway, 1978). The young of farm animals may have little or no contact with parents and are often reared and maintained in same-sex peer groups. While such management practices may be most practical from an economic standpoint, behavioral development can be affected. Price *et al.* (1994) reported that nearly 30% of yearling male domestic sheep (*O. aries*) denied heterosexual experience during their first year of life will initially exhibit little or no interest in estrous females (Table 15.1). In addition, the sexually active rams from the sexually inexperienced group exhibited lower libido than yearlings given sexual experience in their first year (Fig. 15.1).

Effects on Aggressive Behavior

While certain management-related circumstances can increase aggressive behavior (largely defensive) within captive populations, other management techniques reduce aggression. For example, adult rhesus macaque monkeys (*M. mulatta*) reared in social isolation (singly caged) are very aggressive when placed with other adults. However, Reinhardt *et al.* (1987) demonstrated that when previously singly caged adults were paired together in the company of

Table 15.1. Number (%) of sexually inactive yearling rams in four weekly sexual performance tests administered to rams given early heterosexual experience (7–8 months) and rams not exposed to females from weaning (at 3 months) until testing at 22 months of age. (From Price *et al.*, 1994.)

Treatment	Test (Week)			
	1	2	3	4
Sexually-experienced rams (n = 48)	1[a] (2%)	1 (2%)	1 (2%)	1 (2%)
Sexually-inexperienced rams (n = 47)	13 (28%)	10 (21%)	7 (15%)	7 (15%)

[a]This ram was sexually inactive in year 1 (i.e. did not benefit from the early exposure to females).

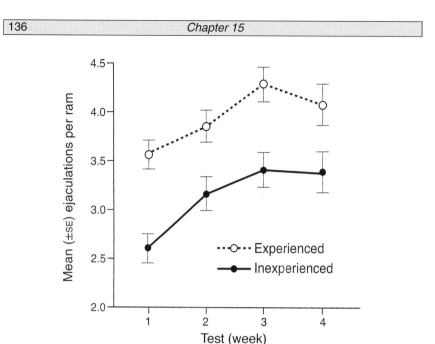

Fig. 15.1. Mean (± SE) number of ejaculations attained by sexually active year-ling rams offered and denied sexual experience as ram lambs (Price *et al.*, 1994).

12–18-month-old individuals, little aggression was exhibited. The adult partners cared for the young as if they were their own kin. In 150 adult pairings conducted in this manner, injury was inflicted in only two cases. In another example, Reinhardt (1989) paired singly caged adult macaques after rank relationships had been established during a period of noncontact familiarization in adjacent wire-mesh cages. Since dominance had already been established when they were placed together, they often engaged in affiliative grooming and/or huddling soon afterwards. Of more than 350 pairs established in this manner, serious aggression occurred in only three pairs.

Effects of Selection and Social Conditioning on Aggression and Reproductive Success in Captivity

Thresholds for agonistic behavior may be influenced by both selection and different opportunities for social conditioning in nature and in captivity. Swain and Riddell (1990) found that hatchery stocks of juvenile coho salmon (*O. kisutch*) were more aggressive than hatchery-reared wild stocks of juvenile salmon. Moyle (1969) reported this same result in brook trout (*S. trutta*). Jonsson (1997) noted that when wild and hatchery-reared Atlantic salmon (*S. salar*) were subjected to pairwise competition tests (e.g. feed competition), the young hatchery-reared fish tended to dominate the young wild fish. He proposed that the greater aggressiveness of farmed over wild salmon parr in competitive interactions may be due in part to unintentional selection of the most aggressive individuals for breeding stock. Wang and White (1994) proposed that a lack of visual isolation and the 'scramble' feeding strategies fish develop in hatchery environments may lead to an increase in aggressive feeding behavior in hatchery-reared salmonids. Rhodes and Quinn (1998) demonstrated the relative importance of hatchery rearing, prior residence, prior winning/losing experience and size (length) advantage on the aggressive behavior of coho salmon (*O. kisutch*). Juvenile hatchery-reared fish were more aggressive and dominant in 42 of 62 pairings (68%) with size-matched fish from the same parental population reared in a stream and in 32 of 41 pairings (78%) with juveniles spawned in nature. The authors postulated that the hatchery-reared fish were more aggressive because they had not previously experienced losing in agonistic interactions with conspecifics. High densities, abundant food and the absence of habitat structure in hatchery environments may greatly reduce the likelihood that hatchery fish will engage in dyadic agonistic interactions with one another, which are so typically observed among wild fish in stream environments, where intense fighting leads to the establishment of dominant–subordinate relationships. Losers in such encounters may become socially inhibited in subsequent agonistic interactions (Chase *et al.*, 1994).

Other studies with fish (Robinson and Doyle, 1990; Ruzzante and Doyle, 1993) reported a reduction in agonistic behavior and/or reproductive success with domestication. Fleming *et al.* (1996) compared the spawning behavior of fifth-generation hatchery-reared and wild Atlantic salmon (*S. salar*) under semi-natural conditions in outdoor tanks. Hatchery-reared males exhibited less aggressiveness and display behavior than wild males. In addition, the spawning behavior of the cultured males was much less intense than that of the wild males. Consequently, the reproductive success of the farmed females was much greater when exposed to wild males than hatchery-reared males. In another study, Fleming *et al.* (1997) compared sea-ranched and naturally reared Atlantic salmon (*S. salar*) from a common genetic background with respect to aggressive behaviors and reproductive success. Although levels of aggressive behavior as adults were similar in the two stocks, sea-ranched males were involved in more prolonged aggressive encounters (i.e. fights) and experienced greater wounding and mortality than wild males (mortality of sea-ranched salmon = 27%; wild = 5%). The authors postulated that the prolonged agonistic interactions in sea-ranched salmon may have been due to inappropriate interpretation or use of contest-resolution signals acquired as juveniles. In addition, sea-ranched males were less able to monopolize spawnings (i.e. dominate the primary courting position in solo spawning). Consequently their reproductive success was only 51% that of the wild males. Wild males may have had more previous experience competing for territories than sea-ranched fish and some wild males may have previously bred as parr. Differences in reproductive success between male sea-ranched and wild fish were probably due to differences in experiences as a juvenile in fresh water (the sea-ranched fish had no river experience) and related to the development of specialized behaviors. Moreover, the reduced spawning success of hatchery-reared stock may be due to the practice of artificial insemination in hatcheries (Fleming and Gross, 1992, 1993).

Berejikian *et al.* (1997) compared the aggressive and reproductive behaviors of wild and hatchery-reared populations of coho salmon (*O. kisutch*) in an artificial stream channel. The wild salmon were captured as adults in their native environment and transported to the designated hatchery for study. The hatchery-reared population consisted of salmon captured as newly emerged fry in a stream 7 km from the source of the wild population and reared to adult stage in tanks in the hatchery. Five males and five females from each of the two populations were placed in two separate sections of the test stream channel and observed for aggressive and reproductive behaviors. Hatchery-reared salmon of both sexes displayed the full range of aggressive and reproductive behaviors shown by the wild-reared salmon. However, wild males dominated captive-reared males in 86% of spawning events. Wild females established nesting territories sooner and constructed more nests per individual than captive-reared females of similar size, suggesting a competitive advantage for wild females. Males from both populations expressed similar levels of aggression toward females from both populations but females exhibited more aggression toward captive-reared males. It was not clear if captive-reared males received more aggression from females because they tended to be less dominant or because they had less striking coloration. Wild males possessed more prominent morphological secondary sexual characteristics and bore the red and black coloration patterns typical of wild fish whereas the captive-reared males were generally dull and brown. Fleming and Gross (1992, 1993) also reported greater differences in the reproductive success of male wild and hatchery-reared coho salmon than between their female counterparts. Berejikian *et al.* (1997) proposed that intrasexual selection is more intense for males than females because of the relatively high level of male–male competition for mates and female choice exhibited in Pacific salmon species.

Interstock Competition and Growth Rate

Ruzzante (1994) proposed that changes in agonistic behavior of hatchery stocks of fish are correlated with selection for growth rate in captivity. If food is limited, larger and more aggressive fish will be better able to compete for existing resources. Fleming and Einum (1997) compared the growth rates of Atlantic salmon (*S. salar*) from a population of wild-genotype fish and a stock of farmed salmon derived from the same source (river) as the wild population but which had been in a hatchery environment for seven generations. Fish from both populations were reared under identical conditions in a hatchery environment. In pure-stock groups, farmed fish grew at a faster rate than wild fish. This was not surprising since the farmed stock had been selected for growth performance. However, mixing fish from the two populations resulted in a decrease in growth rate for the farmed stock but no significant effect on the growth of the wild fish, reflecting the competitive superiority of the latter.

Mate Selection – Natural vs. Artificial

Mate preferences are commonly expressed in reproductively active animals (Bateson, 1983) and stocks may segregate when selecting mates based on a history of domestication. Wong *et al.* (1982) reported that a strain of oriental fruit flies (*D. dorsalis*), cultured in the laboratory for about 330 generations, preferred to mate with members of its own strain when allowed to choose between the latter and wild-type flies. Similarly, wild flies preferred to mate with members of their own strain. Cheng *et al.* (1979a) demonstrated that wild and game-farm (semi-domestic) mallard ducks (*A. platyrhynchos*) prefer to mate with members of their own stock when given a choice. In captivity, mate selection is frequently the choice of the animal breeder as part of the artificial selection process. The period of association between male and female may be very brief, as in controlled breeding, where a receptive female is placed with a male only long enough for copulation to occur. The silver fox (*V. vulpes*) is basically monogamous. On commercial fur ranches, male foxes used to breed multiple females are not left overnight with females in order to prevent pair-bonding (Enders, 1945).

Male–Female Incompatibility

Male–female incompatibility is a significant deterrent to successful reproduction among many animal populations in captivity (Spurway, 1955). Blohowiak (1987) found that when black ducks (*Anas rubripes*) were randomly paired, fewer than a third of the pairs mated. When kept in large groups and allowed to select their own mates, the majority of them bred. Bluhm (1988) reported that captive female canvasback ducks (*Aythya valisineria*) must choose their own mates and be housed with them in visual isolation from other ducks in order to obtain fertile eggs. Gonadal development in unpaired canvasback females was not stimulated by courtship from randomly assigned males. In addition, a water depth of at least 35 cm was necessary for successful copulation. Special diets and large flight pens were unnecessary; a small, bare-walled pen with a nest box was sufficiently 'natural' if an adequate social environment was provided. Kepler (1978) found that in breeding the endangered whooping crane (*G. americana*) in captivity, evidence of compatibility did not ensure mating. Pairs showing no overt aggression toward one another would sometimes fail to breed. It was found that males had to be sufficiently dominant over females to stimulate submissive and copulatory behaviors. Hartt *et al.* (1994) noted that age of California condors (*Gymnogyps californianus*) at pairing may affect reproductive success. Forced cohabitation of potential mating pairs starting early in life may result in mating failure even when male and female are unrelated genetically. Long-term forced cohabitation may lead to a kind of 'assumed kinship' and incest-related mating failure.

Carlstead *et al.* (1999) collected survey data on the behavioral correlates of reproductive success of 60 black rhinoceros (*Diceros bicornis*) at 19 zoos representing 70% of the black rhinoceros in the USA. Fifty-two separate behavioral elements were sampled and only rhinoceros who had previously produced offspring were included in the analysis. Breeding success was found to be related to relative dominance of the male and female. The greatest success was when females were dominant over males, often achieved by pairing older females with younger males. (Females became more dominant with age.) A certain amount of assertiveness and aggressiveness contributed positively to a female's chances of breeding. However, a negative correlation was found between breeding success and ratings of chasing/charging/mouthing. Social interactions of reproductively successful females were less likely to be characterized as 'intense' or 'aroused'. A submissive, adaptable and interactive character was advantageous for males. This example further illustrates the importance of knowing the life history of animals when breeding them in captivity. It is generally assumed that female dominance inhibits the sexual behaviors of males.

Mating Systems in Captivity

The mating system typically employed by the species is an important consider-ation when breeding animals in captivity. Most of our common domestic animals are promiscuous by nature (i.e. have multiple mating partners) since promiscuity constitutes a pre-adaptation for domestication. Captivity may encourage promis-cuity, perhaps because of the greater abundance of potential sex partners, greater densities of animals and lack of privacy. Animal breeders may consciously or unconsciously select for promiscuity by breeding animals that are less discriminat-ing in their choice of mates. Kovach (1974) notes that Japanese quail (*Coturnix coturnix*) are monogamous in the wild but promiscuous on farms or in the labora-tory. Females of some promiscuous species often show greater reproductive success when mated with multiple males. Sullivan *et al.* (1992) reported that Japanese quail produced more fertile eggs when mated every third day with a different male than when mated every third day with the same male.

It is unlikely that polygamous species will become more monogamous under captive breeding. The first breeding colony of the white-toothed shrew (*Crocidura russula monacha*) was established in Israel in 1953 (Hellwing, 1971). Shrews were kept in either monogamous pairs, polygamous breeding groups or sibling pairs. The monogamous pairings were most productive; breeding was rarely successful in the polygamous groupings.

Conclusions

Traditional animal management systems used with captive populations were not necessarily designed with the animals' behavioral biology in mind. Housing systems were sometimes overcrowded, forcing animals to compete for needed resources. Animal caretakers today are more mindful of the psychological and physical needs of captive animals. Even so, the social environment of animals in captivity may dictate important changes in their behavior. Captivity limits the social options available to animals. The sex and age composition of the population may be more uniform than in nature. The relative density of animals is typically greater in captivity than in the wild, often resulting in increased competition and aggression. Being confined, subordinate animals may not have the opportunity to escape the presence of more dominant penmates. The social and physical environments in captivity may result in changes in social organization. Animals that tend to be territorial in nature are sometimes forced to adopt a social hierarchy form of social organization in captivity. Caretakers may socially isolate animals when they are ill or to preclude fighting or breeding. Humans sometimes serve as social companions of isolated animals or surrogate parents for orphaned animals. Lack of choice of mates often results in male–female incompatibility and reproductive failure. Consequently, sexually promiscuous species are favored in the domestication process. Overall, those species which are best able to adapt to the social environment provided by man will constitute the best candidates for domestication.

Chapter 16

Climate and Shelter

General Considerations

Some domestic animals are typically provided with an environment physically similar to the habitat of their wild ancestors. Behavioral and physiological adaptations to such an environment will be readily achieved. Very often, however, the captive environment does not match the ancestral environment and adaptation is challenged. Bartlett (1985) has identified some of the physical characteristics of laboratory environments that compromise the ability of insects to survive and reproduce in captivity.

There is a tendency for animal caretakers to establish captive environments that are believed to support the least adaptive individuals in the population yet still convenient and efficient to maintain (Bartlett, 1993). It is not uncommon for the living space of captive animals to be smaller and less complex than that of their free-living counterparts. In addition, there is typically less temporal variation in the physical components of captive environments. Environmental enrichment is often used to compensate for biologically important deficiencies and to make captive environments more compatible with species-specific behaviors and requirements. The term 'enrichment' is typically applied to 'increasing the physical, social and temporal complexity of captive environments' (Carlstead and Shepherdson, 2000) rather than some measure of the outcome of the change on the animals involved (Newberry, 1995). Changes in physical aspects of the captive environment do not necessarily improve adaptation. Captive animals may actively investigate novel objects when first placed in their living space but do not always retain interest in these objects (Vignes et al., 2001). Enrichment devices should be biologically relevant to the species and designed to achieve specific goals (Newberry, 1995). Feeding devices which require animals to work for their food provide them with perceptual and locomotor stimulation, which may or may not be similar to that required for procuring food in nature. Nevertheless, the stimulation obtained probably helps to promote normal development and good welfare.

©CAB International 2002. Animal Domestication and Behavior
(E.O. Price)

Mellen and Sevenich MacPhee (2001) propose a more holistic approach to environmental enrichment in which enrichment is re-defined for each species and perhaps each individual. They propose that enrichment should be based on an animal's natural and individual histories and the constraints of its living environment and should include defining the physical and social environments of the animal, the role of human caretakers (feeding, cleaning, training, etc.) and diet (type, presentation, variety). It is more than simply adding something to the animal's environment that is believed to improve its welfare.

Wemelsfelder (1997) makes the point that the welfare of animals in captivity may depend on how environmental stimuli are perceived from a functional standpoint. The specific stimulus properties of an environment in terms of variability and complexity may not be as important as what they afford the animal to do. For example, water can be a cool bath, a life-saving drink or a threat to the animal's safety. What is perceived is not merely water, but the functional context of water. Likewise, housing systems can be designed which allow animals to be functionally creative. This may include the use of artificial objects or devices which the animal can manipulate and control (Markowitz and Line, 1989). The ultimate challenge is to create an environment which retains its functional properties over time, one which does not lose its effectiveness.

The next two chapters provide examples of how the physical environment of captive wild and domestic animals differs from that of their free-living counterparts in nature. These reports also include techniques that have been developed or are currently being researched to provide more functionally relevant living environments for captive animals. This chapter begins with a few examples of how temperature, light and day length influence the maturation, reproduction and behavior of captive animals. The adaptive significance of shelter is then discussed, pointing out the possible role of shelter (hiding places) in the development of avoidance responses of animals toward humans and even maturation rates.

Climate – Temperature

The wild ancestors of most domesticated animals lived in temperate regions of the world (Ucko and Dimbleby, 1969) and typically possessed a variety of adaptations to natural temperature fluctuations. Many wild animals escape temperature extremes by burrowing underground, hibernating, estivating, migrating or otherwise seeking out favorable microhabitats. Expression of such temperature-regulating behaviors is sometimes prevented by the captive environment (Geiser *et al.*, 1990) and other behavioral and physiological adjustments have to be employed (Randolph *et al.*, 1990). In contrast, captive animals are sometimes housed in climate-controlled facilities, where temperature and humidity are far more uniform than ever experienced in nature. Constant temperature in the laboratory is responsible for a number of reports of delayed sexual maturation in migratory locusts (*Locusta migratoria*) exposed to long-day photoperiods and reared under crowded conditions (Hasegawa and Tanaka, 1996). Crowded female locusts exposed to cool nights (15°C) and warm days (30°C) during the first two nymphal instars exhibited significantly earlier sexual maturation (i.e. age at first oviposition) than locusts maintained at a constant 30°C. Most of the experimental group had started oviposition by 30 days, whereas a third of the control group had not oviposited by 50 days. The suppression of sexual maturation often reported from laboratory studies may not necessarily occur in the field; free-living locusts may experience low temperatures during the early nymphal instars.

Initial attempts to spawn white sturgeon (*Acipenser transmontanus*) broodstock maintained at constant 15 and 18°C water temperature were largely unsuccessful due to ovarian regression. Normal ovarian development can be obtained in the majority of females with intact ovarian follicles by transferring them to cold water (12 ± 1°C) in the fall or early spring (Webb *et al.*, 2001). There appears to be a temperature-sensitive phase in ovarian development during the transition from vitellogenic growth to oocyte maturation and the degree and timing of sensitivity to environmental temperature are dependent on the female's endogenous reproductive rhythm.

Constant temperature conditions may affect the rate of genetic changes resulting from propagating animal populations in captivity, particularly poikilothermic species (Bartlett, 1984). Economopoulos and Loukas (1986) monitored genetic changes in allele frequencies at the alcohol dehydrogenase locus of olive fruit flies (*D. oleae*) in response to being reared with an artificial larval medium instead of olives under constant temperature regimens of either 17 or 25°C or a daily fluctuating temperature ranging from 17 to 25°C. The magnitude of genetic changes was the same for all three temperature regimens, but the changes occurred more rapidly at the stable temperatures, particularly at 25°C, which favored rapid development and high colony productivity.

There is some evidence that domestication may have reduced the capacity of animals to adapt to extremes in temperature. Richardson *et al.* (1994) compared the offspring of wild-caught house mice (*M. domesticus*) with domestic house mice of the random-bred ICR strain and their reciprocal hybrids for their ability to exhibit nonshivering thermogenesis (NST) in a cold environment, a process that enables small mammals to rapidly increase heat production during cold exposure. They also examined the mice for depositions of brown adipose tissue, an adaptation to living in a cold environment. Wild and hybrid mice together had higher (+18.2%) regulatory NST (maximal NST minus basal metabolic rate) and larger (+21.2%) deposits of interscapular brown adipose tissue than the laboratory strain suggesting that, during domestication, the laboratory strains may have experienced relaxed selection for thermoregulatory abilities since intense cold never occurs in the laboratory environment. Foley *et al.* (1971) demonstrated that neonatal feral (wild) hogs (*S. scrofa*) are more resistant to extreme cold than domestic neonates because of greater heat production and more pelage. Vincent (1960) reported that wild-genotype brook trout (*S. trutta*) could endure higher water temperatures than domestic stock even though reared under the same conditions. Domestic farm animals living outdoors seek protection from cold and wind by using natural vegetation or various types of shelter provided by humans.

Protection against extreme heat is more difficult, however, since avoidance of the direct rays of the sun offers only partial relief. Hansen and Jeppesen (2001a) propose that access to water for swimming can assist farmed mink (*M. vison*) in thermoregulating in response to very high ambient temperatures. Thermo-regulation is less of a problem for domestic stock tolerant of high temperatures (Macfarlane, 1968).

Providing the optimal temperature, humidity and photoperiod can be critical for the survival and reproduction of animals in captivity. Milonas and Savopoulou-Soultani (2000) studied the effect of (laboratory) temperature on the development, fecundity and longevity of *Colpoclypeus florus*, a parasitoid of *Adoxophyes orana*, an economically important pest of peach, apple and cherry orchards in Europe. *C. florus* were reared at constant temperatures of 15, 17, 20, 25 and 30°C with a photoperiod of 16:8 (L:D) hours. The rate of development from egg to adult increased as a linear function of temperature from 15 to 30°C (31.7 ± 0.4 vs. 12.8 ± 0.2 days, respectively) but no adults emerged at 30°C. Fecundity (total number of eggs per female) was greatest at 17°C (57.4 ± 5.1) and least at 15°C (30.7 ± 6.2), only 2° lower. The average number of eggs oviposited per female per day was greatest at 25°C (9.3 ± 0.9) and least at 15°C (2.8 ± 0.3). Longevity of adult females (days) was greatest at 17°C (11.0 ± 1.2) but not significantly different from 15 and 20°C. Longevity was least at 25°C (4.5 ± 0.4 days). Studies such as this one highlight the importance of temperature in rearing invertebrate species such as *C. florus* in captivity to gain information on their potential geographical distribution, population dynamics or for their mass production for field release as a means of biological control of pest species.

Light

Dou *et al.* (2000) found that light is the primary factor regulating the feeding activity of captive-reared Japanese flounder larvae, *P. olivaceus*, an important fish for stock enhancement and sea ranching in Asia (Japan, China and Korea). Tank-reared flounder show a major feeding peak in the morning and a secondary peak in the afternoon throughout larval development with little or no feeding in darkness. Maximum feeding rates occurred at 19°C.

Day Length

Day length can affect growth, reproduction and a host of other biological characteristics. Endal *et al.* (2000) and others have reported that adding additional light enhances the growth rate of Atlantic salmon (*S. salar*) reared in sea cages. Additional light can also increase or decrease the rate of sexual development of salmon depending on timing relative to a critical period of development (see review by Endal *et al.*, 2000).

Reproduction in free-living animal populations is often synchronized to changing day-length. Constant additional light together with good nutrition in the laboratory can facilitate year-round breeding by many rodent species that breed seasonally in nature (Blus, 1971). Constant light–dark cycles may also be a contributing factor to the reproductive failure of many captive animal populations. Price (1966) observed a sevenfold increase in the reproductive success of captive woodland deermice (*Peromyscus maniculatus gracilis*), approximately 15 generations removed from the wild, by delaying the offset of the daily light cycle in the laboratory 1 h per week for 3 successive weeks. Kujawa *et al.* (1999) described a system that controls photoperiod and water temperature for keeping fish prior to artificial spawning. It has been successfully used with a number of commercial and wild European cyrinid and percid fish species, and can be used to alter spawning time relative to the natural environment. More opportunistic species are not as affected by photoperiod and reproduce whenever they are in a positive energy balance (Bronson, 1998).

Ultraviolet-light Exposure

Lack of sunlight or ultraviolet-light exposure in captivity can pose special problems for certain species. Gillespie *et al.* (2000) discuss the deficiencies in vitamin D_3 found in Komodo dragons (*Varanus komodoensis*) kept in captive, indoor exhibits vs. those with daily ultraviolet-B exposure.

Adaptive Significance of Shelter

In nature, natural cover or shelter provides protection from climatic events, privacy for mating and rearing of offspring, and a means of escape from predators and aggressive social partners. An increase in the quantity or quality of cover permits higher population densities by reducing the incidence of intraspecific and interspecific contacts (Jenkins, 1961). Shelter may serve some of these same functions in captivity. Patterson-Kane *et al.* (2001) studied the cage preferences of laboratory rats (*R. norvegicus*) and discovered that paper towels (nesting material) and a tin nest box (coffee can) were consistently preferred over a larger social group (three rats), larger cage space, toys, a plastic tunnel and objects that provided opportunity for chewing (wooden sticks). Bird species that use nest cavities for reproduction in nature may require nest boxes for successful breeding in captivity (Millam *et al.*, 1988). Hansen and Damgaard (1991) reported that the physiological stress level of farmed mink (*M. vison*) increased when they were deprived of nest boxes in their cages. They hypothesized that nest boxes shielded the animals from stress-provoking external stimulation from the animals in neighboring cages. Clark and Galef (1977) demonstrated that the tameness and docility of captive gerbils (*M. unguiculatus*) can be influenced by the physical design of their cages. Laboratory-reared gerbils given access to an enclosed hiding place (e.g. burrow system or shelter) during ontogeny exhibited greater avoidance of a human-like stimulus than gerbils reared in open laboratory cages. Elicitation of escape responses (flight) into a concealed chamber was found to be the 'critical' experience in the development of this avoidance response. Once flight and concealment responses were established, experience in an open-cage environment had little influence on avoidance behavior. In addition, Clark and Galef (1980, 1981) reported that gerbils reared in standard open cages vs. cages with shelters exhibited earlier eye-opening, faster growth, earlier sexual maturity, and a marked decrease in adrenal size, all of which are traits normally associated with domestication (King and Donaldson, 1929; Richter, 1949; Clark and Price, 1981). Harri *et al.* (1998) reported that farmed blue fox cubs (*A. lagopus*) provided with a nest box in their cage were more fearful towards humans than foxes without a nest box. If nest boxes are offered, foxes take advantage of the opportunity to hide and, as a result, do not habituate to the everyday activities of farm staff and visitors. These examples illustrate how environmentally induced developmental events recurring in each generation can play an important role in the development of an important component of the domestic phenotype (i.e. reduced fearfulness of humans).

Nikoletseas and Lore (1981) found that domestic Norway rats (*R. norvegicus*) reared in cages with burrows were more aggressive toward strange intruders than rats reared in standard laboratory cages without shelter. Hubrecht *et al.* (1992) in a study relating the behavior of kenneled dogs (*C. familiaris*) to pen size and housing conditions noted that the provision of a shelter (coop) not only provided a place for the dogs to rest but was also used in play and served as a refuge from other dogs. The shelter provided the dogs with some control (i.e. choice) over their environment. Hansen and Berthelsen (2000) compared the behavior of domestic rabbits (*O. cuniculus*) maintained in conventional cages vs. enriched cage systems with access to shelter. Behaviors in open-field tests and in response to being captured (timidity) were recorded. Rabbits housed in conventional cages, especially females, exhibited more restlessness, excessive grooming, bar-gnawing and timidity than rabbits in the enriched cages. Only a few rabbits used the shelters for resting; more often they were observed resting on the roof of the shelter.

Conclusions

Many captive wild and domestic animals are housed under constant temperature (and humidity). Although unnatural, constant temperature does not constitute a problem for most homeothermic animals providing the temperature is not extreme. In contrast, constant temperature may be very important for cold-blooded (poikilothermic) animals that depend on temperature fluctuations for normal reproduction and maturation. Constant day length is a common feature of many captive environments and may be a key factor in explaining reproductive failure in captive populations of seasonally breeding species.

It is not surprising that captive animals actively use shelter if it is provided. Their wild ancestors may have needed shelter to survive and bear young. Although shelter can shield animals from potentially threatening stimuli and events such as contact with humans and other animals, in captivity it can result in the development of increased fearfulness towards humans and aggression directed at penmates. The traditional use of open-cage designs for laboratory rodents and other captive animals may facilitate habituation to humans and human activity, thus reducing fearfulness and improving animal welfare. Such practices recurring in each generation can play an important role in the development of the domestic phenotype. There is some evidence (in gerbils) that the use of shelter may retard maturation and growth and result in larger adrenal glands. The generality of this interesting finding should be studied because of its potential importance to the domestication process and animal welfare.

Chapter 17

Use of Space

The spatial requirements of free-living wild animals are determined by such proximate factors as the distribution and availability of food and water, the frequency and intensity of social interactions, and the quantity and quality of cover or shelter. In nature, home-range size often varies on a seasonal and annual basis. By comparison, the quantity and quality of space available to captive animals are almost always reduced (Hediger, 1964). Since food, water and shelter for confined animals are normally provided, spatial requirements are dictated primarily to avoid crowding and the need for perceptual and locomotor experience or exercise. The purpose of this chapter is to show how confined animals use space, how a lack of space can influence their behaviors and how environmental complexity can enhance the productivity and welfare of animals. The chapter also demonstrates that preferences for certain physical features of the captive environment are probably based on habitat preferences selected for in nature.

Locomotor Activity

Locomotor activity in most confined animals is largely limited to ambulation and climbing (Dewsbury *et al.*, 1980; Büttner, 1992). Running wheels are sometimes placed in laboratory cages to afford rodents the opportunity for additional locomotor activity. When entering the wheel, mice try to climb but are forced to run due to the rotation of the wheel. Harri *et al.* (1999) compared the behavior of C57BL/6J mice (*M. domesticus*) with and without running wheels in $16 \times 32 \times 12.5$ cm high cages. Experimental mice ran an average of 1.94 km per day ($\pm$ 1.34 km) in the wheels. Individual differences in distance run over days were large (range of 200 m to 4.5 km per 24 h). Mice were most active at the onset of the dark period and sleeping was largely confined to the light period. Experimental mice spent an average of 114 min day^{-1} (7.9%) running the wheel.

Control mice spent more than twice as much time climbing on the lid of their cage than experimental mice (30.7 vs. 12.6 min per 24 h, on average, respectively). In spite of this difference, there was still a weak positive relationship between running time and time climbing (and feeding) on the cage lid for the experimental subjects, suggesting that mice can be broadly categorized as active or less active. Although running the wheel does not totally substitute for climbing, it substituted for climbing proportionally more than any other behavior. Sherwin (1998) reported that the running wheel is preferred over tunnel loops and a complex tunnel system. Lapveteläinen *et al.* (1997) found that running-wheel activity increased over the first 4 weeks of access to the wheel and then leveled off. In the Harri *et al.* (1999) study reported above, the mice had access to the wheel for 3–6.5 weeks before data were collected.

Exploration

Some animals may be motivated to explore and interact with their environment for its own sake (Stevenson, 1983). Exploration may be reinforcing, not just as a means to an end but as an end in itself (Wemelsfelder, 1997). Wheel-running by Norway rats (*R. norvegicus*) trapped as adults was greater immediately following capture than 4 weeks later, whereas rats captured as juveniles continued to show relatively high levels of wheel-running after 4 weeks in captivity (Price and Stehn, 1977). The relatively high levels of running exhibited by adult rats following capture could reflect some combination of a general response to abrupt confinement and an attempt to avoid or escape from an unfamiliar environment (Price and Huck, 1976). The reduced activity of the adults after living in laboratory cages for 4 weeks may represent some combination of habituation to the laboratory environment and becoming conditioned to inactivity. The immaturity of the rats trapped as juveniles could have prolonged these processes and/or may reflect the relatively strong dispersal tendencies of young rodents. Failure to provide captive animals with the opportunity to explore (i.e. be active) may result in an apathy toward environmental stimuli and the development of atypical behaviors (Wood-Gush and Vestergaard, 1989).

Effect of Captive Rearing on Activity

Limiting the movement and physical activity of captive animals can have other nonadaptive or unexpected consequences. Price (1969) found that activity in tilt cages was greater for wild- and semi-domestic-genotype prairie deermice (*P. m. bairdii*) reared in relatively small laboratory cages ($13 \times 28 \times 15$ cm) than for wild-genotype mice reared either in nature (wild-trapped) or in relatively large laboratory cages ($51 \times 35 \times 30$ cm) (Fig. 17.1). Interestingly, confinement in small laboratory cages for 1 month prior to testing did not heighten the activity levels of the outdoor-reared subjects. The wild-caught (field-experienced) animals had access to a perceptually complex and sensory-unrestricted environment during early development, in contrast to the perceptually sterile and restricted conditions that the laboratory-reared subjects experienced.

It is not uncommon for animals reared in enriched environments to exhibit lower levels of locomotor and exploratory activity than those reared in restricted or unenriched environments (see Price, 1969, for references). Melzack (1968) interprets the hyperactivity of stimulus-deprived animals when placed in novel environments as resulting from 'inadequate stimulus filtering causing excessive central nervous system arousal'. Stress associated with the transition from field to laboratory could not account for the reduced activity of the field-reared subjects, since mice reared in large laboratory cages were also less active than the animals reared in small cages. Young and Cech (1993) demonstrated that handling stress (capture and confinement) was less severe (i.e. shorter duration) in captive striped bass (*Morone saxatilis*) that had been exercise conditioned than in unexercised fish. LaRue and Hoffman (1981) also discuss the importance of exercise in the hand-rearing of cranes. Putaala and Hissa (1995) reported that wild gray

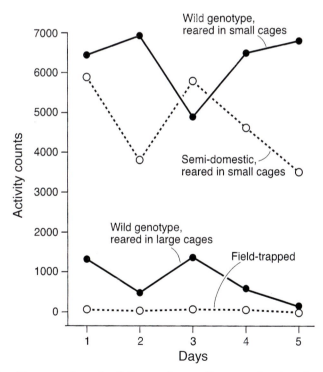

Fig. 17.1. Mean number of activity counts over 5 successive days for populations of wild-genotype prairie deermice reared either in nature, large (51 × 35 × 30 cm) laboratory cages or small (13 × 28 × 15 cm) laboratory cages and a population of semi-domestic-genotype deermice (approximately 20 generations removed from the wild) reared in small laboratory cages (Price, 1969).

partridges (*Perdix perdix*) had heavier hearts and higher glycogen and cytochrome-*c* oxidase activity in the pectoral muscles, suggesting that the wild birds had better flying endurance. Differences were attributed to lack of space in captivity for exercise.

Changes in Activity in Response to Food Deprivation

The spatial requirements of wild and domestic animals in captivity are often determined by fragmented knowledge of the species. Animals with large home ranges in nature are often assumed to require a relatively large area to roam in captivity. This may not be the case inasmuch as home-range size in nature may be dictated more by feeding behavior (i.e. search strategies) and the distribution and abundance of food items than by a need for locomotor stimulation or activity. Price (1976b) tested the hypothesis that wild Norway rats (*R. norvegicus*) would exhibit greater increases in wheel-running in response to food deprivation than domestic rats based on relaxed natural selection in captivity for behaviors that in nature would increase the likelihood of finding food. Domestic rats are typically fed *ad libitum* in small cages and, when they do experience food deprivation, increased activity confers no advantage in finding food. The study demonstrated that wild rats were more active in the wheels than domestic subjects, both in the presence or absence of food, and the proportional increase in wheel-running in response to food deprivation was similar in the two groups (Table 17.1). Wheel-running by the nondeprived control groups did not change. Apparently, increased activity in response to food deprivation is a very conservative trait, which is not readily diminished by domestication. In a similar study, Price (1969) reported that both captive-reared wild and semi-domestic deermice (*P. m. bairdii*) did not show an increase in wheel running in response to total water deprivation.

Table 17.1. Mean daily wheel-revolutions run by first-generation laboratory-reared wild and domestic (Sprague–Dawley strain) Norway rats before, during and after food-deprivation. (From Price, 1976b.)

	Wild	Domestic
Number of subjects	10	10
Pre- and post-deprivation (days 11–15 and 18–22)	1620	381
0–24 h food deprivation (day 16)	2080	521
Mean percentage increase in wheel running (day 16)[a]	28%	37%
24–48 h food deprivation (day 17)	3707	814
Mean percentage increase in wheel running (day 17)[a]	129%	114%

[a]Relative to mean daily pre- and post-deprivation score.

Behavioral Relevance of Space

The space provided in captive environments should be behaviorally relevant, that is, it should allow for the development and expression of a normal complement of basic behavior patterns. Both the quantity and quality (i.e. complexity) of space and opportunities for social interaction may be important. Animals isolated in relatively small enclosures often develop a variety of stereotyped behavior patterns (Fraser, 1968; Mason, 1991a; Lawrence and Rushen, 1993) because of social deprivation and limited perceptual and locomotor experiences (O'Neill *et al.*, 1991). Lagadic and Faure (1987) used operant conditioning techniques to determine preferences for cage size, feeder space and floor types in intensively housed domestic hens (*G. domesticus*). Barnett *et al.* (2001) review the extensive literature on housing systems for domestic pigs (*S. scrofa*). The review emphasizes the complexities associated with designing economically viable housing systems that are ideal for pigs. For example, the use of farrowing crates for sows on commercial pig farms has been criticized because they greatly restrict sow movement, thus jeopardizing her welfare. However, relative to open-pen housing of the sow, farrowing crates reduce piglet mortality resulting from physical trauma inflicted by the sow, thus improving piglet welfare. Weary *et al.* (1998) found that crushing by the sow accounted for 80% of the mortality of healthy piglets that died during the first 3 days postpartum. Edwards and Fraser (1997) reviewed much of the literature on this subject over a 30-year period and reported that the median piglet mortality (from all causes) for litters in farrowing crates was 12.7%, while mortality in more spacious open pens was 24.2%. In contrast, Weber (1997) and Cronin *et al.* (2000) only found a 2% difference in piglet mortality between conventional crates and alternative housing systems for individual sows when comparisons were made on the same farm. Interestingly, Kirkwood *et al.* (1987) reported a relatively high rate of perinatal mortality for the young of wild swine housed at the Zoological Society of London's Regent's Park; 21 of 91 (23%) wild piglets died during the first week following birth. Barnett *et al.* (2001) conclude that in Australia, piglet mortality in alternative farrowing systems must be equivalent to or better than the current industry average of 13.1% for these systems to be considered viable and recommended. In the mean time, the more space-limiting crates will be the industry's farrowing system of choice.

Use of vertical space

Hebert and Bard (2000) studied the use of vertical space by captive orangutans (*Pongo pygmaeus*) in a large indoor habitat equipped with four large molded trees, interwoven vines and a water-flooded floor resembling a continually flooded peat swamp forest. The unique design allowed for a range of species-typical behaviors varying by vertical level with a preference for the upper canopy, which contained many tree limbs for sitting and reclining. The orangutans spent little time (1%) on the flooded floor of the enclosure as planned for by the designers to showcase the arboreal nature of these animals. Reinhardt (2001) noted that the young of a hatchery-reared stock of juvenile Japanese masu salmon (*Oncorhyncus masou*) maintained in captivity for aquaculture purposes for more than 30 years stayed higher in the water column of their artificial stream tanks than fry from either wild or sea-ranched parents. When food pellets were offered from the water surface, the wild stock had to swim upward to intercept the slowly sinking food, while the farmed stock remained close to the surface to feed. Consequently, the farmed fish finished their meal much sooner than the wild fish. The young of sea-ranched salmon were intermediate to the wild and domestic stocks. Staying close to the surface is a successful feeding strategy in hatcheries, where food is traditionally delivered from above, and may have been inadvertently selected for in farmed fish. In natural streams, wild salmon obtain much of their food in the lower part of the water column where the risk of predation may be less than on the surface. Although all three stocks increased their surface orientation after several weeks of surface feeding, the wild stock persisted in remaining lower in the water column, suggesting a strong innate component to surface orientation.

Environmental Enrichment

Even small changes in the physical environment can have important effects on the behavior and welfare of captive animals. Kells *et al.* (2001) reported that young broiler chickens (*G. domesticus*) provided with bales of straw in their rearing houses would peck at the bales, perch on them and cluster around them. Birds with straw bales also walked and ran more in their pens and sat less than birds without bales. Henderson (1970) examined differences in the foraging behavior of domestic laboratory mice (*M. musculus*) representing six inbred strains and their 15 respective F1 hybrids after being reared in either standard or enriched laboratory cages (see p. 36 for an explanation of the rearing and test environments). Enriched animals completed the food-seeking task in faster time than mice reared in unenriched standard laboratory cages in all six pure inbred lines and 11 of the 15 (73%) hybrid lines. Lam *et al.* (1991) reported that when six adult individually-housed male cynomolgus monkeys (*Macaca fascicularis*) were provided a fleece fabric cushion ($0.2 \times 0.2 \times 0.6$ m) in their cages, they showed a significant reduction in time spent engaged in stereotyped behavior (53%, on average). When the fleece was sprinkled with familiar and novel titbits of foods, the monkeys spent a great deal of time foraging (up to 27 min h^{-1}) and rarely groomed the fleece. Grooming time was sixfold greater in monkeys given the fleece without crumbs. Grooming the fleecepad gave the appearance of being therapeutic; the monkeys appeared relaxed after grooming the fleece. In contrast, Line *et al.* (1991) found that passive enrichment devices such as sticks and toys were of limited effectiveness as enrichment objects for adult rhesus monkeys (*M. mulatta*) unless changed on a daily basis. Functionally relevant enrichment devices which promote naturalistic behaviors and which are actively responsive or increase social opportunities appear to be more effective in enriching the lives of captive primates (Markowitz and Line,

1989). They note that one macaque monkey (*M. mulatta*) pressed a control switch 130,000 times in 1 week to obtain food. Similarly, Sambrook and Buchanan-Smith (1997) have proposed that controllability (i.e. capacity of behavior to affect outcome probability) is more important than complexity in environmental enrichment because of its interactive properties. Inglis *et al.* (1997) review a number of publications showing that animals will work to obtain food when the same food is readily available to them in close proximity ('contrafreeloading' experiments). They argue that the contrafreeloading tendency is adaptive because animals willing to work to find food in nature will be more likely to survive when food sources become depleted. Sambrook and Buchanan-Smith (1997) further hypothesize that the acquisition of control is more enriching than the repeated exercise of acquired control. Once control is mastered over a very predictable outcome, there is no incentive for the animal to deviate from its behavior. In nature, animals benefit from learning about their environment, learning how to control their interactions with environmental stimuli, being able to predict environmental events and learning how to adapt to changing environmental circumstances. An element of novelty or unpredictability in these activities retains interest and responsiveness longer than when behaviors routinely result in a predictable outcome. In turn, effective enrichment devices must not only stimulate behavioral interactions but yield unpredictable outcomes to retain the animal's interest. Chamove (1989) suspended a rope between two adjacent cages of chimpanzees (*P. troglodytes*) so that when out-of-sight animals in one cage swung on it, the rope would 'spontaneously' move about in the other enclosure. The shared rope occupied the chimpanzees more than a predictable equivalent.

Enrichment enhances neurological development and function

The standard cage environment of laboratory rodents is relatively barren. Enrichment of the rearing environment of rodents induces a number of structural changes in the developing brain, including increased numbers of neurons, synapses and dendritic branches, especially in the cortex and hippocampus regions (see reviews by Renner and Rosenzweig, 1987; Rosenzweig and Bennett, 1996; van Praag *et al.*, 2000). These changes result in improved learning and memory.

Rodents often develop atypical stereotypic behaviors when reared in standard laboratory cages (Würbel, 2001). Behavioral and pharmacological evidence links cage stereotypies in rodents to housing-induced dysfunction of the basal ganglia, structures implicated in the initiation and sequencing of movements (Hauber, 1998).

Enrichment enhances productivity and reproductive success

Production animals (and the farmer) can likewise benefit from environmental enrichment. Beattie *et al.* (2000) compared the behavior, growth and meat quality of pigs (*S. scrofa*) reared from birth to slaughter in either barren or enriched environments. Barren environments were empty pens with only the minimum recommended space allowance. Enriched environments incorporated extra space, straw in a rack and an area that contained peat for the animals to root in. Enriched pigs spent less time inactive, less time involved in harmful social and aggressive behavior and more time exploring. Food intake was higher and feed conversion ratios lower for the enriched pigs, resulting in faster growth rates and heavier carcass weights at slaughter (21 weeks). Enriched pigs also had greater levels of backfat than their counterparts reared in barren environments and their meat was more tender.

Animals born and reared in confinement may react more favorably to such conditions than animals born and reared in relatively unrestricted environments and then confined later in life. This topic should be investigated more thoroughly since it has particular relevance to the animal welfare issue.

Carlstead and Shepherdson (1994) hypothesized that increasing the physical and temporal complexity of the living environment may improve the reproductive success of captive animals by increasing responsiveness to socio-sexual stimuli. They propose three mechanisms by which this may be achieved: (i) through developmental processes; (ii) modulation of stress and arousal; and (iii) modification of social interactions. In complex environments, animals more readily learn that their behaviors have functional outcomes. Exercising control over their living environment generalizes to novel situations and could influence their willingness to engage in reproductive behaviors and sexual performance. While a degree of acute stress may facilitate reproduction in some species (e.g. cheetahs) by establishing a needed level of arousal, chronic stress is normally detrimental to reproductive physiology and behavior. The authors hypothesize that increased environmental complexity can reduce stress by offering greater opportunity for behavioral coping responses (e.g. hiding, fleeing or diverting attention away from the stressor). Furthermore, they propose that increased environmental complexity could enhance reproductive success by stabilizing social relationships, reducing aggression and increasing affiliative and play behaviors. For example, scattering food in a woodchip litter substrate is known to reduce aggression among stumptailed macaques (*Macaca arctoides*; Anderson and Chamove, 1984) and chimpanzees (*P. troglodytes*; Bloomsmith *et al.*, 1988). While the authors' hypothesized effect of environmental enrichment on reproductive success seems reasonable, support is largely indirect and anecdotal at this time.

Space Effects on Social Organization

The social structure of animal populations may also change when space becomes limited (Calhoun, 1962; Lott, 1991). Sundell *et al.* (1998) proposed that the artificial confinement of juvenile salmonids at high densities in rearing tanks inhibits their natural territorial behaviors and imposes schooling behavior. Butler (1980) reported that the social organization of wild house mice (*M. musculus*) changes from territoriality to what resembles a dominance hierarchy if available space is significantly reduced. The inability of subordinate animals to escape from more dominant conspecifics may result in the formation of more highly polarized social hierarchies than would otherwise occur in nature. Sandnabba (1997) demonstrated an association between high levels of aggression in mice (*M. musculus*) and their motivation to enlarge their home range or territorial area and gain exclusive dominance. Nevison *et al.* (1999) suggested that enrichment (increasing complexity) of laboratory cages may increase a naturally selected territorial response in some strains of laboratory mice, as reflected in higher levels of testosterone, corticosterone and IgG titers (indicator of peripheral immune responsiveness).

Density-related Effects on Behavior

Animals are often housed at greater densities in captivity than in the wild, a fact that is frequently used to explain the higher incidence of agonistic interactions and social unrest among captive populations relative to their free-living counterparts. Höhn *et al.* (2000) described the more frequent agonistic interactions among male gray kangaroos (*Macropus giganteus*) in a 4.2-ha outdoor enclosure at the Neuwied Zoo in Germany than in a comparable free-living population in the Grampians National Park in Victoria, Australia. de Waal (1994) compared the behaviors of a captive population of chimpanzees (*P. troglodytes*) in their outdoor (summer) and indoor (winter) quarters in the Arnhem Zoo (The Netherlands). The 20-fold decrease in space during the winter resulted in increases in the frequency of aggression, submissive greetings and social grooming, while the incidence of social play declined. Severe aggression was not appreciably changed by the reduction in space, due largely to the appeasement behaviors and calming gestures frequently accompanying agonistic interactions. Hall (2001), working with production broiler chickens (*G. domesticus*), describes an experiment that compared the effect of two levels of stocking densities (34 and 40 kg m^{-2}) on broiler chicken welfare and behavior. The trial monitored 121,900 birds housed in eight flocks under commercial conditions. Behavioral data were collected on 384 birds between the ages of 23 and 37 days. At the higher density, the birds' resting and preening behaviors were reduced, largely due to disruption by other birds. In addition, walking (steps taken) and ground pecking decreased but time spent feeding and drinking, time spent preening and pecks directed at other birds were not affected. Mortality was greater for birds held at the higher stocking density, especially during the first week, and the incidence of leg problems, contact dermatitis and carcass bruising increased. It was concluded that the birds' welfare was adversely affected at the higher stocking density.

Structure of Environment Can Affect the Outcome of Competitive Interactions

Fleming and Einum (1997) compared the aggressiveness and dominance of wild-genotype Atlantic salmon (*S. salar*) to the aggressiveness of farmed salmon in a stream-like test environment and a tank environment typical of hatchery facilities. The farmed population was derived from the same river system (in Norway) as the wild population and was seven generations removed from the wild. Fish from both populations were reared under identical conditions in a hatchery environment. Wild fish dominated farmed salmon in the stream-like test environment in 21 of 30 contests, while farmed salmon dominated their wild counterparts in only three of 30 contests (six contests were unresolved). Conversely, the farmed salmon tended to dominate the wild fish in the tank environment, but the difference was not statistically significant. The fact that the expression of aggression and dominance in this study was context-specific may explain why reports on the relative dominance of wild and farmed fish in aggressive interactions have been equivocal (Ruzzante, 1994).

Environmental complexity can influence the social dynamics of captive populations. Stolba (1981) found that the social activities of fattening pigs (*S. scrofa*) increased as their living environment became more barren. Juvenile steelhead (*O. mykiss*) reared in an enriched environment containing a combination of in-water structure, overhead cover and underwater feeders became more dominant (i.e. were more competitive) and exhibited faster growth rates than juveniles reared conventionally with no in-water or overhead structure and hand-fed by scattering food across the water's surface (Berejikian *et al.*, 2000). This result confirmed other studies showing that manipulation of rearing parameters such as fish density and food presentation can affect competitive ability. The present study also showed that structural modifications of the hatchery-rearing environment, under conditions of equal ration, density and flow, can also affect the competitive ability of salmonids. In the enriched treatments, the defensible food resources (underwater food inlets) and visual isolation from conspecifics provided by submerged structure probably reduced the frequency of aggressive interactions (reduced 'intruder pressure') and facilitated the expression of territorial behaviors. Rearing fish in more naturalistic hatchery environments could result in fish that behave more like wild fish and which integrate more readily into natural stream environments when released in nature.

Substrate Preferences

In a study on substrate preferences (Peake, 1999), juvenile hatchery-reared lake sturgeon (*Acipenser fulvescens*) were given a choice of sand, gravel, rock or a smooth plastic bottom. All of the fish were strongly attracted to the sand substrate, a result previously described for free-living wild lake sturgeon (Kempinger, 1996). Preference of wild sturgeon for a sandy substrate has been associated with food acquisition (Kempinger, 1996) but in Peake's study, food rations were always distributed uniformly over the four substrate choices. Kellison *et al.* (2000) compared the behavioral responses of hatchery-reared and wild summer flounder (*P. dentatus*) to different substrates (sand and mud). Substrate type had no effect on time spent by the two stocks swimming in the water column, buried in the substrate or stationary (not buried). However, hatchery-reared fish spent significantly more time swimming in the water column than wild fish, regardless of substrate type. Swimming in the water column, a behavior which probably increases susceptibility to predation, may have been encouraged in hatchery-reared fish by them having been fed fish pellets dropped into the water from above (i.e. fish were rewarded for swimming to the surface). Such behavior may confer a selective advantage in the hatchery environment.

Substrate Effects on Aggressive Behavior

Mork *et al.* (1999) examined the effect of substrate (of the test environment) on the aggressive behaviors of wild and hatchery stocks of Atlantic salmon (*S. salar*). The wild stock were first-generation captive reared and the hatchery stock had been selected for growth rate for five generations. Data on aggressive behaviors were obtained for pure and mixed groups of salmon on fiberglass and for mixed groups on a divided substrate (fiberglass vs. river cobblestones). Each test group consisted of ten individuals; the mixed groups contained five wild and five hatchery fish. The investigators also collected data on the vertical distribution of the two strains of fish. Levels of aggression did not differ between the pure groups of wild and hatchery fish on the fiberglass substrate or in mixed groups on the divided substrate. In the mixed groups on fiberglass, the wild salmon were more aggressive toward farmed fish than other wild fish and they were more aggressive toward farmed fish than vice versa. Total aggression in mixed groups on fiberglass was greater than in mixed groups with a divided substrate. In addition, the farmed salmon tended to prefer the open water column in tanks with fiberglass substrate, while the wild fish tended to reside at the bottom, a result also observed by Olla *et al.* (1994). Increased pelagic swimming may be a result of selection for adaptation to the more open aquaculture environment, while wild fish may be more attracted to situations providing refuge (Olla *et al.*, 1994).

Light Intensity and Color Preferences

The physical aspects of the rearing environment can sometimes have important effects on measures of fitness. Tamazouzt *et al.* (2000) compared the relative effects of light intensity and tank wall color on the survival and growth of Eurasian perch larvae (*Perca fluviatilis*) for 15 days after hatching. Three light intensities (250, 400 and 800 lux) and four wall colors (black, dark gray, light gray and white) were used. The highest survival rates were found for the lowest light intensity (250 lux) and the light gray walls (17%). Growth rates were best in strongly illuminated tanks with either light gray or white walls, while the poorest growth rates were recorded in the black tanks with 250 lux illumination.

Conclusions

Space for locomotor behavior and exploration is typically limited in captivity and may represent a form of stimulus deprivation. Confinement often leads to the development of hyperactive behaviors when exposed to novelty and atypical stereotyped behaviors. Running wheels have been successfully used to compensate for the lack of space for laboratory rodents. In recent decades, animal caretakers have been keenly interested in enriching the physical environments of captive animals. Environmental enrichment has been found to enhance brain structure and function and to improve animal productivity and reproductive success. These benefits are attained through the positive effects of enrichment on: (i) developmental processes; (ii) modulation of stress and arousal; and (iii) the modification of social interactions. Enrichment research has focused on promoting naturalistic behaviors and enabling captive animals to acquire control over their environments. Enrichment techniques that allow animals to gain repeated control over aspects of their environment in ways that yield unpredictable outcomes appear to be more beneficial and sustaining than simple repetition of acquired control.

Limitations of space in captivity frequently preclude the expression of territoriality; populations may shift to the dominance hierarchy form of social organization. Social strife and disruption of daily behavior patterns associated with increased population density can jeopardize the welfare of animals. The frequency and outcome of competitive interactions can be influenced by the structure of the physical environment and environmental preferences.

Chapter 18

Behavioral Development in Captive Animals

The development of the domestic phenotype mirrors the domestication process. Genes and environment interact to facilitate the adaptation of captive animals to their environment. Genetic changes occurring over generations in captivity provide a continually evolving template for behavioral development, while the captive environment helps to shape and mold the final product. This chapter begins by discussing the topic of degeneracy. Are domestic animals degenerate or simply adapted to a different environment? Are certain behavioral patterns typically lost when animals are domesticated? Why do domesticated animals exhibit some behaviors more frequently than their wild counterparts? The chapter also addresses rates of development during domestication. Accelerated physical maturation and the retention of juvenile behavioral characteristics (neoteny) are given special attention. Examples are given to show how the social environment and certain aspects of the physical environment can influence the development of agonistic behavior. Evidence is presented to support the hypothesis that domestication has attenuated defensive more than offensive agonistic behaviors. This is followed by reports showing that domestication has buffered some species against genetic change and changes in the captive environment. The chapter ends by proposing the single greatest impact of domestication on behavior.

The Notion of Degeneracy

The belief that domestic animals are 'degenerate' is largely derived from reports that many stocks of domestic animals have lost their biological integrity or are incapable of surviving the rigors of life in the wild (e.g. Lockard, 1968). Darwin (1868) believed that in captivity biological traits are lost through disuse. Hemmer (1990) proposed that domestic animals have experienced a kind of regressive evolution, in which their behavior is less influenced by environmental stimuli than that of their wild counterparts. Lockard (1968), in speaking of the domestic Norway rat, stated: 'It is at least a waste of time, if not outright folly, to experiment upon the degenerate remains of what is available intact in other animals' (p. 739). Although it is true that intense artificial selection and modern husbandry practices may have produced populations of domestic animals with reduced potential for survival in nature (Vincent, 1960; Foley *et al.* 1971; Eibl-Eibesfeldt, 1975; Wood-Gush and Duncan, 1976), it is also true that most domestic animals do not live in nature and have inherited and/or acquired adaptations that facilitate adaptation to the captive environment. Whether or not domestic animals are con- sidered degenerate depends upon which environment, natural or captive, serves as the reference habitat (Boice, 1973, 1980). While it is reasonable to assume that most wild animals will probably have greater fitness than their domestic counterparts in nature, the converse is probably true in captivity (Kruska, 1988).

Supposed loss of behavioral patterns

Accompanying the degeneracy hypothesis is the claim that domestic animals have 'lost' many of the biologically adaptive traits of their wild ancestors (Kavanau, 1964). As stated earlier, artificial selection for large breast size in domestic turkeys (*M. gallopavo*) has made it very difficult for males to copulate naturally (Etches, 1996). Artificial selection of laying hens (*G. domesticus*) for nonbroody behavior has resulted in strains of chickens that normally do not incubate eggs or brood chicks. Although these and other relatively extreme examples of behavioral, physiological and structural changes in domestic animals imply that domestication has produced inferior animals, there is little evidence that domestication has resulted in the loss of behaviors from the species repertoire or that the basic structure of the motor patterns for such behaviors has been changed (Scott and Fuller, 1965; Hale, 1969; Miller, 1977). In nearly all cases, behavioral differences between wild and domestic stocks are quantitative rather than qualitative in character and best explained by differences in response thresholds. Likewise, the characteristic physiological traits and perceptual capabilities of most species remain highly stable under domestication (Kretchmer and Fox, 1975; Heaton, 1976). The most extreme differences are due to intense artificial selection.

Absence of key stimuli or experiences

The absence of certain key stimuli in the environment of captive animals can result in a failure to express certain behavioral patterns. Whereas burrows constructed by wild and domestic Norway rats (*R. norvegicus*) are similar in every respect, Boice (1977) noted that domestic Norway rats seldom initiate burrowing without an object (e.g. stone) to dig under. Wild rats were less dependent upon such stimulation. Similarly, Huck and Price (1976) reported that a wild stock of

Norway rats would exhibit climbing behavior even when denied early climbing experience; male domestic Norway rats would not climb unless they had early climbing experience. Hale and Schein (1962) reported that broodiness could be induced in female turkeys (*M. gallopavo*) of a 'nonbroody' strain by letting eggs accumulate in the nest.

A lack of key social experiences, particularly early in life, can influence adult behavior. Hannah and Brotman (1990) studied the tendency of captive female chimpanzees (*P. troglodytes*) to show inappropriate maternal behavior, particularly with their first-born infant. Following a hunch that abnormal maternal behavior may result from a lack of exposure to young infants, ten infant-naive female chimpanzees were given the opportunity to adopt unrelated infant chimpanzees or to observe lactating female chimpanzees with infants, or a combination of both, prior to giving birth themselves. A control group of eight pregnant infant-naive animals were denied these experiences. With the exception of one animal, all female chimpanzees given previous exposure to infants successfully reared their young. All eight females lacking previous experience with infants either handled them incorrectly or did not allow them to suckle, requiring the infants to be removed and hand-reared. Beck and Power (1988) found that female lowland gorillas (*Gorilla gorilla gorilla*) hand-reared by humans from birth are less likely to reproduce in captivity than mother-reared female gorillas (35% vs. 71%, respectively; Table 18.1). Lack of early social experience with other gorillas was believed to be the salient factor. Female gorillas without first-year social experience were less likely to copulate with males (73% vs. 83%), conceive (27% vs. 63%) and give birth to young (27% vs. 57%, respectively). Differences in conception rates and offspring production were statistically significant. Reproductively unsuccessful females were less likely than successful females to exhibit sexual solicitation behaviors and to copulate with males. Interestingly, there was no difference in the reproductive success of hand-reared and mother-reared males. Furthermore, the maternal competence of female gorillas that produced offspring

Table 18.1. Reproductive success of mother-reared and hand-reared gorillas born in captivity in North America. (From Beck and Power, 1988.)[a]

	Total number	Number (%) that reproduced	Reproductive years	Number of infants	Infants per reproductive year
Females					
Mother-reared	14	10 (71%)	57	17	0.30
Hand-reared	17	6 (35%)	127	14	0.11
Males					
Mother-reared	4	2 (50%)	5	2	0.40
Hand-reared	13	7 (54%)	68	25	0.37

[a]Data do not include partly mother-reared animals.

was not influenced by whether she was born in the wild or in captivity or whether she was hand-reared or mother-reared.

Animal management systems employed in captivity often limit opportunities for social learning. Cade and Fyfe (1978) reported that the one requirement most necessary for the development of full reproductive competence in captive-reared peregrine falcons (*Falco peregrinus*) is close association with siblings during the nestling period and possibly beyond. Relative to sibling interactions, being reared by adult falcons or human beings is relatively unimportant. Falcons reared in social isolation or with minimal contact with conspecifics early in life experienced the poorest reproductive success. Mellen (1992) demonstrated that hand-rearing of domestic cats (*Felis catus*) can be detrimental to their subsequent reproductive success. Twenty-one female kittens were divided equally into three treatment groups: (i) human-reared alone; (ii) human-reared with a single sibling; and (iii) mother-reared with a single sibling. Only two of seven human-reared-alone females successfully copulated as adults, compared to three of seven human-reared-sibling and six of seven mother-reared-sibling subjects. In addition, many of the cats in the human-reared-alone group were very aggressive toward human handlers and other cats, whereas the cats in the mother-reared treatment were calm and friendly toward human caretakers and appeared playful and calm with other cats. The human-reared-sibling animals were generally intermediate in behavior to the cats in the other treatments. Goldfoot (1977) found that rhesus monkeys (*M. mulatta*) require a complex social environment for normal repro-ductive behaviors to develop. Monkeys raised only with peers exhibited relatively high rates of play activity but were still deficient in sexual behavior. Goldfoot suggested that normal rearing eliminates excessive fear of peers and provides opportunities to develop affectional relationships (i.e. 'to learn to trust'). For rhesus monkeys, both mother and peers provide the needed early experience for the establishment of normal sexual relationships as adults. In Mellen's (1992) study, the human-reared-alone cats could be described as lacking 'trust' in both their human caretakers and the males with whom they were paired. Through inter-actions with mothers, kittens may learn not be fearful of other cats larger than themselves such as male sexual partners.

Coppinger and Coppinger (2001) discuss the fact that captive wolves (*C. lupus*) are better at problem-solving tasks and observational learning than dogs (*C. familiaris*) but have difficulty in learning operant conditioning tasks. In contrast, dogs readily learn to perform certain tasks for simple rewards. 'Dogs have been described as just smart enough to do a job and just dumb enough to do it' (Coppinger and Coppinger, 2001, p. 49). The superior performance of dogs in operant conditioning trials may reflect the high priority humans place on trainability and learning simple obedience-related behaviors, where food rewards or tactile stimulation provide the necessary incentives. In addition, dogs may be less stressed than wolves from handling and exposure to the artificial, highly structured circumstances typically associated with operant learning tasks. Mendl *et al.* (2001) discuss how chronic stress or very high levels of stress hormones, as well as novel events, can interfere with memory formation and retrieval in emotionally reactive animals. Frank and Frank (1982b, 1985) and Frank *et al.* (1989) argued that the decreased problem-solving ability of dogs, relative to wolves, was due to relaxed selection related to the 'buffering effect' of human care-takers. The argument is that people not only solve problems for dogs but can intervene when they make mistakes. Topál *et al.* (1997) examined a closely related hypothesis that the decreased problem-solving ability of domestic dogs was not due to inferior cognitive abilities resulting from domestication but rather their strong attachment to and dependence on their owners. He reached this con-clusion by observing 28 dog–owner pairs representing 14 breeds of dogs. Sixteen dog–owner pairs were described as having a 'companion relationship' (dogs were kept in the house as an integral part of the family) and the remaining 12 pairs were described as having a 'working relationship' (dogs were kept outside as a guard or for some other purpose). In the problem-solving task, the owner was asked to enter a 2×2 m fenced enclosure with the dog. The owner sat down behind the dog and remained motionless and silent. The experimenter was located just outside the enclosure opposite the dog and owner. Ten small food dishes containing meat were placed under the fence between the experimenter and the dog so that the dishes had to be pulled out from under the fence to get at the food. The handles on five of the dishes faced the experimenter and the other five faced the dog. At the start of the test, the experimenter pulled out a dish in his direction every 30 s and ate the food. After all of the experimenter's food had been eaten (2.5 min), the owner was asked to encourage the dog to get the food by manipulating the food dishes. After another 2.5 min, the test was terminated. During testing, the observer recorded the number of glances made by the dog toward its owner, the latency for the dog to first manipulate a food dish and the number of food items eaten with and without the owner's encouragement. Interestingly, dogs that glanced at their owners the most tended to start manipulation later and consumed a smaller number of food items. It was also noted that these same dogs followed their owners more closely in a second test, where the dog and owner were placed in an unfamiliar furnished room. Furthermore, dogs viewed anthropomorphically by the owner on an attitude questionnaire showed more dependent behavior in the experimental situations. It is as though they were waiting for help from their human companion. On the basis of these results, the authors concluded that the inferior problem-solving ability of dogs is not due to domestication but to their strong social attachment to humans. Serpell (1996) also has reported a relation-ship between the behavior of companion animals and owner attachment levels.

Enhancement of Behavior

Behaviors are not always attenuated by domestication; in some cases thresholds are lowered and frequencies of expression increased. Numerous claims have been made (e.g. Hale, 1969) that the sexual behavior of domestic animals is exagger-ated relative to that of their wild counterparts. Villavaso and McGovern (1986) reported that males of a laboratory strain of boll weevils (*A. g. grandis*) were more than twice as competitive as wild males for wild females and that laboratory males were nearly 3.5 times as attractive as wild males to wild females. Raina *et al.* (1989) found that males of a population of the corn earworm (*Heliothis zea*) maintained in the laboratory for over 120 generations were less discriminating than wild males in their response to female sex pheromone that had been experimentally modified. Domesticated males had also lost the requirement of low light intensity to respond to the pheromone. Künzl and Sachser (2000) demonstrated that male domestic guinea pigs (*C. porcellus*) courted females more frequently than their wild counterparts (*Cavia aperea*) maintained under identical conditions. These studies do not necessarily prove that domestic animals have more libido (sexual motivation) than wild conspecifics. Domestic animals may simply be less inhibited by the test conditions, unfamiliar animals or the presence of humans than their wild counterparts. The relatively high-density conditions in which many domestic populations are housed may facilitate hypersexuality because of the constant close proximity of potential sexual partners and the relatively large amount of 'discretionary time' available to them. Studies that have controlled for stimulus novelty have provided no evidence that domestic animals have higher libido than their wild counterparts (Estep *et al.*, 1975; Price, 1980).

Domestication enhances vocalization frequency in dogs and other animals

Barking by domestic dogs (*C. familiaris*) is much more frequent than by wild canids and is elicited in a greater variety of contexts (Cohen and Fox, 1976). Wolves (*C. lupus*) seldom bark in nature or in captivity (Scott and Fuller, 1965; Harrington and Mech, 1978). Schassburger (1987) notes that barks comprise only 2.3% of all wolf vocalizations. Wolf barks are lower pitched and less variable in structure and social context than the barks of dogs (Feddersen-Petersen, 2000). Some dog barks are low in sound frequency; some are very high.

Dogs bark in a number of contexts, such as alarm, food solicitation, in response to confinement, while playing and to get attention, while wolves bark primarily in the context of becoming alerted and in agonistic/territorial interactions (Feddersen-Petersen, 2000). The density of dogs is typically much greater and the territories smaller than those of wolves, offering more opportunity for vocal communication between animals and for barking to be socially facilitated (Fox, 1971).

Dogs can learn to bark to achieve some desired end result, such as to get attention or announce their presence to humans and other animals. Since dogs rely heavily on humans to fulfill many of their needs (e.g. food, exploration/ activity, access to and from house), soliciting information or attention may account for much of the barking by dogs (Yin, 2002). If people are relatively non-discriminating in the kind of barks they respond to, contextual cues may not be readily learned by dogs. Morton (cited in Coppinger and Feinstein, 1991) believes that dogs sometimes bark when they are experiencing indecision or conflicting motivational states (i.e. approach–avoidance). Coppinger and Feinstein (1991) describe a livestock-guarding dog that barked continuously for 7 hours 'with no other dog in sight or ear shot'. Could the persistent barking of dogs in certain situations be a form of out-of-context displacement activity or stereotyped behavior, serving no obvious function?

Domestic dogs may bark more than wolves due to various selective processes. Manwell and Baker (1984) proposed that humans selected barking dogs for their guarding ability. Feddersen-Petersen (2000) hypothesized that barking was an adaptation to living with man. Certain sounds were selected for human-related activities such as sport hunting and working. Coppinger and Feinstein (1991) make a case for the idea that the frequent barking of dogs in a large number of contexts is merely a product of pedomorphosis (i.e. neoteny). Up to a certain stage in development, young juvenile wolves and other wild canids frequently vocalize, including barking. They suggest that the persistent barking of dogs, relative to their wolf ancestors, reflects the fact that the behavioral development of the dog has become 'stuck' in adolescence. Put simply, barking is what juvenile canids do. Cohen and Fox (1976) proposed that dogs bark more than wolves because of relaxed selection for 'silence' necessary for wild predators to remain undetected by their prey.

Case (1999) notes that adult dogs also whine more than adult wolves. Whining may be a soliciting behavior retained into adulthood in domestic dogs to get attention from humans.

Rood (1972) noted that domestic guinea pigs (*C. porcellus*) are generally more vocal than their wild relatives (*C. aperea*). It is believed they were named guinea *pigs* because their squeals resemble those of domestic pigs (*S. scrofa*; Weir, 1974). Wood-Gush (1959) discussed selection by man for more frequent crowing in domestic chickens (*G. domesticus*). In the country of Turkey, cocks are selected for the length of each crow and contests are held to determine the longest continuous crow, some up to 30 s in length. The Russian Yurlov Crower and the German Bergische Kräher breeds of chickens are famous for their long crowing (Romanov and Weigend, 2001). Many ancient peoples were impressed by the crowing of the cock. In Roman times, the cock's crowing was believed to herald premonitions of future events.

Enhancement of reproduction

Hale (1969) and Setchell (1992) maintained that in a number of species, domestication has improved reproductive efficiency. Better nutrition, reduced energy expenditure and, in some cases, reduced stress in captivity have contributed to the improved reproductive performance of domesticated animals. With improved nutrition, farmed female red deer (*C. elaphus*) can reach puberty a year earlier than is typical in the wild (as yearlings rather than as 2-year olds; Hanlon, 1999). In the wild, lactating red deer hinds have about a 50% chance of calving the following year (Kay and Staines, 1981). With good nutrition, the conception rates of captive lactating females can exceed 85%. Cavalli *et al.* (2001) found that under adequate and stable rearing conditions, captive freshwater prawns (*M. rosenbergii*) were able to spawn up to five times during a 6-month period, while spawning in wild populations is typically three to five times per year (Rao, 1991).

Domestication has extended the breeding season for some species. Seasonal changes in testicular size and plasma concentrations of reproductive hormones have been reduced in domestic mallard ducks (*A. platyrhynchos*) relative to their wild counterparts (Haase and Donham, 1980). The breeding season of wild female mallards (*A. platyrhynchos*) is restricted to a few months in the spring whereas the breeding season of domestic strains is greatly extended (see also Prince *et al.*, 1970). Wild male mallards produce sperm from March until May or June, while domestic mallards produce some sperm in all months of the year (see review by Haase and Donham, 1980). Leopold (1944) reported that in free-living wild × domestic hybrid turkeys (*M. gallopavo*) the peak of hatching is 1 month earlier than in pure wild turkeys. Trut (1999) reported that 40 years of selection for tameness in silver foxes (*V. vulpes*) has resulted in some females breeding out of season, in November–December or April–May, rather than the normal late January to late March (in Siberia). Interestingly, no offspring born to an extraseasonal mating has survived to adulthood. In contrast, the estrous cycles of female dogs (*C. familiaris*) are normally 6–7 months apart without regard to season (Christie and Bell, 1971). Their wild ancestors, wolves (*C. lupus*), breed only once a year, in the winter months (Scott and Fuller, 1965; Mech, 1970) like the foxes in Siberia. In addition, the monogamous mating system characteristic of free-living wolves and the associated psychosocial inhibitions to breeding observed in wolf packs have been largely replaced by promiscuous sexual relations in domestic dogs (Scott and Fuller, 1965). The shift from monogamy to polygamy in domestic dogs may reflect man's intervention as a dominant figure in their social structure and the practice of artificial selection. Dogs that accept any mate chosen by man will have a selective advantage over dogs who are more selective in their mating proclivity. In addition, the parental care normally provided by (wolf) pack members in nature is no longer needed in captivity since man can provision the mother and readily intervene in the care of the pups.

Developmental Rates

Heterochrony refers to alterations or shifts in rate of development (Gould, 1977; Richardson, 1995) and can be conveniently categorized into three patterns: direct development (acceleration in somatic development), progenesis (acceleration in gonadal development) and neoteny (retardation in somatic development) (Wakahara, 1996). Developmental rates can be either accelerated or reduced by selective mechanisms (Cairns, 1976). It is not surprising, then, that many heterochronic shifts in development have been observed among domesticated animals.

Accelerated sexual maturation

Certain aspects of sexual maturation are accelerated in domestic animals. Trut (1999) reports that foxes (*V. vulpes*) selected for tameness for 40 years reach sexual maturity approximately 1 month earlier than nondomesticated foxes. Clark and Price (1981) demonstrated that two different strains of domestic female Norway rats (*R. norvegicus*) became fertile about 14 days earlier than female first-generation captive-reared wild rats reared under identical conditions (Table 18.2). Copulatory behavior was first evident in domestic males about 7–10 days earlier than in captive-reared wild males (Table 18.2). On the other hand, domestic and wild males did not differ in age at first appearance of spermatozoa in the testes. In nature, selection for early puberty in male rats may be limited because of the inability of very young males to compete successfully with older, more experienced males for estrous females (Steiniger, 1950). In the laboratory, a single male is usually paired with one or more females of similar age, and often at weaning, thus eliminating male–male competition for females. Under these conditions,

Table 18.2. Mean (± SD) values for measures of sexual maturation and fecundity in male and female first-generation captive-reared wild and two strains of domestic Norway rats. (From Clark and Price, 1981.)

Variable	Stock	Number of rats	Age (days)	Body weight (g)
Males				
Mature spermatozoa	Wild	24	42.1 ± 1.9	111.5 ± 17.3
first obtained	Long–Evans	21	42.7 ± 2.8	156.4 ± 14.9
	Sprague–Dawley	27	41.7 ± 1.7	178.0 ± 26.4
First copulation	Wild	23	64.6 ± 20.5	185.5 ± 62.8
	Long–Evans	19	58.7 ± 3.2	258.4 ± 27.4
	Sprague–Dawley	24	53.9 ± 5.5	275.3 ± 42.6
Females				
Vaginal opening	Wild	26	42.5 ± 5.7	81.1 ± 9.2
	Long–Evans	25	37.1 ± 2.9	105.9 ± 13.0
	Sprague–Dawley	26	37.5 ± 2.7	126.0 ± 15.5
First estrus	Wild	26	45.7 ± 6.9	86.6 ± 9.6
	Long–Evans	25	37.9 ± 3.3	110.7 ± 16.5
	Sprague–Dawley	26	37.8 ± 2.8	127.7 ± 16.2
First conception	Wild	21	55.7 ± 22.6	103.2 ± 23.3
	Long–Evans	25	41.3 ± 2.9	125.9 ± 17.6
	Sprague–Dawley	25	42.0 ± 4.5	151.8 ± 29.3
Length of first estrus	Wild	11	4.5 ± 1.0	
cycle (days)	Long–Evans	9	4.7 ± 0.9	
	Sprague–Dawley	12	4.3 ± 1.7	
Litter size at first	Wild	20	6.4 ± 1.0	
parturition	Long–Evans	25	9.7 ± 1.8	
	Sprague–Dawley	24	10.4 ± 2.8	

early maturation of copulatory function would confer an important selective advantage for the domestic male rat.

A similar trend has been noted in wild and domestic fowl. Kawahara (1972) indicated that the age at first egg for wild Japanese quail (*C. japonica*) bred in laboratory cages for the first three generations averaged 110 ± 2.7, 82 ± 2.3 and 81 ± 1.5 days, respectively. Domesticated quail (controls) averaged 50 ± 0.3 days.

In some cases, reports that domestic animals reach puberty earlier than their wild counterparts may be based on the absence of socially dominant conspecifics or environmental factors (e.g. photoperiod), which inhibit reproductive development and mating. In spite of reports that wolves do not reach puberty until nearly 2 years of age (Mech, 1970; Asa and Valdespino, 1998), evidence is accumulating (e.g. Haase, 2000) that wolves (*C. lupus*) hand-reared in captivity can reach sexual maturity by 9 months of age, which is within the age range reported for most breeds of domestic dogs (*C. familiaris*; Scott and Fuller, 1965; Fox, 1978). Leopold (1944) reported that first-year wild turkeys (*M. gallopavo*) seldom breed in nature whereas first-year domestic turkeys readily engage in mating. Bauer and Glutz von Blotzheim (1968, cited in Haase and Donham, 1980) noted that greylag geese (*Anser anser*), the ancestor of most domestic geese, pair during their second fall but do not become sexually mature until their third spring (2 years of age), while domestic geese start to breed in their first year. Controlled studies are needed to confirm that these differences in the timing of puberty are genetically based rather than a result of social or climatic factors.

Accelerated growth and maturation rates

Growth rates of captive animals are typically greater than for their wild counterparts. Improved nutrition and reduced activity in captivity directs proportionately more energy toward growth-related phenomena. Faster growth rates are often associated with undesirable characteristics. For example, the mobility of captive animals may be impaired. Leg abnormalities develop in hand-reared whooping cranes (*G. americana*) when chicks gain more than 20% of their body weight per day (Kepler, 1978). Food is routinely withheld from such fast-growing chicks. Thorpe (1991) observed that faster growth rates of cultivated fish relative to wild fish (a desired objective) are often accompanied by accelerated maturation, which is undesirable in the fisheries industry. Crozier (1998) demonstrated that faster-maturing Atlantic salmon (*S. salar*) were genetically distinct from their parent population. Eggs were collected from a wild population and resulting fry were reared in either a hatchery or the wild. Allozyme variation at five polymorphic loci were examined in both populations over time. No genetic changes were found in the wild population over time or with respect to the parent population. However, the fish that smolted in their first year in the hatchery were genetically different from the parent population. Fish smolting in the hatchery in year 2 did not differ from the parent population. Most hatchery operations select for large size (growth rate). By so doing, selection for smolting in year 1 (early maturation) is indirectly practiced. Hatchery managers can decelerate maturation by reducing food intake to deplete fat stores by the end of winter of their first year (Thorpe *et al.*, 1990; Rowe *et al.*, 1991) or by selection for late maturity (Thorpe, 1991). In livestock operations, the developmental events associated with maturation in males are frequently controlled by castration, permitting growth rates to be maximized without a corresponding effect on maturation rate.

Accelerated behavioral development

Behavioral development in captivity may be influenced by early experiences related to the physical characteristics of the rearing environment. Clark and Galef (1982) demonstrated that maturation in the relative darkness of a burrowlike shelter delayed the expression of adult-like patterns of exploration in the Mongolian gerbil (*M. unguiculatus*) in the days following eye-opening. Animals of comparable age and development reared in standard open laboratory cages were more active in the test enclosure. In nature, it may be maladaptive for young rodents to leave their natal burrows and disperse too quickly (Daly, 1973). Prolonging the onset of sensitivity to visual stimuli by rearing in a dark environment is the norm for this species in nature.

Retarded development

Whereas, in general, the development of reproductive capabilities has been accelerated by domestication, other developmental processes have been prolonged. Miller and Gottlieb (1981) reported that domestication has decelerated the rate of development of behavioral arousal in the newly hatched domestic mallard (Pekin) duck (*A. p. domestica*). The sensitive period of primary socialization was extended in a population of silver fox (*V. vulpes*) selected for tameability toward humans (Belyaev *et al.*, 1984/1985; Trut, 1999). The window of social bonding starts 1–2 days earlier and ends about a month later in the selected foxes relative to unselected farmed-reared controls (Fig. 18.1).

Developmental events are often coupled. Sundell *et al.* (1998) reported that hatchery rearing of brown trout (*S. trutta*) suppressed the physiological and morphological changes normally associated with smoltification (parr–smolt transformation), maturational changes associated with their first migration to the sea. In addition, the migration tendency of the hatchery-reared fish was weak. In one river studied, less than 2% of the hatchery-reared fish migrated compared with an estimated 60% of the wild trout.

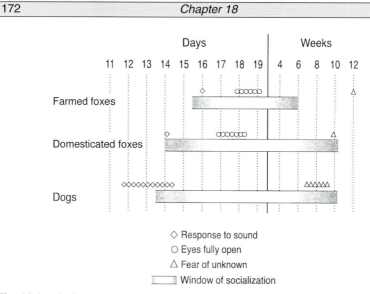

Fig. 18.1. In foxes bred for tameability (i.e. domesticated), the window for socialization begins earlier and ends later than for farmed foxes not subjected to selection and is similar in duration to the sensitive period for socialization in domestic dogs (Trut, 1999).

Neoteny

The retention of juvenile characteristics into adulthood has also been cited as an effect of domestication (Fox, 1968). This process has been referred to as neoteny, a type of pedomorphosis (Sheldon, 1993; Goodwin *et al.*, 1997). Neoteny can affect the whole developmental process (global effects) or be restricted to certain developmental events (local effects), and more than one process may be affected simultaneously (mosaic heterochrony; Sheldon, 1993). Domestic dog breeds (*C. familiaris*) exhibit considerable morphological diversity, much of it thought to be the result of both global and local pedomorphosis (Goodwin *et al.*, 1997). There are reports in the literature (Dechambre, 1949; Clutton-Brock, 1981; Morey, 1994; Trut, 1999) that certain anatomical characteristics of the juvenile wolf (*C. lupus*) (e.g. shortening of the jaws and facial region) are retained into adulthood in the dog, perhaps as a result of selection to preserve the greater esthetic appeal of the young animal. However, Wayne (1986) found that the ratio of palate to overall skull length was the same in adult wolves, dogs and coyotes, in spite of sizable differences between neonates and adults in all three groups. Dogs do not retain the juvenile skull dimensions of wolf puppies. Rather, dogs' heads look shorter because they are wider than wolves. Skull width is not a pedomorphic trait (Coppinger and Coppinger, 2001). The overall reduction in growth in some dog breeds appears to be the result of progenesis (earlier completion of development) in addition to neoteny, since small breeds achieve adult weight earlier than large breeds, as well as grow more slowly (Goodwin *et al.*, 1997).

Coppinger and Smith (1983) proposed that behavioral neoteny was selected for in the domestication of the wolf (*C. lupus*) and is a prerequisite to its successful domestication. Their argument is based on the assumption that tameability is necessarily linked to neoteny (i.e. that neoteny is a prerequisite for the development of a placid temperament in the adult animal). The idea that behavioral neoteny may accompany domestication is further supported by the fact that traditional animal management practices allow for the retention of juvenile social behaviors by reducing the selective advantages of aggressive potential in securing needed resources. In captivity, where food and water are normally provided and mating is largely controlled by humans, survival and reproductive success are not necessarily contingent on the attainment of high social status. Selection for early sexual maturation, before the animal's aggressive potential is fully developed, is also made possible by mating and management systems, which largely eliminate competition for the opportunity to breed. Multi-sire mating programs are the exception rather than the rule for most of our common domestic animals (see Table 9.2).

An alternative to the selection hypothesis is that behavioral neoteny in domesticated animals may be environmentally induced. First, young animals may be conditioned to retain their juvenile behaviors. Offering positive rewards for juvenile behaviors (e.g. care-soliciting, playfulness, submissiveness to humans) may, in effect, retard the development of more independent adult activities, or alternatively, may mask their expression. This latter effect could be achieved by raising the threshold for elicitation of adult behaviors or by sufficiently lowering the threshold for juvenile behaviors to allow for the continued expression of many juvenile characteristics throughout adulthood. Rearing animals in physical isolation from older, socially dominating conspecifics may further discourage the development of normal adult-like patterns of agonistic behaviors (Price, 1978) and facilitate the retention of many juvenile social behaviors. The proclivity of dogs (*C. familiaris*) to socialize to humans and the sensitivity of dogs to their social environment (Ginsburg and Hiestand, 1992) may facilitate the retention of juvenile social behaviors in this species. The intense submissive and conciliatory behaviors exhibited by dogs toward human social companions represent a most important behavioral adaptation to living with man. It is also important to note that intense submissive behaviors are an important adaptation of subordinate wolves living in packs.

Behavioral neoteny in dogs

It is commonly believed that the social behaviors of the adult domestic dog (*C. familiaris*) are relatively juvenile compared to the wolf (*C. lupus*), its wild ancestor (Fox, 1978; Frank and Frank, 1982a; Coppinger *et al.*, 1987; Ginsburg and Hiestand, 1992; Morey, 1994). A case can be made that neoteny or pedomorphosis may characterize the behavioral development of the dog because of selection for tractability or the capacity to be closely controlled by humans, a trait frequently associated with juvenile animals. Coppinger *et al.* (1987) have pointed out that behaviors commonly associated with the adult wolf may be lacking in different degrees in various dog breeds. Dog breeds selected for herding livestock exhibit the predatory behaviors of eye/stalk/chase/(sometimes) bite sequence but refrain from attack. Dog breeds selected for protecting livestock (from predators) lack even these predatory behaviors and will sometimes engage them (livestock) in play. The ease with which most dogs can be trained to perform various tasks may reflect a high degree of plasticity relative to the more structured (i.e. less variable) adult-like behavior of their wild relatives. Adult dogs engage in play behaviors more frequently than adult wolves (Case, 1999).

A similar 'mesomorphic remodeling theory' proposes that the dog becomes reproductive during what is the juvenile or mesomorphic period of its wolf-like ancestor and that there are traits present in the dog not found in either wolf pups or adult wolves (Coppinger and Schneider, 1995). 'Dogs use bits of behavior, mix them or delete them in ways not available to the adult wolf.' This theory is attractive because the juvenile period represents an unstable time in the life of the animal, when it shows considerable flexibility and is highly responsive to learning, traits that to a great extent characterize the domestic dog.

The behavior of the dog provides the most compelling case for behavioral pedomorphosis accompanying the domestication process. Even so, more work is needed to determine the extent to which neoteny or the 'mesomorphic remodeling theory' best describes the behavioral development of the dog and the extent to which their 'juvenile' behaviors represent developmental processes controlled by their rearing environment. It is also important to determine the extent to which changes in response thresholds may explain the behavioral differences between wolves and dogs. In actuality, all of these processes may be involved to varying degrees in different dog breeds.

It is not clear that behavioral pedomorphosis has had a significant role in the domestication of any other species. From this standpoint, it is difficult to argue that neoteny is a unifying principle associated with domestication. For many species (e.g. livestock) the retention of juvenile characteristics may be more apparent than real. Species raised for meat are normally slaughtered before reaching maturity; adult behavior patterns are seldom observed except in breeding stock (Beilharz and Zeeb, 1982).

Interaction of Genes and Environment

Both genetic and experiential contributions to behavioral development are illustrated by Wecker's (1963) account of the development of habitat preferences in wild and semi-domestic prairie deermice (*P. m. bairdii*). Wild-genotype deermice preferred a field over a woodland environment, even when reared in a woodland enclosure or born and reared in the laboratory. A semi-domesticated stock of the same subspecies, originally trapped in the same area and approximately 12–20 generations removed from the wild, preferred the field habitat only if given early experience in a small field enclosure. After 15 years of laboratory breeding, the semi-domestic stock had 'lost' the innate tendency to prefer field over woodland stimuli (presumably through relaxed selection) but it had not lost the psychological bias toward a grassland habitat once this preference was reinforced by early field experience. It is assumed that in the evolution of habitat preferences, prairie deermice were originally selected for sensitivity to field stimuli, whereby early exposure led to a field preference (i.e. a kind of 'habitat imprinting'). Further selection on the most field-sensitive members of the population could have led to an innate preference for field habitat through a process of 'genetic assimilation' (Waddington, 1961). Wecker speculated that the loss of the innate field preference resulted from 12–20 generations of relaxed selection in captivity and represents 'genetic assimilation in reverse'.

Development of Agonistic Behavior

Social experience and physical aspects of the rearing environment can play a surprisingly important role in the development and expression of agonistic behavior in captive animals (Lore and Flannelly, 1981; Adams and Boice, 1989). Banks and Popham (1975) found wild-caught brown lemmings (*Lemmus trimucronatus*) dominant over lemmings conceived and reared in the laboratory. Price (1978) compared the agonistic behavior of wild-caught and first-generation laboratory-reared wild male Norway rats (*R. norvegicus*) in a resident–intruder context designed to encourage the expression of territorial behaviors. Experimental subjects were housed in 122 × 30.5 × 33 cm enclosures for at least 1 month prior to their exposure to hybrid (wild × domestic) intruders. The field-trapped animals were highly defensive in this context and were significantly more aggressive toward the intruders than their laboratory-reared wild counterparts (Table 18.3). Novelty of the laboratory environment was an unlikely explanation for this

Table 18.3. Aggressive behaviors displayed by field-trapped and first-generation laboratory-reared wild Norway rats toward wild × domestic intruders. (From Price, 1978.)

Variable	Field-trapped	Laboratory-reared
(a) Number (%) of trials (*n* = 16) in which certain aggressive behaviors were exhibited by experimental subjects		
Broadside displays	11 (69%)	2 (12%)
Attacks	11 (69%)	3 (19%)
Fights	9 (56%)	4 (25%)
(b) Mean frequency of aggressive behaviors per test trial		
Broadside displays	5.2	0.1
Attacks	2.2	0.2
Fights	2.0	0.4

difference, since the wild-caught rats had been in captivity for a minimum of 3.5 months prior to testing and both stocks were placed in the test chambers a minimum of 1 month before exposure to intruders. A more parsimonious explanation was the difference in early social experience. It was hypothesized that social interactions of the field-trapped rats prior to capture allowed for the full development of their aggressive potential and when denied the option of escape in the test chambers, well-developed defensive behaviors were exhibited. In contrast, wild rats born and reared in the laboratory were probably denied certain social experiences critical for the full development of their aggressive potential. Relative to their free-living wild counterparts, the frequency and intensity of aggressive interactions is much reduced for rodents reared from a very young age in small stable social groups (often littermates) in laboratory cages (Barnett *et al.*, 1979; Lore and Flannelly, 1981; Nikoletseas and Lore, 1981; Boice and Adams, 1983). Exposure to strange conspecifics is a rare event and when it occurs, investigatory or fear responses are more likely evoked than aggressive behaviors. This phenomenon was illustrated in a study by Boreman and Price (1972) comparing the agonistic behaviors of captive-reared wild, hybrid and domestic Norway rats (*R. norvegicus*). Wild rats were the first-generation captive-reared offspring of wild-caught adults and the domestic stock was derived from cross-breeding four different strains of laboratory rats. Four animals from each of the three stocks (12 animals total) were placed together and observed for spontaneous and competitive interactions in a large 7.6 × 3.6 × 2.7 m unfamiliar room over a 13-day period. Four replications were conducted. The most obvious difference observed between the wild and domestic rats was the uninhibited manner in which domestic rats initiated social interactions. This behavior was in sharp contrast to the highly inhibited nature of the wild stock. Spontaneous interactions initiated by the domestic rats had the appearance of being play-motivated and their uninhibited approach to wild rats resembled a form of 'bullying'. In this context, the wild rats were clearly dominated by their domestic counterparts (Table 18.4). A similar study (Price *et al.*, 1976), in which food-deprived wild (captive-reared) and domestic rats (*R. norvegicus*) competed for food in cross-strain round-robin pairings, yielded similar results. Domestic subjects spent more time eating and were dominant over wild rats in 23 of 36 trials in which dominance could be determined on the basis of

Table 18.4. Mean (± SD) percentage cross-strain interactions won in spontaneous and competitive interactions between wild, domestic and wild × domestic hybrid Norway rats. (From Boreman and Price, 1972.)

Replication	Spontaneous interactions			Competitive interactions		
	Wild	Hybrid	Domestic	Wild	Hybrid	Domestic
1	0.29 ± 0.20	0.59 ± 0.31	0.79 ± 0.13	0.32 ± 0.11	0.56 ± 0.06	0.48 ± 0.15
2	0.25 ± 0.21	0.58 ± 0.17	0.67 ± 0.10	0.29 ± 0.06	0.36 ± 0.19	0.70 ± 0.13
3	0.27 ± 0.05	0.64 ± 0.30	0.51 ± 0.47	0.39 ± 0.05	0.38 ± 0.11	0.68 ± 0.02
4	0.17 ± 0.29	0.27 ± 0.47	0.85 ± 0.31	0.35 ± 0.19	0.42 ± 0.05	0.62 ± 0.03

fights won (or lost), and successful and unsuccessful takeovers of the food stimulus (slice of apple). Again, the lack of social inhibitions of the domestic rats provided the best explanation for their competitive advantage. The socially uninhibited nature of the domestic rats predisposed them to dominate the more inhibited wild subjects from the start. Wild rats responded to unfamiliar conspecifics with the same neophobia they typically exhibited toward strange inanimate objects or changes in diet. In addition, when wild rats experienced a decisive defeat (usually by a domestic competitor), they were noticeably more inhibited and submissive in the next competition trial. Domestic rats defeated in competition would recover more quickly and were not necessarily more submissive in subsequent trials. Because the reduced emotional reactivity of domestic Norway rats pervades so many different behavioral traits, it appears to be the most important single change in their behavior during the domestication process.

If field-reared (wild-caught) and domestic Norway rats both dominate laboratory-reared wild rats, how does the aggressive potential of these two groups compare? It is known that rearing domestic laboratory rats in groups in outdoor enclosures enhances their aggressive nature (Nikoletseas and Lore, 1981; Boice and Adams, 1983; Adams and Boice, 1989). Perhaps the instability of the social environment under these circumstances and/or the experience of defending nest sites from intruders results in lowering their threshold for aggressive behavior. In addition, there are other circumstances in which the aggressive behaviors of domestic rats can be just as intense as for their wild-caught counterparts (Blanchard *et al.*, 1975; Adams, 1976; Takahashi and Blanchard, 1982). Clearly, the developmental plasticity of this species may explain its successful status in both nature and the laboratory.

Attenuation of Defensive Behaviors

As noted in the previous section, the uninhibited manner in which domestic rats engage in social interactions stands in sharp contrast to the naturally more cautious and ambivalent demeanor of their wild counterparts (Boreman and Price, 1972; Price, 1978). This observation suggests that it is the defensive behaviors of the domestic rat that have become attenuated by the domestication process (Blanchard *et al.*, 1986). Selection for reduced defensive behaviors may represent an adaptation to minimize stress when living in an environment where escape from human handlers and aggressive conspecifics is not possible. Associated with changes in defensive behaviors, there appears to be a corresponding decrease in the frequency and intensity of submissive behaviors. The 'loss of social inhibitions' ascribed to domestic animals by Lorenz (1965, p. 94) reflects higher thresholds for defensive aggression. In a study comparing the social interactions of wild and domestic house mice (*M. musculus*), Smith *et al.* (1994) found that social encounters were more frequent among domestic mice than their wild counterparts; subordinate domestic mice were more likely to interact with dominant conspecifics. Stahnke (1986) found that domestic guinea pigs (*C. porcellus*) showed less agonistic behavior patterns towards other members of their social group than wild guinea pigs (*C. aperea*) and they engaged in more amicable social relationships. In guinea pigs, the vocalization pattern 'purr' functions as an important stabilizing pattern used mainly by the alpha male. This pattern was used frequently by domestic alpha males but was seldom heard in the wild cavies. Künzl and Sachser (2000) also found domestic cavies to be less aggressive than their wild counterparts and related the reduced reactivity of domestic guinea pigs to decreased sensitivity of their pituitary–adrenocortical and sympathetic–adrenomedullary systems (i.e. lower serum cortisol, adrenaline and noradrenaline titers).

Domestication Effects on Genotype–Environment Interactions

Domestication can affect the extent to which developmental events are canalized or buffered against both genetic changes and changes in the animal's environment. The relative effects of genotypes and environments may vary. Devlin *et al.* (2001) demonstrated that a growth-hormone transgene increased the size of wild but not domesticated rainbow trout (*O. mykiss*). Trout eggs from a slow-growing wild stock and a domesticated stock selected for growth rate were injected with a salmon gene construct overexpressing growth hormone (construct *OnMTGH1*). By 14 months of age, the transgenic wild trout were 17.3 times heavier than their nontransgenic controls (167.6 ± 14.9 g vs. 9.7 ± 0.6 g, respectively) whereas the transgene did not enhance the growth of the domestic stock (282.7 ± 26.2 g vs. 269.4 ± 5.9 g, respectively, for the transgenic and nontransgenic fish). Likewise, wild trout treated with exogenous growth hormone protein grew 2.7 times faster than untreated controls whereas treated domestic trout displayed a modest increase of only 9% (McLean *et al.*, 1997). It was concluded that previous genetic selection for a trait may limit the extent to which the trait can be enhanced by genetic manipulation. The domestic trout had been previously selected for rapid growth and were less sensitive to further attempts to increase body size.

Domesticated rodents may adapt more readily to a relatively small, impoverished environment (e.g. laboratory cages) than their wild counterparts. Huck and Price (1975) found that growth rates and the development of behavior in a stock of domestic laboratory rats (*R. norvegicus*) were more highly buffered against perceptual and locomotor deficits than growth and behavior in a stock of first-generation laboratory-born wild Norway rats. Rats of both stocks were reared in either empty or enriched 0.61 × 0.61 × 0.91 m cages. Enriched cages were partitioned into three horizontal tiers connected by wooden ramps and contained the following items: a 35.6-cm activity wheel, two large nest boxes with nesting material, a 30.4 × 30.4 m wooden maze, a teeter-totter, small blocks of wood, marbles, golf balls, three bells suspended by chains, a 0.61-m climbing chain, a horizontal 0.61-m-long rod and a 12.7 × 17.7 cm mirror attached to one wall. At about 100 days of age, the rats were weighed and tested for open-field behaviors. The scores of enriched and unenriched wild rats were more divergent than the scores of enriched and unenriched domestic rats for nine of ten variables, with open-field defecation being the only exception. Statistically significant interactions were obtained between genotype and rearing environment in each case; see Fig. 18.2 for several examples. A climbing behavior test administered after the open-field test (Huck and Price, 1976) revealed a similar result. The climbing behaviors of enriched and unenriched wild rats were more divergent than those of enriched and unenriched domestic groups. Although it was not clear if the domestic rats were less responsive to experiences gained in the enriched environments or if they simply had fewer experiences, the results support the general conclusion that, of

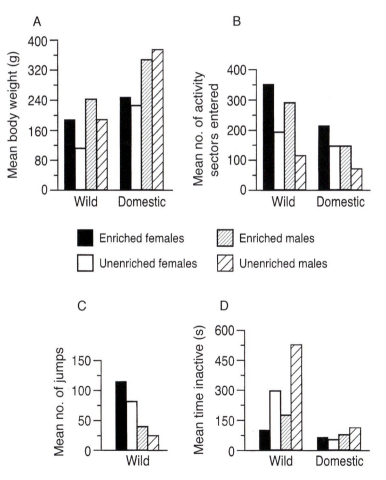

Fig. 18.2. Mean body weights and open-field behaviors for wild and domestic Norway rats reared in enriched and unenriched cages: (A) body weight at 95 ± 5 days of age, (B) total number of activity sectors entered, (C) frequency of jumping (only wild rats jumped), and (D) time inactive. Open-field tests were 15 min in length and were administered on 5 consecutive days (Huck and Price, 1975).

the two strains tested, the physical and behavioral development of the laboratory rat were more highly buffered against changes in its living environment.

Reduced Emotional Reactivity: the Single Most Important Effect of Domestication on Behavior

The information presented in this and preceding chapters points to the conclusion that the single most important effect of domestication on behavior is reduced emotional reactivity or responsiveness to fear-evoking stimuli (i.e. environmental change). This characteristic is observed in virtually all populations of domestic animals and pervades a wide variety of behavioral responses to both the social and physical environments (e.g. intraspecific social interactions, reactions to the presence of humans, responses to novel objects and places). Reduced responsiveness to fear-evoking stimuli is seen as an adaptation to living in a biologically 'safe', predator-free environment with: (i) limited opportunities for perceptual and locomotor stimulation; (ii) frequent invasions of personal space, with little opportunity to escape from dominant conspecifics; and (iii) frequent association with humans, who are prone to cull untamed and intractable individuals. Available information supports the hypothesis that individuals less reactive to fear-evoking stimuli experience reduced levels of stress in captivity, greater reproductive success, greater productivity (e.g. growth rate, animal products) and are handled by humans with greater ease. For example, Malmkvist and Hansen (2001) review selection experiments demonstrating that mink (*M. vison*) selected for reduced fear of humans also show less fear of novel objects, less fear of unfamiliar mink placed in their cage and increased willingness to mate, all traits associated with the domestic phenotype. It is not surprising that one biological trait can be so important to the domestication process. Consider the importance of emotional reactivity to the fitness of animals living in nature.

Conclusions

If one believes that animal degeneracy must be measured against the environment in which the animal typically resides, one cannot make a good case for the notion that domestic animals are degenerate. They are simply adapted to a different environment than their wild counterparts. Behaviors thought to be lost during domestication are usually expressed when the right key stimuli or experiences are provided. In nearly all cases, behavioral differences between wild and domestic stocks are quantitative rather than qualitative in character, and best explained by differences in response thresholds. Sexual and certain communicatory behaviors are exhibited more frequently in captivity than in the wild, perhaps because of artificial selection or the close quarters in which captive animals typically live and the 'extra' time available for such activities. Once a species has become adapted to its captive environment, reproductive and growth rates are typically enhanced through improved nutrition, reduced energy expenditure and, in some cases, reduced stress in captivity. Sexual maturation may have been accelerated in males of many species by the long-term effect of captive breeding systems that have eliminated male–male competition for access to sexually receptive females. In other cases, maturation may have been retarded during domestication. The domestic dog appears to have retained into adulthood many of the juvenile traits of its wolf-like ancestor. Such a compelling case for neoteny has not been found in other domesticated species.

Response thresholds for agonistic behavior in domesticated rodents are readily modified by early social experience and certain features of the captive environment. Response thresholds for defensive agonistic behaviors have been raised during domestication, resulting in animals that exhibit fewer submissive behaviors or social inhibitions than their wild counterparts. Lastly, evidence points to the fact that the single most important result of domestication on animal behavior is a reduction in emotional reactivity. This effect is seen in all populations of domesticated animals and pervades a variety of behaviors (e.g. social interactions, response to humans, response to novel objects and places) characterizing the domestic phenotype and enhancing economically important traits, such as growth rate.

Chapter 19

Reintroductions and Feralization

There is a long-standing tradition of releasing captive-reared wild animals in nature for the sporting industries (hunting and fishing). In more recent years, captive-reared wild animals have been reintroduced in nature to initiate new or supplement existing wild populations of rare or endangered species. Domestic animals (e.g. livestock, companion animals) have been purposefully or inadvertently released in nature, establishing feral populations in different parts of the globe. Some of these released animals have survived and reproduced but many more have died without leaving progeny. This chapter reviews concepts and factual information relative to the native ability of captive-reared wild and domestic animals to adapt to life in the wild, and management techniques employed to improve their survival and reproductive success in nature. Studies identifying the factors responsible for the failure of captive-reared wild and domestic animal populations to become established in nature may lead to a better understanding of the genetic changes accompanying the domestication process and the effect of the captive-rearing environment on behavior and physiology.

Feralization Defined

As in the case of domestication, the process of feralization has meant different things to different people. Some definitions assert that feral animals are merely free-living populations of animals that originated from domestic stock (Pullar, 1950; Shank, 1972). Others suggest that in addition to their free-ranging status, feral animals must be unowned, not intentionally cared for by humans and not dependent on humans for breeding (McKnight, 1976; Baker and Manwell, 1981). Those with a more evolutionary viewpoint describe feral animals as undergoing the domestication process in reverse (Letts, 1964; Hale, 1969; Brisbin, 1974; Price,

©CAB *International* 2002. *Animal Domestication and Behavior*
(E.O. Price)

1984). This latter definition implies that feral animals are no longer exposed to artificial selection by man or natural selection imposed by the captive environment. If one accepts the thesis that domestication involves genetic change, it is only logical to postulate that feralization will involve evolutionary processes as well. Consequently, the process of feralization, like domestication, is seldom achieved in a single generation. Daniels and Bekoff (1989a) propose a more ontogenetic approach to feralization. In their view, domestic animals undergo feralization when they fail to become socialized (or become desocialized) to humans and behave as untamed, wild animals. Under these conditions, an animal can become feral during the course of its lifetime.

Wolfe and Bekoff (unpublished observations) maintain that the term 'feral' should be restricted to animals that have reached some endpoint in the feralization process, in much the same manner that we use the term 'domestic' to refer to animals that have reached some endpoint in the process of domestication. Although the gene pools of feral populations will not necessarily be identical to those of their wild progenitors, it seems appropriate from a biological perspective to use 'feral' to refer to free-living populations that have become adapted to the environment in which they are living and have become emancipated from man's direct care and control. That is not to say that feral (or wild) populations have static gene pools; change is continually occurring in nature.

Examples of Feralization

Animals may become feral for a variety of reasons. Domestic animals may be deliberately released in nature to establish populations for human food and recreation or to avoid euthanasia. Others may escape from captivity and establish free-living populations. Captive-reared wild or semi-domestic animals may be released in the wild to replenish (restock) natural populations that have been depleted because of loss of suitable habitat, disease or human harvesting for food or sport. The common denominator for feral and restocked animals is that they or their ancestors lived in captivity at some point in their past. Van Vuren and Hedrick (1989) have pointed out that feral animals may be sources of genetic variation with potential commercial, scientific, historical or esthetic value. Important variants may include primitive traits absent in modern breeds and novel or rare adaptations. It is clear that many of our common domestic animals will undergo feralization if given the opportunity. McKnight (1964) has discussed the distribution and ecology of feral horses, burros, cattle, sheep, goats, swine, dogs and cats in North America. Others have reported on the behavior and ecology of feral fowl (McBride *et al.*, 1969; Brisbin, 1974; Duncan *et al.*, 1978), feral dogs (Scott and Causey, 1973; Nesbitt, 1975; Daniels and Bekoff, 1989b), feral goats (Shank, 1972; O'Brien, 1984), feral pigs (Graves, 1984) and feral horses (McCort, 1984; Berger, 1986). Occasionally, reports are made of domestic Norway rats living in landfills (Minckler and Pease, 1938). King and Donaldson (1929) noted five attempts to feralize the albino laboratory rat (*R. norvegicus*) at sites ranging from

Massachusetts to an island group in the Gulf of Mexico (Tortugas). For various reasons, all attempts failed. Although many attempts to feralize domestic stocks of animals have failed, Baker and Manwell (1981) note that in Australia and New Zealand domesticated animals have been just as successful as exotic wild animals in establishing self-perpetuating free-living populations. Although the scarcity of large predators in these countries (except possibly for the dingo in Australia) may account for their success, the notion that domesticated stocks of animals are poorly equipped to survive and reproduce in nature is not a 'safe' generalization.

Feral dogs

Beck (1973), Brisbin (1977), Brisbin and Risch (1997) and Coppinger and Coppinger (2001) discussed free-ranging populations of feral dogs still dependent on man for their breeding or subsistence and, thus, subject to certain selective mechanisms accompanying the domestication process. Brisbin (1977) used the term 'pariah' to describe these populations and has illustrated how pariah dogs can serve as a filter or bridge between domestic and feral (or wild) populations. Coppinger and Coppinger (2001) describe how study of these primitive/pariah dogs and their anthropologic relations with humans can provide a reasonable hypothesis for how the dog's wolf-like ancestor was domesticated. One interesting fact is that most of the present-day feral/primitive dogs resemble the Australian dingo (*Canis dingo*) in general appearance more than they resemble the typical wolf (*C. lupus*; Brisbin and Risch, 1997). There may be certain gene combinations and phenotypic traits of dogs that are favored by natural selection in most natural environments and produce a rather generic feral animal. Daniels and Bekoff (1989a) maintain that feralization of the dog (*C. familiaris*) is largely attained through emancipation from the social environment and food provisioning of humans. The implication is that the domestic behaviors of the dog are largely induced by social attachment and dependence on humans. This is a reasonable hypothesis considering the sociobiology of wolves and humans and the relatively strong social attachments that typically develop between dogs and their human caretakers. The feralization of other less socially dependent species may be less dependent on emancipation from human influence.

Rate of Feralization

One could gain a better understanding of the process of feralization by monitoring domesticated populations over generations following release into a suitable natural environment. The rate of feralization will, of course, depend on the ontogenetic processes involved and how completely the population is divorced from human influence. Populations of domestic animals (e.g. dogs or cats) living on the fringe of captivity that rely on humans for food or shelter (Coman and Brunner, 1972; Beck, 1973; Iverson, 1978) will probably revert to the wild (feral) phenotype at a slower rate than animals living entirely on their own, or may never become totally free from human intervention.

Wild animals that are highly pre-adapted for domestication should, theoretically, be highly pre-adapted for feralization unless, of course, significant changes in the gene pool have been made through artificial selection. A high degree of pre-adaptation to captivity implies that the captive environment is relatively similar to the natural environment, or that the species is highly adaptable. Under either of these circumstances, there should be relatively few changes in the gene pool during domestication, and reversion to the wild phenotype should be readily achieved. Pearce-Kelley *et al.* (1995) found that when a population of captive-bred snails (*Partula taeniata*) maintained in the laboratory for six generations and fed an artificial diet were released into the wild, they exhibited patterns of feeding and microhabitat choice similar to those observed in wild snails. Of course, the ecological conditions at the release point should be compatible with requirements for survival. Hundreds of thousands of domesticated Japanese quail (*Coturnix coturnix japonica*) were released at multiple sites in the midwestern USA in the late 1950s without success (Labisky, 1959). Yet, this species was successfully introduced to the Hawaiian islands (Moulton *et al.*, 2001).

Interbreeding of Domestic and Wild Animals in Nature

Feralization may be facilitated by the interbreeding of domestic and wild animals (Moav *et al.*, 1978; Baker and Manwell, 1981). Although, in theory, the resulting hybrids will not be well adapted to either nature or captivity, they are more likely to survive and reproduce in nature than their domestic counterparts since the hybrid population should have more genes in common with their wild ancestors. Hybridization of domestic and wild animals can facilitate feralization by enlarging the range of genetic and phenotypic variation on which natural selection can operate. Of course, hybridization is dependent on the ability and desire of the wild and domestic stocks to interbreed. Cheng *et al.* (1979a) have reported difficulties in hybridizing domestic and wild mallard ducks (*A. platyrhynchos*).

Genetic Considerations for Releases

Preservation of the natural gene pool in captivity

Animal populations whose gene pools have been significantly modified during domestication may initially experience high mortality when placed in a natural environment (Vincent, 1960; Leider *et al.*, 1990). Hence, the diversity of the 'natural' gene pool must be protected when breeding animals in captivity for the purpose of re-establishing (stocking) free-living populations of animals in their natural habitats (Ballou, 1984). This is particularly important when breeding rare and endangered species in captivity, where lost genes may be irretrievable.

Potential problems must be acknowledged when rearing animals in captivity for restocking. Protection of the natural gene pool and genetic diversity will be difficult to achieve in relatively small, isolated populations where inbreeding and genetic drift are likely to occur and where natural selection in captivity may be severe (Bartlett, 1985; Philippart, 1995). In effect, the biologist is faced with somewhat conflicting objectives: (i) to maximize reproduction and, by so doing, possibly accelerate the domestication process; and (ii) to minimize domestication in order to protect and preserve 'natural' gene pools. Small-bodied taxa tend to have shorter generation times and so lose genetic variability more quickly than larger species. To compensate for this, small species must be kept in larger numbers in captivity than larger taxa (Soulé et al., 1986). It may be difficult to obtain an adequate number of rare animals to establish an effective captive breeding program, especially when the species is poorly pre-adapted for captivity and survival and/or reproductive rates are relatively low.

Changes in the gene pool of a population resulting from domestication can be reduced by: (i) continued introduction of genes from wild populations; (ii) prolonging the generation interval; (iii) minimizing mortality in captive populations; (iv) minimizing or preventing genetic bottlenecks; and (v) reducing the heritability of traits by equalizing family sizes (Frankham and Loebel, 1992). In equalizing family size, between-family selection is reduced by restricting reproduction to equal-size full-sib families with each male–female pair parenting a single family (i.e. two progeny from each family will become breeding adults; Allendorf, 1993). The rate of gene frequency change with 'familial selection' is approximately one-half that of ordinary selection. It is most often recommended when population size is small enough that genetic drift is a major concern. In species with very high fecundity (e.g. many fish, amphibians and insects), where hundreds of progeny can be produced by individual matings and the intensity of selection can be considerable, equalizing progeny number should be considered even when populations sizes are large. In species with relatively low fecundity (e.g. birds and mammals), rapid genetic change caused by adaptation to captivity is less likely to occur because low fecundity limits the potential intensity of selection.

Genetic changes may occur in captivity in spite of efforts to preserve the wild gene pool. Perez-Enriquez et al. (1999) found that even though a hatchery population of red seabream broodstock was genetically similar to the wild population, the progeny produced for release were genetically different, presumably due to a reduced number of contributing parents. Tessier et al. (1997) suggested that management practices contributed heavily to changes in allele frequencies between wild and hatchery populations of Atlantic salmon (S. salar).

Genetic and demographic effects of reintroductions

Failure to protect the natural gene pool not only jeopardizes the survival and reproductive success of animals released in nature, but significant changes in the gene pool of natural populations can occur indirectly from nongenetic causes such as competition or disease (Utter, 1998) or directly by hybridization between released and wild animals (Evans and Willox, 1991; Garcia-Marin, 1991; Hindar *et al.*, 1991; Waples, 1991; Cagigas *et al.*, 1999). Garcia-Marin *et al.* (1999) studied the genetic introgression of hatchery-reared brown trout (*S. trutta*) on a native trout population in the Spanish Pyrennees. Using gene correlation matrices between individuals, they estimated introgression at 10% after 2 years. In extreme cases, native populations could be displaced by domestic stocks. Cultured salmon released in nature have a varying degree of impact on natural populations (Allendorf, 1983). These impacts may be mediated through genetic introgression (Hindar *et al.*, 1991) or competition (Fleming and Gross, 1993). Hilborn and Eggers (2000) believe that the release of hatchery-reared pink salmon (*Oncorhynchus gorbuscha*) in Prince William Sound, Alaska, has replaced rather than augmented wild salmon production. As hatchery production increased, wild production declined.

In reality, it is difficult to estimate the impact of hatchery fish on wild salmonid production because there are so few areas that can be considered adequate controls. It is estimated that between 25 and 40% of the salmon in the North Atlantic and more than 90% of the salmon in the Baltic Sea are cultured fish (Hansen *et al.*, 1993; Jonsson and Fleming, 1993). If supplementation of wild populations is the goal, the production of interspecies sterile hybrids is one way to avoid introgressive hybridization and consequent contamination of wild gene pools in the case of accidental escape of hybrids into the wild. Tave (1993) describes the production of sterile triploid striped bass × white bass (*M. saxatilis* × *Morone chrysops*) by chromosomal manipulation. Another approach to lessen the risk of gene-pool contamination is to use only naturally reared fish as broodstock in each generation (Waples, 1999). Of course, this procedure would not eliminate natural selection in captivity on the broodstock or on the resulting offspring prior to release. Using a relatively large sample of naturally reared fish for broodstock reduces the risk of a founder effect and the likelihood of inbreeding but at the same time exposes a larger portion of the population to the captive environment, which could affect the remaining natural population demographically or genetically. Alternatively, the risk of genetic change due to captive rearing is reduced if animals are released early in their life cycle (e.g. as fry or parr rather than smolts in the case of fish; Waples, 1999). However, doing so reduces the survival benefit of the captive environment and exposes the animals to competition with native animals at a young age.

The fact that most hatchery-reared fish are released into the wild as juveniles raises the question of the extent to which this management procedure reduces the rate of domestication (Waples, 1999). Does natural selection in the wild during the post-juvenile stage offset or negate any selective advantage in any segment of the population prior to release (Fig. 19.1)? To what extent do hatcheries merely prolong the survival of some fish that would be lost prior to reproduction in the wild anyway? A 5% survival of released chinook (*Oncorhynchus tshawytscha*) and coho salmon (*O. kisutch*) is considered very good, even for fish reared for a year in the hatchery, fed extensively and released at a relatively large size (McNeil, 1991;

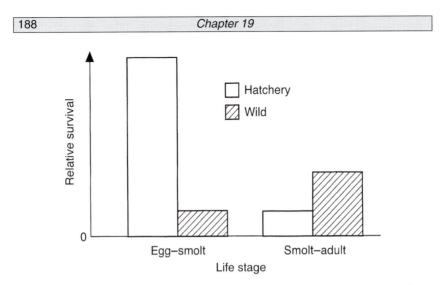

Fig. 19.1.   The survival rate of hatchery-reared fish is considerably greater than that of their wild counterparts prior to release in streams (egg–smolt stage) but lower than wild fish following release (smolt–adult stage) (Waples, 1999).

Hilborn and Eggers, 2000). Weiss and Schmutz (1999) examined the growth, survival and movement of released hatchery-reared brown trout (*S. trutta*) in two streams and the demographic effect of these releases on existing indigenous wild trout populations. After 12 months, 1–19% of the hatchery-reared fish and 13–52% of the wild fish had survived, dependent on stream and strain of the hatchery-reared fish. There was no significant change in the population size or biomass of the wild trout populations due to stocking in either of the two streams. The authors make the important point that as restocking programs become more successful (i.e. stocked animals experience enhanced growth and survival rates), the potential for negative demographic and/or genetic interactions between stocked and native animals will probably increase.

Habitat Considerations for Releases

Availability of habitat

The availability of suitable habitat is frequently a critical determinant of success in reintroductions (Ellis *et al.*, 1978; Griffith *et al.*, 1989; Wilson and Stanley-Price, 1994; Balmford *et al.*, 1996). Kleiman (1996; Kleiman *et al.*, 1994) has been a strong advocate for not attempting releases if suitable, stable, unsaturated habitat is not available. Van der Meeren (2000) reported that hatchery-reared lobsters (*Homarus gammarus*) are particularly susceptible to predation in the hours following release (10% in the first hour), before the lobsters have found shelters in which to hide. Lobsters released on a rocky substratum with numerous hiding places were less subject to predation than those released in areas where the bottom was sandy.

Kenward *et al.* (2001) noted that predation by buzzards (*Buteo buteo*) on released, captive-reared ring-necked pheasants (*P. colchicus*) was greatest in areas with little shrub cover for the pheasants and open deciduous canopies providing perching and nesting sites for the buzzards.

In many instances, the improvement of wild habitat may be a more cost-effective way to enhance population numbers in the wild than captive breeding and restocking (Snyder *et al.*, 1997). Loftin (1995) asserts that habitat protection has had a greater effect on increasing free-living populations of certain endangered or threatened species (e.g. golden lion tamarins, whooping cranes) than the release of captive-reared animals.

Acclimation to release site

A relatively short period of acclimation to the actual release site/habitat may improve the success of releases. Van der Meeren (2000) found that lobsters (*H. gammarus*) acclimated to ambient seawater temperature in open trays for at least 1 hour prior to release found shelters more quickly than lobsters released directly into the water from their transportation boxes. Jonssonn *et al.* (1999) compared the post-release recovery rates of 1-year-old hatchery-reared brown trout (*S. trutta*) released either directly into stream locations after transport from the hatchery or held in small enclosures in the stream for 6 days prior to release. After 2 months in the stream, acclimatized fish were captured in higher numbers, had gained significantly more weight and were in better body condition than nonacclimatized fish. Cresswell and Williams (1983) reported a greater percentage of recaptured brown trout by keeping them in stream enclosures at low flow rates for 24 h prior to release. Acclimation to release sites may be more important when releasing fish in streams than in standing water because streams tend to be harsher environments (White *et al.*, 1995). Linley (2001) compared the growth and ocean survival of hatchery-reared coho salmon smolts (*O. kisutch*) given 2–5 weeks of exposure to salt water in estuarine net-pens prior to release vs. smolts released directly into salt water from their freshwater rearing pens. Fish released from net-pens were consistently larger at return than those released directly from fresh water, and short-term estuarine rearing increased ocean survival by an average of 35%. Short-term estuarine rearing may help counter the effects of osmotic stress and loss of swimming stamina during seawater acclimation, factors that increase susceptibility to predation, delay foraging or both.

A short period of adaptation to the release site (i.e. 'hacking') is traditionally employed when releasing captive-reared birds of prey into the wild (Temple, 1978). Castro *et al.* (1998) reported that providing captive-reared golden lion tamarins (*Leontopithecus rosalia*) with food provisioning for 6–18 months following release in nature was a more cost-effective and time-effective way to improve post-release survival than pre-release training designed to develop foraging techniques. Post-release food provisioning allowed the animals time to become acquainted with their new environment and fine-tune their behaviors in preparation for living on their own. Meyers and Miller (1992) found that captive-reared bald eagles (*Haliaeetus leucocephalus*) were more reluctant to leave their release site than translocated young wild-reared eagles.

Acclimation to the release site may also give individuals the opportunity to return to normal physiological status after the stressful events of handling, transportation and exposure to a novel environment. Stress is known to increase the vulnerability of fish to predation and disrupt nearly all components of feeding behavior (see review by Schreck *et al.*, 1997). Releasing captive-reared animals in nature immediately following a series of stressful events greatly decreases their chances for survival.

Importance of Age and Size When Released

The importance of age and size of released individuals is often underestimated. Young animals theoretically have less time and opportunity than older animals to become conditioned to the captive environment and associated management techniques but tend to experience greater mortality than older animals when released in nature. Older animals may have greater difficulty than younger individuals in shedding nonadaptive habits acquired in captivity and learning new behaviors important to survival in their new environment. Older animals may also be in poorer physical condition. Sarrazin and Legendre (2000) developed a model for predicting the relative efficiency of releasing juvenile or adult animals and then applied that model to the reintroduction of Griffon vultures (*Gyps fulvus*) in southern France. They concluded that releasing adults was more efficient than releasing juveniles in restoring diminished populations of relatively long-lived species. Kristiansen *et al.* (2000) demonstrated that the natural mortality of released Atlantic cod (*G. morhua*) decreased with increasing size at release and became minimal when the fish reached a size at which they were not preyed upon by their usual predators.

Competition of Released Animals with Wild Stock

In many instances, released animals must compete with wild residents for food, space, mates, etc., and their competitive ability may determine whether or not they survive and reproduce. Deverill *et al.* (1999) compared the aggressiveness and growth rates of hatchery-reared and wild-reared brown trout (*S. trutta*) when placed in an artificial stream environment inhabited by resident wild trout obtained from the same population as the nonresident wild trout. Hatchery-reared trout initiated more aggressive acts than stocked wild trout (34 vs. 22%, respectively) when placed in the artificial stream, but resident wild trout initiated the most aggressive interactions (44%) and exhibited the highest growth rates, supporting a prior-residence effect on aggressive potential. Hatchery-reared trout had the poorest growth rates, perhaps due, in part, to the change in diet from a high-energy commercially prepared food (pelleted trout diet) to a relatively less energy-dense natural food (bloodworms) or by an initial period of reluctance to accept unfamiliar food items. Crowding, the manner of feeding hatchery-reared salmonids and the lack of visual isolation in rearing tanks may facilitate the development of a 'scramble' feeding strategy and aggressive feeding behaviors (Wang and White, 1994).

Breeding success of size-matched adult male wild- and captive-reared coho salmon (*O. kisutch*) was compared in indoor flumes simulating natural spawning channels (Berejikian *et al.*, 2001). Wild salmon dominated captive-reared males very quickly after pairing and controlled access to spawning females in 11 of 14 (79%) paired trials. A similar study by Berejikian *et al.* (1997) found wild-reared salmon to be dominant in 86% of all spawning events. The decidedly poorer breeding success of the captive-reared fish may be due to deficiencies in the hatchery environment, which fail to produce appropriate behavior, nuptial coloration and the development of certain secondary sexual characteristics (e.g. snout length and hump size).

Biological Handicaps of Captive-reared Animals Released in Nature

Animals reared in captivity for release in nature may lack some of the natural structural, physiological and behavioral responses to environmental stimuli that are normally acquired (often early in life) by their free-living counterparts (Shrimpton *et al.*, 1994). Released animals must be able to find food and shelter, develop anti-predator skills, interact appropriately with conspecifics and orient (disperse, navigate and, in some cases, migrate) in a structurally complex environment (Kleiman, 1989; Box, 1991). These activities may require greater physical conditioning than in captivity. McDonald *et al.* (1998) compared the anaerobic capacity and swimming endurance of wild and hatchery-reared Atlantic salmon (*S. salar*) of the same age and derived from the same parent population. Wild fingerlings reared in a natural stream had a higher anaerobic capacity than hatchery fingerlings, and wild fish showed a small but significant glycogen recovery following exercise, which was not observed in the hatchery stock. Furthermore, once the hatchery fish had reached the yearling stage, their sprint swimming endurance was also inferior to the wild-reared yearlings.

The relatively large body size of some captive-reared animals can be partially explained by better nutrition and lack of exercise. Large size may constitute a handicap for released animals if it impairs mobility and agility. Pyörnilä *et al.* (1998) examined the muscle fiber types and their proportions in the breasts and leg muscles of hand-reared and wild gray partridges (*P. perdix*) to determine if differences could explain in any way the poor survival of captive-reared birds following release in nature. The relative number of fibers and the cross-sectional areas of fibers were larger in captive-reared birds but these differences *per se* could not account for their higher mortality.

Rearing animals in captivity does not always lead to poorer growth and survival when released in nature. Elsey *et al.* (1992) reported that captive-reared alligators (*Alligator mississippiensis*) released in coastal Louisiana (USA) wetlands exhibited faster growth rates than wild stock of comparable size. The captive-reared alligators studied in this investigation were all hatched from eggs collected in the wild and reared in a ranch environment.

Other studies have shown that behavioral deficits may result in slower growth and higher mortality for captive-reared animals released in nature (Hessler *et al.*, 1970; Schladweiler and Tester, 1972; Kraus *et al.*, 1987; Roseberry *et al.*, 1987; McCall *et al.*, 1988; Hesthagen *et al.*, 1999). Sharpe *et al.* (1998) documented the high mortality experienced by captive-reared ruffed grouse (*B. umbellus*) following release in nature. The first four 'tamed' birds died within 7 days of release and at least three of them were killed by predators. Munakata *et al.* (2000) reported that the relatively low survival of hatchery-reared honmasu salmon parr (*Oncorhynchus rhodurus* × *Oncorhynchus masou*), once released in nature, may be due in part to their inexperience in feeding on natural prey. During the first week, stomach fullness of the released parr was considerably lower than for the wild parr, and stomach contents included weeds and pebbles, implying that the foraging behavior of the released fish was sub-optimal. After 2 weeks, the stomach fullness of the released fish was similar to the wild stock. Plasma levels of growth hormone increased in the released fish within 1 day of release and were higher than in the wild fish for a 3-month period. The dramatic growth-hormone response in the released fish may have been a response to increased levels of physical activity or exposure to natural environmental cues.

Steingrund and Fernö (1997) reported that in laboratory experiments, captive-reared cod (*G. morhua*) used a pursuit strategy to capture live gobies (*Gobiusculus flavescens*), whereas wild cod used a more efficient stalk, wait-and-lunge strategy for the highly mobile gobies. The pursuit strategy of the captive-reared fish resembled the high-speed attack used in competing with other reared fish for food pellets placed in their rearing tanks. When captive-reared fish were released in nature they had to relearn their prey-catching approach, since highly mobile prey could quickly disappear into hiding places. In addition, highly visible pursuits placed the reared cod in greater danger of detection by their own predators. The failure of the pursuit strategy in natural habitats probably explains the different diets of newly released cod relative to their wild counterparts (Nordeide and Salvanes, 1991). After 4 months, the diets of reared and wild cod were similar, indicating that the reared cod had changed their feeding strategy.

Dieperink *et al.* (2001) compared the avian predation rate on emigrating wild and hatchery-reared sea trout (*S. trutta*) in a fjord in the western Baltic Sea. Fifty wild-caught and 50 farmed smolts were intraperitoneally tagged with radio-transmitters and predation was recorded by signal interception in an estuarine breeding colony of cormorants and herons near the outlet of the stream where the smolts were released. Predation rates on similar-size wild and hatchery-reared sea trout were 40 and 77%, respectively, in the first days in the marine environment. Maynard *et al.* (1995) proposed that hatchery-reared fish are more vulnerable to predation since they are less exercised, less able to endure sustained swimming, more surface oriented and less experienced with predators than wild fish. Kanid'hev *et al.* (1970) reported that naive hatchery chum salmon (*Oncorhynchus keta*) were preyed upon up to 30% more than similar-size wild salmon. This difference dropped to 12–18% shortly after release, and after 4 weeks, there was no difference in predation rates of wild and hatchery-reared fish, suggesting that the survivors had learned to avoid predation.

Preconditioning as Preparation for Release

Habitat preconditioning

It is generally held that the survival of released animals can often be improved by preconditioning (i.e. training) individuals to respond appropriately to important environmental stimuli prior to their release (Mydans, 1973; Spivey, 1973; Suboski and Templeton, 1989; Box, 1991; Kleiman, 1996). This is not always the result obtained. Beck *et al.* (1991, cited in Castro *et al.*, 1998) noted that 41% of captive-reared golden lion tamarins (*L. rosalia*) introduced into a Brazilian forest died in the first year. Neither free-ranging experience in a naturalistic wooded area prior to release nor pre-release training in foraging techniques significantly improved post-release survival.

Biggins *et al.* (1998) reared 282 black-footed ferrets (*M. nigripes*) either in small cages or large outdoor enclosures with different prey species and subsequently released them in natural prairie dog colonies in three states (USA) over a 5-year period to determine the effects of early experience on survival at 1 and 9 months post-release. Cage-reared animals were given live hamsters (*M. auratus*) as prey once per week and dead prairie-dog (*Cynomys leucurus*) carcasses. Outdoor pen-reared animals were exposed to prairie dogs living in the pens and the carcasses of dead prairie dogs. Outdoor pens were of different sizes (18–280 m square). Minimum post-release survival rates were higher for the ferrets reared in outdoor pens (30% for 1 month and 20% for 9 months) than for ferrets reared in cages (11% for 1 month and 2% for 9 months). Cage-reared ferrets were generally more active above-ground than pen-reared animals, which may have made them more susceptible to predators such as coyotes (*C. latrans*). Biggins *et al.* (1998) also compared ferrets: (i) born and reared in outdoor pens; (ii) born in cages and trans-ferred to outdoor pens (with dam) at 60–90 days; and (iii) transferred to outdoor pens at >90 days. Post-release survival was greatest for those animals spending the most time in outdoor pens prior to release. In a similar study, Biggins *et al.* (1999) compared the post-release survival of black-footed ferrets: (i) reared in cages (1.5 m^2) and fed live hamsters and dead prairie dogs; (ii) reared in cages and fed both live hamsters and prairie dogs; and (iii) reared in outdoor pens (approximately 80 m^2) containing live prairie dogs and prairie dog burrows starting at 2 months of age. Recovery rates indicated that outdoor-pen-reared ferrets were more prone to remain at the release sites (quality habitat) than cage-reared animals. In spite of a 10-day period of preconditioning to the release site in cages, 77% of the cage-reared animals moved >7 km within a 12-h period and 50% had dispersed >21 km in 72 h. Many of these dispersing ferrets were tracked to areas without prairie dog colonies, where food and water were scarce and where they probably suffered high mortality from starvation and predation. Outdoor-pen-reared ferrets may have used burrows (at the release sites) as refuges more effectively because of having been reared in them. Exposure to prairie dogs in cages during rearing did not translate into a post-release survival advantage.

A common rearing practice in many salmonid hatcheries is to transfer juvenile fish from concrete raceways to large rearing ponds with earth and gravel bottoms prior to release in nature. Pond rearing is believed to acclimate fish to an environment more closely resembling natural streams (Maynard *et al.*, 1995). Tipping (2001) examined the post-release survival of sea-run cutthroat trout (*Oncorhynchus clarki*) reared in ponds for either 1 month or 4–7 months. Adult returns (recoveries) were 31% greater for the group housed in ponds for the longer period. In addition to habitat conditioning, rearing fish in ponds provided exposure to avian predators. A longer period of pond rearing provides more anti-predator conditioning and selection for anti-predator behaviors prior to release (Maynard *et al.*, 1995).

Preconditioning prey species to predator stimuli

A number of investigators have studied the efficacy of preconditioning captive-reared animals to predator stimuli prior to their release in nature (see review by Griffin *et al.*, 2000). Kellison *et al.* (2000) demonstrated that hatchery-reared summer flounder (*P. dentatus*) were more susceptable to predation by blue crabs (*Callinectes sapidus*) than wild flounder captured in nature as juveniles. In a subsequent experiment they found that hatchery-reared fish exposed to a caged blue crab for 24 h prior to predation tests were less susceptible to crab predation than nonconditioned fish, but still had higher losses than wild-born fish. It is not known exactly how this preconditioning reduced susceptibility to predation. Johnsson and Abrahams (1991) and Berejikian (1995) found that hatchery-reared strains of steelhead (*O. mykiss*) were more susceptible to predation than wild strains. In a subsequent study, Berejikian *et al.* (1999) showed that hatchery-reared juvenile chinook salmon (*O. tshawytscha*) learned to associate the odor of a natural predator, cutthroat trout (*O. clarki*), with chemical alarm substance extracts from chinook salmon tissue produced by damage to epidermal cells (i.e. injured fish stimulus). Fish treated with predator odor and chinook salmon alarm extract spent significantly more time motionless than fish treated with predator odor and green swordtail (*Xiphophorus helleri*) extract, which does not contain a natural chemical alarm substance, or fish treated with predator odor and distilled water. Survival rates of conditioned and unconditioned fish in field tests were equivocal. They concluded that chemically based anti-predator conditioning is a promising technique for improving the post-release survival of chinook salmon.

In a similar study, Brown and Smith (1998) tested the ability of hatchery-reared rainbow trout (*O. mykiss*) to acquire chemical recognition of a natural predator, northern pike (*E. lucius*). Hatchery-reared rainbow trout do not innately exhibit anti-predator responses to pike odor. However, they can be readily conditioned to exhibit anti-predator behavior by pairing trout chemical alarm skin extract with water in which the predator had been living. Trout showed increases in anti-predator responses with only one pairing of trout skin extract and pike odor (Fig. 19.2). Conditioned fish still exhibited anti-predator behaviors 21 days after conditioning when presented with pike odor only. It is not known if salmonids can be conditioned to respond to the visual cues of a predator using chemical alarm signals.

Mirza and Chivers (2000) not only confirmed that fish can be readily trained to avoid the odor of predators but that this learning enhances survival.

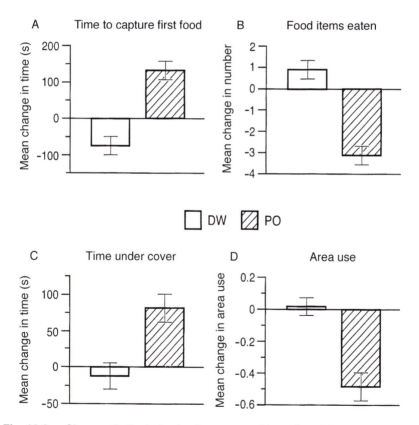

Fig. 19.2. Changes in the behavioral response of juvenile rainbow trout following injection of 15 ml of distilled water (DW) into the test tank or 15 ml of water containing the odor of a predator (pike) (PO) that the fish had previously experienced in a single pairing with trout-skin extract (a chemical alarm signal). All differences were statistically significant ($P < 0.05$) (Brown and Smith, 1998).

Hatchery-reared predator-naive brook trout (*Salvelinus fontinalis*) exposed to the odor of predatory chain pickerel (*Esox niger*), paired with chemical alarm substances released by injured trout, exhibited a greater ability than untrained trout to avoid predation in staged encounters with pickerel in both a laboratory test area and a field stream enclosure. Treatment differences in predation were greatest when both trained and untrained fish were exposed to the predator together.

Thompson (1966, cited in Brown and Smith, 1998) and Healy and Reinhardt (1991, cited in Brown and Smith, 1998) reported that conditioning salmon fry (*O. tshawytscha* and *O. kisutch*, respectively) to avoid a novel model predator by pairing the model with an electric current did not increase their survival relative to untrained controls when released into a natural stream.

Ellis *et al.* (1978) described a procedure for training masked bobwhite quail (*Colinus virginianus ridgwayi*) to exhibit appropriate anti-predator behaviors once released in nature. For a 2-week period prior to release, the birds were pursued and harassed by humans, dogs and a trained hawk. During training, the birds rapidly improved in general mobility, covey coordination and ability to hide and avoid 'predators'. The dogs were particularly useful in simulating mammalian predation; the quail quickly learned when to freeze and when to flush.

McLean *et al.* (1999) demonstrated that New Zealand robins (*Petroica australis*) can be trained to avoid predators by exposure to a model predator either in the wild or in the laboratory. Studies on free-living birds involved exposing them to a simulated attack by a stuffed predator (ferret or cat), or a stuffed robin in the aggressive or mobbing posture accompanied by robin alarm and distress calls. Studies in the laboratory used the same birds trained in the wild and involved exposing the birds to a pursuing predator model. Predator-trained birds became more cautious or fearful in the presence of the model predator. Birds avoided the model predator more after aviary training than after training in the wild. The results suggested that captive-reared, predator-naive birds can be readily trained to avoid predators prior to release in nature. Slaugh *et al.* (1992) found that the survival of captive-reared chukar partridges (*Alectoris chukar*) in nature is greater if they are given predator conditioning starting at 4 weeks of age. The latter consisted of twice-weekly exposures to a dog and a hawk model passing over the birds in their outdoor rearing pen. During these conditioning sessions, the young birds learned to associate the alarm calls of nearby adult birds with predator stimuli. In addition, birds imprinted on adult birds by visual exposure shortly after hatching with little or no subsequent contact with people showed greater fear of humans than birds reared conventionally with regular human contact.

These and other studies with vertebrate species have demonstrated that preconditioning captive-reared animals to avoid predation once released in nature may be most successful if techniques are designed to take advantage of a species' natural defense mechanisms and predisposition to learn (Griffin *et al.*, 2000). Animals can quickly improve their anti-predator responses through experiences with predators, and cues from conspecifics can facilitate that learning process. Of course, anti-predator training with no real predation could result in habituation to predator cues and reduce post-release survival.

Castro *et al.* (1998) cited earlier work reporting that both captive-reared and wild-born golden lion tamarins (*L. rosalia*) responded similarly to the presentation of an aerial predator model (stuffed hawk) and exhibited mobbing behavior toward an arboreal–terrestrial predator (rubber snake). However, the captive-born animals scanned the ground more than wild animals after being exposed to the hawk and spent less time on the ground after seeing the snake. The captive-reared animals may have been conditioned to avoid the open 'understory' area of their enclosures whenever threatened, and exhibited this response regardless of the stimulus perceived. Wild-born tamarins showed more context-appropriate responses to predator stimuli, presumably based on experiences in the wild.

It is also likely that individuals of prey species learn to recognize predators and show appropriate anti-predator behaviors by observing the responses of conspecifics or individuals of other species to key predators (Griffin *et al.*, 2000). Such social learning may be particularly evident in group-living prey populations and species bearing altricial young requiring relatively prolonged parental care or with delayed maturation. Griffin *et al.* (2000) point out that the considerable time and effort required to precondition captive-reared animals to recognize, avoid and respond appropriately to predators following release in nature may be cost-effective if the trained individuals serve as effective models to their offspring and the learned behaviors are culturally transmitted over subsequent generations. Mathis *et al.* (1996) demonstrated intra- and inter-specific cultural transmission of anti-predator behaviors in fathead minnows (*Pimephales promelas*) and brook sticklebacks (*Culaea inconstans*). The release of hatchery-reared fish in high numbers may attract predators but may also facilitate rapid observational learning of anti-predator behaviors. It appears that animals learn anti-predator behaviors very quickly (see references cited in Griffin *et al.*, 2000).

Preconditioning to natural diets

Survival and reproduction of captive-reared birds following release in nature may be facilitated by preconditioning them in captivity to the diet they are likely to encounter in the wild. Liukkonen-Anttila *et al.* (1999) assessed the nutritional status of captive-reared gray partridges (*P. perdix*) reared on a commercial diet and abruptly shifted to a natural diet. Birds shifted to a natural diet ate less than birds in the control group on the commercial diet, and lost body mass for the first 7–10 days and then started to regain mass. Experimental birds also had lower metabolized energy coefficients than birds in the control group. No differences were seen in blood metabolites. Weights of gizzards increased over time in response to the change to a natural diet, perhaps due to the need for more effective grinding ability. However, captive-reared partridges still had smaller gizzards than wild partridges after 6 weeks. The authors concluded that a 6-week period of preconditioning to a natural diet before release may be insufficient to prepare gray partridge for utilizing nutrients from foods available in the wild.

Commercial diets for birds are typically low in fiber content compared with natural diets. Accordingly, captive gallinaceous gamebirds fed commercial diets have shorter intestinal tracts and lighter gizzards than their wild counterparts (Putaala and Hissa, 1995; Millán *et al.*, 2001). Some released partridges may effectively starve to death before their digestive system adapts to the new diet (Putaala and Hissa, 1993, cited in Liukkonen-Anttila *et al.*, 1999).

Preconditioning to foraging techniques

The development of efficient foraging techniques is one of the more important survival skills acquired by young free-living animals (Yoerg, 1994). Preconditioning captive-reared animals to locate and recognize appropriate food items may be critical for their survival following release. This may be particularly important for species occupying specialized feeding niches or utilizing specialized foraging techniques. Certain predators fall into this category. Vargas and Anderson (1999) demonstrated the importance of early exposure to prey on the development of predatory behaviors in captive-reared juvenile black-footed ferrets (*M. nigripes*). Ferrets exposed to live hamsters (*M. auratus*) during their development killed significantly more prey during individual predatory tests than ferrets fed dead prey (Fig. 19.3). Vargas (1994, cited in Biggins *et al.*, 1999) demonstrated that cage-reared black-footed ferrets exposed to live prairie dogs (*C. leucurus*) were more proficient at killing prairie dogs than cage-reared ferrets exposed to live hamsters only. Furthermore, ferrets reared in cages where food was hidden in different places at different times were more adept at killing hamsters than those raised in cages where food was always placed in the same location (Vargas and Anderson, 1999). This latter point addresses the role of predictability (of the captive environment) on behavioral development. Food-getting in captivity is usually highly predictable from both a spatial and temporal standpoint. Food-getting is highly unpredictable for free-living predators, and success may require a variety of specialized hunting skills learned from parents and actual predatory experience. Csermely (2000) released hand-reared long-eared owls (*Asio otus*) into the wild after ascertaining they could capture a live laboratory mouse (*M. m. domesticus*). High mortality in the released birds was believed to result from lack of experience in recognizing and hunting a variety of suitable prey species.

The release of captive-reared herbivores (e.g. ungulates) in nature may be less problematic because of their ready acceptance of a variety of plant species and relatively unspecialized feeding techniques. Species that exist on fruits, nuts and seeds can be preconditioned to accept these natural food items while in captivity, but training them to utilize natural foraging techniques can be more difficult. Scientists in the Golden Lion Tamarin Conservation Program have demonstrated that attempts to precondition tamarins to various feeding techniques prior to release had little effect on post-release survival rates (Castro *et al.*, 1998). Provisioning these animals following release, while they were learning to forage naturally, proved to be more effective.

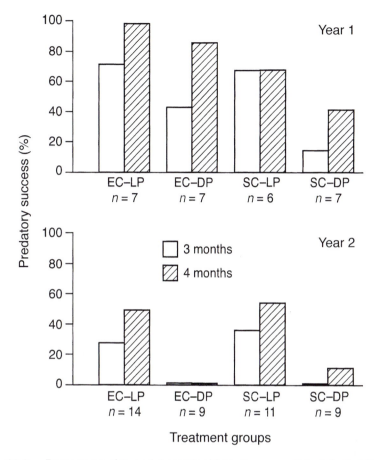

Fig. 19.3. Percentage of 3- and 4-month-old captive-reared black-footed ferrets that killed prey (hamsters) in individual predatory tests in years 1 and 2. Ferrets were reared in either enriched or standard cages and fed a diet of either live or dead prey. EC–LP, enriched cage–live prey; EC–DP, enriched cage–dead prey; SC–LP, standard cage–live prey; SC–DP, standard cage–dead prey (Vargas and Anderson, 1999).

Effect of Hand-rearing on Survival in Nature

There is some evidence that animals reared in captivity by their parents (or surrogate parents) are better equipped to survive in nature than hand-reared cohorts. Hill and Robertson (1988) hypothesized that effective anti-predator behaviors are not as well developed in hand-reared ring-necked pheasants (*P. colchicus*) as in wild pheasants because of a lack of reinforcement of survival strategies provided naturally by the mother bird. Black (1994) reported that parent-reared nene geese (*Branta sandvicensis*) were more vigilant than hand-reared birds after viewing a predator. Dowell (1990) indicated that parent-reared gray partridges (*P. perdix*) showed better organized anti-predator behavior, with more specific responses to their own species' alarm calls. Van Heezik and Seddon (1998) found that parent-reared houbara bustard chicks (*Chlamydotis undulata*) were more active than hand-reared birds, possibly because the parent-reared birds had to move about to keep up with their mothers. However, this did not improve their survival following release in nature. Approximately half of both the parent-reared and hand-reared bustards were killed by predators in their first 6 months in the wild. It was noted, however, that the hens used for parent-rearing were themselves hand-reared. Brittas *et al.* (1992) found no difference in the survival of ring-necked pheasants (*P. colchicus*) reared (fostered) by domestic chickens (*G. domesticus*) vs. hand-reared (by machine) following their release in nature. However, the surviving foster-reared birds were in better body condition and experienced greater reproductive success in the field than the surviving hand-reared birds. The authors speculated that the foster-reared birds may have been taught better foraging skills by their chicken mothers, which they passed on to their offspring.

Horwich (1989) demonstrated the importance in restocking programs of training captive-bred sandhill cranes (*G. canadensis*) to behave like wild cranes. Cranes in the wild receive intense and prolonged parental care. Accordingly, Horwich imprinted crane chicks on realistic crane models using brooding calls and by feeding the young cranes natural foods with crane-like puppets. Humans dressed in crane costumes also fed natural foods to the young birds. Upon release, all of the birds showed an interest in wild cranes within days and had formed a continuous association with wild cranes within 30 days. Ellis *et al.* (2000) found that Mississippi sandhill cranes (*Grus canadensis pulla*) hand-reared by costumed humans survived as well as parent-reared birds following release in the wild. At the end of the first year post-release, 77% of 56 hand-reared and 68% of 76 parent-reared cranes in their study area were known to be alive. Costumed humans using taxidermy mounts of crane heads would occasionally lead hand-reared chicks afield to help them gain experience in foraging for insects and to use a natural marsh (Fig. 19.4). The hand-reared cranes were also exposed to a live adult bird in an adjacent pen, and sandhill crane brood calls were played via a tape recorder during hatching and when the costumed caretaker was interacting with chicks. Interestingly, parent-reared birds showed greater survival if released with hand-reared cranes. Although not formally tested, parent-reared cranes appeared generally less wary than hand-reared birds, perhaps as a consequence of being habituated to noncostumed people and automobiles at the rearing site. The authors speculated that once released, the parent-reared birds benefited from the greater wariness of the hand-reared birds when exposed to natural predators.

California condors (*G. californianus*), puppet-reared in an attempt to avoid social imprinting to human caretakers, were found to exhibit considerably less fear of people following release into nature than parent-reared birds (Meretsky *et al.*, 2000). In repeated releases over a 12-year period, puppet-reared condors have consistently shown a tendency to approach humans and human structures, even after conditioning to avoid humans. Some have accepted food handouts from people and some have vandalized human property, ranging from tents and

Fig. 19.4. Hand-reared crane chicks were occasionally led afield by a costumed caretaker and given opportunities to forage for insects and to use natural marsh habitat (Ellis *et al.*, 2000).

sleeping bags to vehicle windshield wipers and roof shingles. These chronic misbehaviors have led Meretsky *et al.* (2001) to recommend that only parent-reared birds be released in the future.

Effect of Rearing with Wild Conspecifics

Hessler *et al.* (1970) and Haensly *et al.* (1985) reported increased survival of semi-domestic captive-reared ring-necked pheasants (*P. colchicus*) if reared with wild-genotype birds prior to release. Game-farm birds became more alert and flightier when reared with wild birds. Ellis *et al.* (1978) and Carpenter *et al.* (1991) described improved post-release survival of captive-reared masked bobwhite quail (*C. v. ridgwayi*) by housing them with wild-caught male Texas bobwhite quail (*Colinus virginianus texanus*) in 'following' pens in dense vegetation starting at 2 weeks of age.

Parasite and Disease Transmission

An additional concern is parasite and disease transmission between captive-reared animals and their free-living counterparts (Starling, 1991; Cunningham, 1996).

This concern is particularly acute if natural populations are threatened or endangered. The first appearance of the monogenean parasite, *Gyrodactylus salaris*, and the bacterium, *Aeromonas salmonicida*, in Norwegian waters was linked to the introduction of salmon from other regions (Håstein and Lindstad, 1991; Johnsen and Jensen, 1991). Both pathogens have resulted in high mortality of wild salmon. Chew (1990) noted that a North American parasite, *Bonamia*, devastated European populations of the European oyster (*Ostrea edulis*) when colonized North American populations of this same oyster species were reintroduced to Europe. Jones (1966) reported that parasite-free populations of caribou (*R. tarandus*) have been established on the Aleutian Islands by hand-rearing calves captured on the Alaska mainland.

Conclusions

Like domestication, feralization is a process that may involve both genetic changes over generations and experiential events during each generation. Feralization proceeds at different rates depending on the number of pre-adaptations of the population to their respective environment. Free-living animals are feral when they become adapted to the environment in which they are living and have become emancipated from man's direct care and control.

All things considered, preparing most captive-reared animals for release in nature is a costly, labor-intensive and complicated task. To the extent that a species must learn, often early in life, to select an appropriate habitat (including shelter), learn to avoid predators, learn to develop efficient foraging and social behaviors and learn adaptations to an ever-changing physical environment, it will be increasingly difficult to provide captive-reared animals with the necessary skills to survive and reproduce in nature. In addition, body structures and physiological systems must have sufficient time to adapt to natural diets and changeable climatic conditions before release. Consequently, the specific requirements of each species must be carefully studied and appropriate preconditioning procedures developed before large-scale releases are justified on either practical or ethical grounds. It appears that these criteria have been reached with only a few species.

To whatever extent possible, animals reared in captivity for release in nature should be reared in large outdoor enclosures with natural shelters, given a natural diet that they must search for and an appropriate mix of social partners. Anti-predator conditioning is more problematic, particularly if the species is subject to both aerial and ground predation. Rearing with nontamed wild adults will facilitate the development of fear responses to novel stimuli. The resulting increase in reactivity will probably generalize to predators but the development of appropriate responses to different predators may require actual exposure to predators likely to be encountered in the wild. Predator species require experience hunting for and killing natural prey prior to release. Both prey and predator species will benefit from acclimation to the release point.

The above steps apply primarily to rearing animals in captivity to replenish depleted natural populations on a long-term basis. If the goal of the captive-rearing program is to provide animals for hunting, fishing or other uses for the short-term, many of the above steps can be de-emphasized or circumvented, since most released animals will be caught or killed in the first few days or weeks after release. In either case, concerns regarding the introduction of 'foreign' genes or diseases into natural populations will depend on the existence or size and distribution of the natural population at the release point and the potential mobility of the species, which could affect the intermixing of natural and captive-reared populations. This concern is minimal if wild stock is used to produce the captive-reared animals.

Chapter 20

Welfare and Ethics

The welfare of captive animals has both physical and psychological parameters. Feelings of well-being are usually high if animals can sustain physical fitness (i.e. freedom from disease, injury and incapacity) and avoid mental suffering (Webster, 2001). No great cognitive powers are needed to feel hunger or pain, so it may be that the capacity to feel emotions is widespread in the animal kingdom (Dawkins, 2000). Some species may experience 'higher feelings' such as friendship or grief. Scientists are just beginning to understand the nature of emotions, their function and mechanisms of control.

Humans cannot totally shield their children from physical and mental suffering. Likewise, animal caretakers cannot guarantee that the animals under their care will never suffer. Nevertheless, it is an ethical obligation for animal caretakers to strive to develop and implement animal management systems that enhance physical and mental well-being and minimize suffering. This goal cannot be reached without the input of science-based studies on how animals respond to various management systems (e.g. Barnett *et al.*, 2001).

The first part of this chapter discusses selected animal welfare issues as they relate to the process of domestication and the management of captive animals. It begins by discussing welfare issues associated with bringing animals into captivity and subjecting them to routine handling procedures. Welfare problems in nature and captivity are then compared and guidelines proposed for identifying welfare-promoting animal management practices and behaviorally relevant housing systems. This first section ends with a discussion of atypical behaviors exhibited by captive animals.

The second part of this chapter focuses on ethical issues associated with rearing wild and domestic animals in captivity. Factors influencing the quality of care provided for captive animals are presented. The welfare impact of artificial selection for specific characters is discussed as well as the balance that must be maintained between welfare, production efficiency and profitability. The chapter

©CAB *International* 2002. *Animal Domestication and Behavior*
(E.O. Price)

ends by addressing ethical issues associated with releasing captive-reared animals in nature and the role of zoos in conservation programs.

Bringing Wild Animals Into Captivity

Capture, restraint, handling and transportation of wild animals from their native environments into captivity constitutes a host of potentially stressful events (Martin, 1975; Grutter and Pankhurst, 2000; Moore *et al.*, 2001). Once in captivity, wild-caught animals are typically exposed to novel places, unfamiliar conspecifics, presence of humans and periodic handling, changes in diet, and restriction in the quantity and quality of space for locomotion and associated activities. In short, the wild-caught animal is abruptly forced to inhabit a very different environment from that in which it had been living, an event which can cause suffering and challenge the animal's adaptability. As noted in previous chapters, some species can make this transition more successfully than others.

Capture and Confinement Stress

The stress of handling and confinement can challenge the animal's physiological homeostasis and psychological well-being (Wedemeyer, 1997). Pankhurst (2001) reported that spiny damselfish (*Acanthochromis polyacanthus*) could be stalked and chased in their native habitat for up to 60 min without a significant increase in plasma cortisol concentrations. In contrast, capture and confinement in the laboratory for either 2 or 6 h resulted in significant increases in cortisol and reductions in gonadal steroids. Suleman *et al.* (2000) determined the early effects of capture and confinement on wild African green monkeys (*Cercopithecus aethiops*) by monitoring changes in body weight and pathological changes in two stress target organs, the adrenal gland and stomach. By day 45 post-capture, body weight losses ranged from 33 to 50% of their weights at capture. This weight loss occurred in spite of being fed familiar locally available grains *ad libitum* and maintaining a good appetite, suggesting a catabolic effect on body weight mediated by hypercortisolimia. Compared with wild control monkeys euthanized with a rifle, pathological lesions were noted in both adrenal glands (acidophilic, hyperplastic and hypertrophic cells) and stomach (gastric ulcers) after 1 and 45 days in captivity. High levels of circulating cortisol and a twofold enlargement of the adrenal glands between day 1 and day 45 post-capture were also found. High plasma cortisol levels were correlated with a decrease in the capacity of the cellular immune response by day 45 (Suleman *et al.*, 1999).

Hennig and Dunlap (1978) noted that the duration of the tonic immobility response to manual restraint in wild anoles (*Anolis carolinensis*) decreased over days in captivity and was partially ameliorated by providing a more natural housing environment (i.e. foliage rather than an empty terrarium). Boice and Williams (1971) also showed decreased susceptibility to tonic immobility in frogs (*Rana pipiens*) with length of time in captivity. To the extent that duration of tonic immobility in these species is related to relative levels of stress or fearfulness of humans, adaptation to the laboratory environment, including man, is a gradual process and may require many days or months depending on the species.

Ha *et al.* (2000) described the mortality associated with the translocation of both captive-bred and wild-caught pigtailed macaque monkeys (*Macaca nemestrina*) from the Washington Regional Primate Research Center's Primate Field Station (Medical Lake, Washington, USA) to the Tulane Regional Primate Research Center (Covington, Louisiana, USA). The move affected both the physical and social environments of the animals by moving them from indoor to outdoor quarters and from small, single-male groups to large, multi-male groups. Wild-caught animals experienced significantly lower survival rates during the 250 days following the move to Tulane than did wild-caught animals that remained in Washington (Fig. 20.1). Survival of captive-reared animals transfered and not transfered to Tulane did not differ. The wild-caught monkeys were clearly not as successful (i.e. flexible) as captive-born animals in adapting to a change in their living environment even after prior experience in captivity. Baker *et al.* (1998) documented indices of stress experienced by brushtail possums (*Trichosurus vulpecula*) when transfered from the wild to captivity. Females lost weight for 6 weeks following capture and males were still showing signs of maladaptation (hormone and immune responses) at 20 weeks.

Capture and confinement stress can have a negative effect on reproductive performance. Wild-caught female primates imported during pregnancy frequently lose their infants (Hertig *et al.*, 1971). Alibhai *et al.* (2001) found that immobilization of wild female black rhinoceroses (*D. bicornis*) for dehorning, ear notching and radio collaring resulted in reduced fertility (calves born per year). Hartley *et al.* (1994) reported that reproductive failure in red foxes (*V. vulpes*) was associated with stress-related elevation of plasma cortisol concentrations. Persistent handling of pregnant blue foxes (*A. lagopus*) results in an increase in plasma concentration and adrenal production of cortisol and reduces maternal and fetal body weights (Osadchuk *et al.*, 2001).

Handling stress can affect the timing of reproduction in fish but the specific response may differ in different species (Schreck *et al.*, 2001). For example, in rainbow trout, *O. mykiss*, handling or disturbance delayed ovulation if experienced during early ovarian development, whereas the same treatment accelerated ovulation if experienced late in the vitellogenesis period (Contreras-Sanchez *et al.*, 1998).

Self-imposed food deprivation often occurs in response to capture and confinement. In a review of the behavioral response of fish to stress, Schreck *et al.* (1997) point out that stress induced by handling frequently results in loss of feeding behavior for different periods of time depending on the severity of the stress and the physiological state of the fish. Resumption of feeding behavior by Pacific salmon following a stressful event is often correlated with their return to pre-stress physiological status. DiLauro (1998) described the failure of wild-caught Atlantic sturgeon (*Acipenser oxyrinchus*) to feed once brought into captivity. After 2 months of

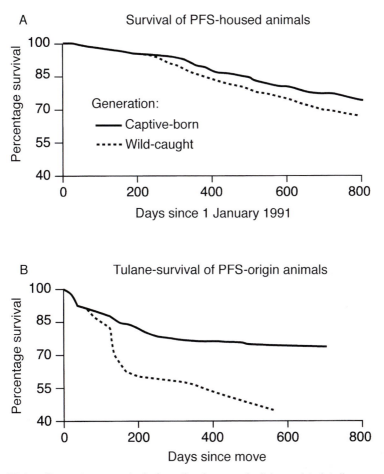

Fig. 20.1. Percentage survival of captive-born and wild-caught pigtail macaque monkeys when transfered from indoor enclosures at the Primate Field Station (PFS) of the Washington Regional Primate Center to outdoor enclosures at the Tulane Regional Primate Center (Louisiana) compared with monkeys that remained at the Primate Field Station. Survival of captive-born animals was identical in the two locations, while wild-caught animals moved to Tulane experienced much higher mortality than wild-caught animals remaining at the PFS (Ha *et al.*, 2000).

refusing to eat a variety of natural and artificial foods presented to them, the care-takers force-fed the fish under anesthesia. After three to six force-feedings spaced 3–4 days apart, the sturgeon began feeding voluntarily. In seeking an explanation for the loss of appetite, the authors ruled out nonoptimal oxygen and temperature levels, ammonia stress and intraspecific behaviors. Handling and disturbance stress associated with removing the fish from their native habitat and relocating them to a new environment were the most likely causes of this self-imposed fasting.

Recovery from Stressful Events

Being captured and handled by humans is not only stressful for the wild animal but the captive environment may impede the animal's natural recovery mechanisms. Milligan *et al.* (2000) noted that when trout are subjected to 'angling stress' (caught on a hook and line) and released, they seek refuge in the current and continue swimming rather than seeking refuge in still water. An experiment was designed to test the hypothesis that moderate exercise following an exhaustive event facilitates metabolic recovery. Rainbow trout (*O. mykiss*) were manually chased around a circular tank for 5 min, after which they were assigned to either a swimming recovery or recovery in still water for a 6-h period. Swimming after the exhaustive exercise had a significant impact on the recovery of metabolic and acid–base status. For example, muscle glycogen and lactate levels had both returned to pre-exercise levels within 2 h in swimming fish, compared to more than 6 h in the still-water fish. A most interesting finding was that plasma cortisol levels did not increase at all in fish placed in a current after exercise, while fish placed in still water showed a fivefold increase in cortisol levels. The researchers concluded that the prolonged recovery associated with exhaustive exercise in trout is due to elevations in plasma cortisol concentrations caused by post-exercise inactivity. It is not known whether a similar response to exhaustive exercise exists in other vertebrate taxa, but this knowledge may prove useful in designing captive environments for fish that assist them in their recovery from stressful events.

Tolerance to Handling and Confinement Stress

A number of studies have reported strain and individual differences in the physiological response to handling by humans. Weil *et al.* (2001) noted that growth rates of rainbow trout (*O. mykiss*) were greatest for fish that showed the fastest decrease in post-stress cortisol concentrations following a standardized handling treatment. Salonius and Iwama (1993) reported that wild coho (*O. kisutch*) and chinook salmon (*O. tschawytscha*) exhibited significantly higher plasma cortisol levels in response to handling than hatchery-reared counterparts derived from the same genetic stock. Furthermore, coho salmon collected as eggs in the wild, reared in a hatchery for 5 months and then transferred to a natural riverine environment as fry showed a cortisol response to stress similar to wild fish and higher than the response of their hatchery-reared sibs. Cleary *et al.* (2000) examined the effect of capture and handling stress on plasma steroid levels and the gonadal condition of three populations of snapper (*Pagrus auratus*): (i) wild-caught (trapped); (ii) reared in captivity for 5 years after being captured in nature as juveniles; and (iii) hatchery born and reared. Compared to the latter, both wild-caught and hatchery-reared wild fish showed higher levels of plasma cortisol, higher incidence of ovarian atresia and poorer indications of post-capture recovery of plasma gonadal steroid levels following capture and confinement. These studies suggest that the rearing environment can have an important effect on responsiveness to stressful events.

Interestingly, Pottinger and Carrick (1999) found a relatively high heritability (0.41) for confinement-induced plasma cortisol response in farmed rainbow trout (*O. mykiss*), suggesting that stress responsiveness in trout can be manipulated by selective breeding. Male and female trout were segregated into high- (HR) and low-responding (LR) individuals on the basis of their plasma cortisol response to a 3-h period of confinement (six or seven fish in a 50-l tank) imposed at monthly intervals for 5 months. Progeny groups (full-sib families) were obtained from the pairing of HR with HR and LR with LR fish. Randomly selected fish (US) served as a control. In spite of the fact that there was no difference in baseline cortisol rates in the three lines, the cortisol responses of progeny groups reflected the responsiveness of their parents (HR > US > LR; 178, 126 and 81 ng ml^{-1}, respectively). This demonstration that responsiveness to confinement stress is subject to modification by selective breeding identifies a useful tool for optimizing the performance of fish under intensive rearing conditions and possibly improving animal welfare. It is not clear why the heritability of the cortisol response to confinement stress was so high in this case, since traits critically important for survival generally have low heritabilities (i.e. a large proportion of the genetic variability is already exploited by selective pressures). Perhaps the confinement conditions routinely imposed in the hatchery from which the test fish were obtained were within the adaptability range for these fish, and thus their response to confinement had not been subjected to significant evolutionary pressures. Frequent exposure to stress during a sensitive developmental period may 'down-regulate' responsiveness (Salonius and Iwama, 1993; Pottinger and Pickering, 1997).

Further research is needed to test the hypothesis that natural selection in captivity purges stress-sensitive animals from populations of animals undergoing domestication. Pankhurst and Van Der Kraak (1997) claim that even though stress continues to be a problem in domestic stocks of fish, anecdotal evidence suggests that stress effects on reproduction decline with each cultured generation. While this proposal has intuitive appeal, data are needed from different taxa to confirm its validity. Unlike reproduction, no single endocrine cascade is responsible for the control of growth, making the assessment of the specific effects of handling and confinement stress on growth more difficult (Pankhurst and Van Der Kraak, 1997).

Training Animals to Handling Procedures Can Reduce Stress

Training captive animals to engage in various handling procedures can potentially reduce the distress experienced by both animal and handler. Adult rhesus monkeys (*M. mulatta*) have been trained to extend a leg out of their home cage for blood collection (Fig. 20.2), using simple positive reinforcement techniques (Reinhardt, 1991, 1996). Inappropriate handling of semi-wild ungulates such as farmed red deer (*C. elaphus*) can induce severe acute stress, trauma and capture myopathy (Hanlon, 1997). Animals familiar with the stockperson and the layout

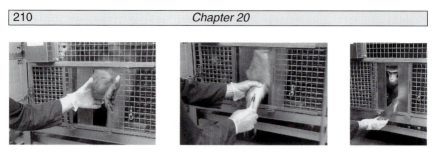

Fig. 20.2. Left to right: an adult male rhesus monkey extends his leg out of the cage, blood is collected and he receives a food reward (Reinhardt, 1991, 1996).

of the handling facilities are less prone to panic and experience severe stress. New stock and calves at weaning are particularly difficult to handle because of their inexperience (Hanlon, 1999).

Welfare Problems in Nature and Captivity Compared

A common error made in comparing the welfare of wild and domestic animals is to assume that captive-reared animals, as a whole, necessarily experience better (or worse) welfare than their free-living wild counterparts. Animals in captivity have their basic needs (e.g. food, water, shelter) provided, are not forced to contend with predators and are normally given medical attention for injuries, disease and parasites. Admittedly, these facts stand in sharp contrast to the suffering often experienced by animals in nature. However, there are important trade-offs associated with life in captivity, particularly with species reared in intensive management systems. Animals may be periodically subjected to various handling procedures, physical restraint and transportation, which may be stressful (see review by Grandin, 1997). While the availability of food and water is typically assured, time normally spent searching for food and water in their natural environment is now available to the animals in a stimulus-deprived environment, providing little opportunity for locomotor activity and perceptual stimulation. Young mammals are normally weaned at an earlier age than in nature so the mother can devote more of her energies to her next pregnancy. In addition, the young are typically abruptly and totally separated from their parent(s) at weaning, a practice that can cause distress and reduced welfare to both young and adults (Heller *et al.*, 1988; Worobec *et al.*, 1999; E.O. Price *et al.*, unpublished observations). Subordinate animals may not be able to escape from aggressive cagemates and thus experience severe social stress. Does the absence of predators improve the quality of life of captive animals? To what extent do animals view human caretakers as quasi predators? Domestic animals typically live much longer than their wild counterparts. What special problems does that represent? The point being made is that for most animals, life in captivity is a very different experience from life in the wild and the welfare of captive animals should be assessed independently of what their wild counterparts experience in nature. Moran (1987) expresses this sentiment very aptly in this statement, 'The (suffering in the) natural world should not be used to justify cruelty or abuse of animals (in captivity), but neither does it serve as a reliable basis for the argument that all captivity is cruel by comparison.' As discussed in an earlier chapter, the pre-adaptations of animals, including their developmental plasticity, are critical to their well-being in the captive environment.

Animals born and reared in captivity, such as the captive-reared offspring of wild-caught animals, have the opportunity to develop adaptations to captivity early in life, which may improve their long-term well-being relative to wild-caught counterparts. Studies are needed to further examine the hypothesis that wild animals born and reared in captivity experience better welfare than wild animals reared in nature and subsequently brought into captivity. The species considered and the animal management systems employed in captivity will undoubtedly influence the outcome of such comparisons.

While domestication can facilitate adaptation to the captive environment, it offers no guarantee that the animal will experience good welfare in captivity. Similarly, it cannot be assumed that animals which have just begun the domestication process will experience poorer welfare than populations which have been maintained in captive environments for many generations (Vinke, 2001). Diet, housing and other management practices can swing the welfare pendulum either way.

A Framework for Identifying Welfare-promoting Animal Management Practices

Hughes and Duncan (1988) provided a theoretical model, which, although simplified, serves as a general framework by which animal caretakers can better assess which animal management practices are likely to promote good welfare and which are not. Hughes and Duncan broadly classify behaviors into those: (i) internally motivated; (ii) externally motivated; and (iii) motivated by both external and internal mechanisms, all the while acknowledging the fact that all behaviors are ultimately controlled by both internal and external factors. Internally motivated behaviors should be accommodated; they are not optional to the animal. Examples include the behavioral need to lie down, stretch, seek food when hungry and possibly to engage in locomotor activity for its own sake (not driven by curiosity). Conversely, providing captive animals with the opportunity to engage in specific behaviors motivated by external cues should be optional. While the animal's response to external stimuli may be adaptive or may give the animal something to do, specific externally motivated behaviors are often not essential to the animal's well-being. For example, animals do not normally seek out the opportunity to exhibit aggressive behavior for its own sake. Motivation to exhibit aggressive behavior is not related to time since the last aggressive encounter. It is a means to an end, not an end in itself. Animals may choose to respond aggressively to conspecifics once they are encountered in the course of their daily activities. There is no evidence that animals suffer if not given the opportunity to participate in aggressive social interactions. Between these two extremes are behaviors motivated by both internal and external stimuli. The importance of these behaviors to the animal's welfare is more difficult to assess. For example, there are obviously both internal and external stimuli that motivate animals to engage in sexual behavior. Hormonal changes prime them to be attracted to potential mates and respond with species-specific courtship and/or mating behaviors once a sexually receptive mating partner is perceived. If a healthy adult male perceives a sexually receptive female and cannot gain access to her, he is likely to experience some degree of distress (i.e. frustration). But will he suffer if a suitable sex partner is not perceived in the first place? Probably not. Obviously, all behaviors cannot be placed neatly into these three categories. Rather, many behaviors will fall somewhere between categories. Nevertheless, the rationale used in this classification scheme may be useful to animal managers seeking more welfare-friendly management practices. As the mechanisms underlying motivation become better understood, it will be possible to more intelligently assess and provide for the behavioral needs of captive animals.

The 'Need' for Perceptual and Locomotor Stimulation

While there are specific behavioral needs of captive animals that are more or less easily identified, satisfying the animal's internally driven recurring need to be active and to engage in perceptually stimulating events is more problematic. Re-creation of natural environments in captivity is the logical solution to the problem but this course of action is seldom possible. The lives of free-living animals are enriched by countless forms of stimulation periodically changing in dynamic ways. In captivity, the need for perceptual and locomotor stimulation can be met in different ways, provided that alternative behaviors can be substituted for many of those performed in the wild without jeopardizing the animal's welfare (Swaisgood et al., 2001). Finding acceptable forms of stimulation for captive animals may be limited only by the housing systems employed and the creativity of animal caretakers. Mason et al. (2001) tested the motivation of captive mink (*M. vison*) to work for the opportunity to access: a water pool measuring about 1.5×0.5 m and filled with 0.2 m of water; a raised platform reached by a 2-m vertical wire tunnel; novel objects changed daily; an alternative nest site (box of hay); toys for manipulation and chewing (e.g. tennis ball); a plastic tunnel; and an empty chamber, each located in a separate compartment. Doors providing access to the seven compartments were weighted (a series of 0–1.25-kg weights) to determine the 'costs' mink were willing to 'pay' to access the various compartments. The animals worked the hardest to access the water pool, a result not totally unexpected considering that wild mink typically live in or near aquatic environments. An interesting follow-up study showed that when denied access to the water pool, urinary cortisol levels (produced in response to stress) increased to an average of 34% over baseline. This effect did not differ significantly from the increase in cortisol levels observed (50%) in response to food deprivation. In contrast, cortisol levels did not increase in response to blocking access to the alternative nest site or the empty compartment. Not only was the water pool the most valuable resource out of those presented to the mink, but they appeared to be frustrated when access to the pool was denied. It was concluded that preventing captive mink from swimming causes frustration. While this may be the case for caged mink that have been given prior access to a water pool, it does not necessarily mean that being given the opportunity to swim is an absolute requirement for captive mink. Hansen and Jeppesen (2001b) found no difference in the levels of stereotyped behavior exhibited by mink reared with and without the opportunity for swimming. In contrast, mink reared in large cages exhibited less stereotyped behavior than mink reared in small cages. What other forms of activity could serve as an effective substitute for swimming or vice versa? How specific are the animals' requirements?

Behavioral Relevance of Captive Housing

Determining how free-living wild or feral animals spend their time can offer important clues to the kind of management techniques and housing conditions which might best serve their behavioral needs in captivity. Newberry and Wood-Gush (1988) used this approach to gain a better understanding of the requirements of young pigs (*S. scrofa*). They studied the development of behavior of young domestic piglets living in a 2.3-ha enclosure (Edinburgh Pig Park) containing a running stream, grassy slopes, gorse bushes (*Ulex europaeus*) and an area of woodland comprised mainly of pine trees (*Pinus sylvestris*). The population consisted of five adult sows, an adult boar and a variable number of juvenile (weaned) pigs that had been born in the Pig Park. Young piglets used considerable space in locomotor activities, especially between 2 and 6 weeks of age, suggesting that the availability of space during this period is particularly important. Interestingly, none of the piglets developed leg or foot disorders during the course of the study although these are relatively common under intensive housing conditions. Coprophagy was not observed in the young piglets in the park, while rooting in feces is commonly seen in confined young piglets. Redirected sucking behavior was not observed in the Pig Park piglets. Weaning occurred naturally between 6 and 10 weeks of age, which gave the piglets ample time to adjust to eating solid foods and to lose their motivation to suck. By contrast, piglets weaned at 4 weeks or earlier in production units typically exhibit relatively high rates of sucking and mouthing of other pigs and inanimate objects (Fraser, 1978). Stolba (1981) and Stolba and Wood-Gush (1981, 1984) had earlier studied the 'natural' behavior of pigs in the Pig Park and subsequently designed a captive housing system for pigs ('Edinburgh Family Pen System of Pig Production'), which attempted to accommodate the behaviors observed in the Pig Park in a semi-commercial, welfare-friendly production system. Three versions of the 'Family Pen System' were studied (Kerr *et al.*, 1988), each involving alterations to the previous design to improve welfare and productivity. Specific areas in the Family Pen System were constructed for specific functions (Fig. 20.3) in keeping with behavioral

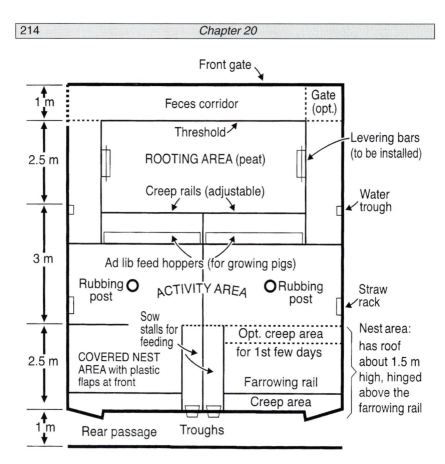

Fig. 20.3. Third version of the 'Family Pen System'. Sows and their offspring are able to perform a variety of activities due to the high level of pen complexity, and welfare is improved (Kerr *et al.*, 1988).

observations in the Pig Park and observations on wild boar. Levels of aggression were low in the Family Pen System due to keeping families (familiar animals) together throughout their life and by not re-grouping or mixing pigs, as is commonly practiced in commercial pig units. The Family Pen System was designed to allow sows to remove themselves from physical and visual contact with their piglets without altering nursing frequency. Consequently, a very high 93% of the sows conceived at a lactational estrus, negating the need for early artificial weaning of piglets at the traditional 21–24 days postpartum. Not only did the piglets avoid the stress of early weaning, they were not exposed to stressors associated with being moved to unfamiliar living quarters and changes in group composition. Farrowing rates (litters per year), litter sizes and reproductive failure in the Family System were comparable to the best commercial pig farms. Mortality rates of piglets were relatively high in the Family System (18.5% from birth to 21 days, due largely to crushing by the sow), growth rates were average and labor requirements were greater than in commercial enterprises. Overall, the Family Pen

System represents a welfare-friendly alternative to more conventional pig rearing and management practices. Further refinement in design may reduce losses in productivity and profitability. Acceptance of such welfare-friendly farm-animal rearing systems will largely depend on consumer willingness to pay the added costs.

The physical location of enclosures can have important behavioral implications for welfare. Rekilä *et al.* (1996) noted that farmed blue foxes (*A. lagopus*) penned in the front of sheds were less fearful of humans than those penned in the rear. They proposed that foxes in the front pens were more frequently exposed to humans and farm activities, and thus had become more habituated to people and disturbance or they were less fearful because of higher levels of stimulation.

In nature, socially subordinate silver foxes (*V. vulpes*) seldom breed (Nimon and Broom, 2001). On fox farms, breeding vixens are typically housed singly but in close proximity to other foxes. Killing and injury of cubs is a common problem. Bakken (1993a,b) reported that infanticidal behavior in fox vixens (*V. vulpes*) often was related to the social status of neighboring foxes. Vixens of low social status were more likely to kill their cubs than vixens of high social status, and the incidence of infanticide was greatest when vixens of low social status were housed in a pen adjacent to a female of high competitive ability. Since fox farmers seldom know the social status of their animals, neighbor-induced infanticide is likely to continue.

Designing a captive environment that simulates living conditions in the wild is not necessarily in the best interest of the animals involved. Early fox farmers initially reared and maintained their animals in groups in relatively large, earthen-floored pens with underground dens. Unfortunately, aggression was common, young mortality was high in the damp underground nests, nearly all animals got intestinal parasites and the foxes did not habituate to the presence of humans (Forester and Forester, 1973, cited in Ahola *et al.*, 2000). This led to the practice of housing foxes singly after weaning in wire-mesh cages located in outdoor sheds. A recent study (Ahola *et al.*, 2000) revisited this topic and found that, although a larger space benefitted the physical endurance of silver foxes (*V. vulpes*), there was no clear correlation with physiological indices of welfare. Group-reared foxes experienced social tension, and the foxes housed in enclosures had a greater fear level toward humans than foxes reared in cages. Korhonen *et al.* (2001) reported a similar finding in blue foxes (*A. lagopus*). Providing 4.2 times more space than in standard cages, with or without an earthen floor, did not significantly improve the animals' welfare based on the behavioral and physiological indices used.

Designing a behaviorally relevant cage for farmed foxes presents an interesting dilemma from a welfare standpoint. Group housing in large enclosures may result in increased social tensions and fear of people. Individual housing in small cages restricts movement and activity. Provision of nest boxes may preclude habituation to humans and thus increase fearfulness. Increasing the physical complexity of cages also does not significantly improve welfare (see review by Nimon and Broom, 2001). Interestingly, when given a choice, farmed blue foxes (*A. lagopus*) prefer cages that provide the best view of their immediate surroundings

(Mononen *et al.*, 2001). This makes sense when one considers that their wild counterparts live in arctic areas with open landscape and few places to hide from predators. Since it appears that the welfare of farmed foxes is not significantly improved by modification of their cage environment, new directions should be pursued, such as artificial selection of foxes for tolerance and adaptability to existing housing conditions and the presence of man. A similar approach has been suggested for farmed mink (*M. vison*), whose welfare also has not been significantly improved by new cage designs (Malmkvist and Hansen, 2001).

Environmental Control as a Form of Enrichment

As discussed in Chapter 17, simple enrichment devices that permit animals to acquire some degree of control over their environment have the potential to improve welfare. Hanson *et al.* (1976) trained one group of rhesus monkeys (*M. mulatta*) to control the onset of white noise, while an experimental group had no control over the noise. The monkeys with control of the noise exhibited lower levels of cortisol (stress hormone) than the controls. When the privileged monkeys were denied control of the noise, their cortisol levels rose higher than those of the control group, even though noise exposure was the same for both groups. It would appear that the benefit of control may be relative to expectations and that a net loss in control can have a negative effect on welfare.

The consequences of a stressor can be altered by psychological factors such as predictability. Weiss (1970) conducted four experiments to investigate the effects of stressor predictability on a variety of stress responses such as stomach ulceration, plasma corticosterone concentration and changes in body weight. Rats that received unpredictable electric shocks showed greater stress reactions and more stress-induced pathology than animals that received the same shocks but could predict their occurrence by an auditory signal. Stomach ulceration was not much more severe for signaled animals than for the nonshocked controls, suggesting that the shock itself was only a minor determinant of stomach ulceration in comparison to the psychological variables. Weiss (1971) points out that having control over the shock was not itself the significant psychological factor but rather the frequent relevant feedback from responses.

Effects of Early Positive Handling on Welfare

Captive animals' fear of humans and unfamiliar objects, places and conspecifics represents one of the greatest challenges to their welfare. Early positive handling can reduce the stress experienced by animals when in contact with humans (Hemsworth *et al.*, 1981; Fig. 20.4). Collette *et al.* (2000) describe the effects of an early handling experiment with orange-winged Amazon parrots (*A. amazonica*) on their subsequent responses to handling and their immune responsiveness. At 10 days post-fledging, the birds were restrained for 10 min either by being held

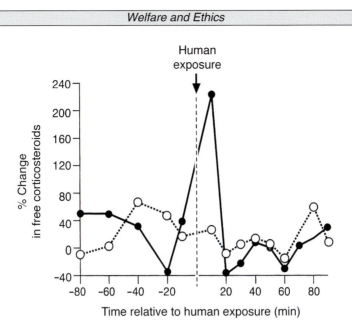

Fig. 20.4. Percentage change in free corticosteroid concentrations in negatively (solid circles) and positively (open circles) handled gilts before and after a 2-min exposure to a human (Hemsworth *et al.*, 1981).

while perching ('tame/handled' group) or by being restrained in a towel ('untamed/nonhandled' group). The delayed-type hypersensitivity response to phytohemagglutinin-P injection was greater in nonhandled chicks, as were serum corticosterone levels. Heterophil:lymphocyte ratios did not differ for handled and nonhandled birds, but nonhandled birds tended to exhibit reduced antibody titers to a killed Newcastle disease virus challenge. It was concluded that early handling can reduce the distress experienced by young parrots during routine handling. Neonatal handling is also known to enhance tameness, alter the stress response and improve immune competence in domestic rats (*R. norvegicus*) and chickens (*G. domesticus*) (see reviews by Liu *et al.*, 1997 and Collette *et al.*, 2000).

Atypical Behaviors

Captive animals often exhibit atypical behaviors not seen in their free-living counterparts. For example, captive wild rodents often exhibit repetitive wire-gnawing (Würbel *et al.*, 1996), jumping up and down on a vertical surface (Ödberg, 1986) or backflipping behavior (Callard *et al.*, 2000) in laboratory cages. Such atypical behaviors are generally viewed as a sign of poor welfare, particularly if the behaviors are repetitive and stereotyped (Mason, 1991a; Duncan *et al.*, 1993). It appears that stereotyped behaviors in captive animals typically develop as a consequence of frustrating experiences or stimuli imposed on animals genetically predisposed to develop such behaviors (Würbel and Stauffacher, 1998;

Schoenecker and Heller, 2000). Intense stimulation rather than understimulation may play an important role in stereotypy development (Cooper and Nicol, 1991). Although one could argue that stereotyped behaviors provide an outlet for locomotor and perceptual stimulation or general arousal in an otherwise impoverished environment and thus may be beneficial to the animal, there is little evidence to support this hypothesis (Nicol, 1999). Stereotyped and other atypical behaviors often become insensitive to changes in the environment directed at 'satisfying' the original motivational basis for the behavior. They can also result in self-injury, reduced reproductive success and increased mortality.

Effects of early weaning and social isolation

There is increasing evidence that many stereotyped behaviors reflect feeding rather than housing problems (Rushen and de Passillé, 1992). Würbel and Stauffacher (1998), working with laboratory mice (*M. musculus*), demonstrated that premature weaning and low weaning weight results in the development of relatively high levels of stereotyped wire-gnawing on the lids of their cages. ICR mice were weaned at 17 (premature) or 20 days (normal weaning age) following birth. Of the 20-day-old mice, the heaviest and lightest were selected for study. All mice developed stereotyped wire-gnawing. At 80 days of age, mice weaned prematurely and the lightest mice weaned at 20 days exhibited significantly more wire-gnawing

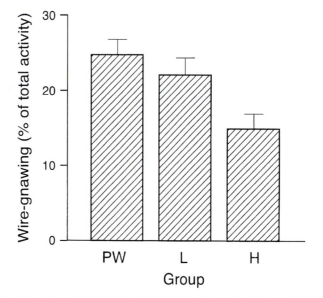

Fig. 20.5. Levels of stereotypic wire-gnawing for ICR mice at 80 days of age. Mice weaned 3 days prematurely (PW) at 17 days and mice with low weaning weights (L) for their litter exhibited significantly ($P < 0.05$) more wire-gnawing than mice with relatively high weaning weights (H) (Würbel and Stauffacher, 1998).

than the heaviest mice weaned at 20 days (Fig. 20.5). Callard *et al.* (2000) reported that captive-born roof rats (*R. rattus*) began exhibiting stereotyped backflipping behavior almost immediately following weaning. Nicol (1999) noted that two-thirds of 14 horses that developed stereotyped behavior did so within 1 month of weaning, even though they were weaned at different ages. Both the nutritional and the social environment of the foal is substantially altered during this period.

Stress early in life could predispose animals to stereotypy by affecting the persistence of behaviors exhibited at that time. Jeppesen *et al.* (2000) studied the development of stereotyped behavior in mink pups (*M. vison*) weaned at 6, 8 or 10 weeks of age and subjected to two different housing conditions (traditional cages or three adjoining traditional cages). All subjects were socially isolated at 5 months of age. It was found that early weaning, small cages and individual housing all promote the development of stereotyped behavior in farmed mink. Mason (1996) also reported higher frequencies of stereotyped behavior among early-weaned female mink. Hubrecht *et al.* (1992) found that a higher percentage of single-housed dogs (*C. familiaris*) exhibited repetitive behaviors than group-housed animals. Mason (1991b) hypothesized that adult stereotypy levels may resemble a 'behavioral scar' representing the level of stress at the onset of stereotypy development. Amir *et al.* (1981) have pointed out that stressful experiences may lead to the massive release of endogenous opioid peptides, which sensitize the dopaminergic pathways that mediate the behavioral responses to stress. The 'scarring' effect may be a result of this process (Cabib, 1993; Würbel and Stauffacher, 1997).

Nutritional considerations

Feeding practices may have a greater effect than housing practices on the performance of atypical behaviors by horses (*E. caballus*; Marsden, 1993). Feeding horses a high-concentrate, low-forage diet consistently results in a greater incidence of oral-based stereotypical behaviors such as wood chewing and crib-biting (Haenlein *et al.*, 1966; Willard *et al.*, 1977; Johnson *et al.*, 1998). Concentrate diets are known to alter cecal fermentation and increase cecal acidity. Oral stereotypies that develop in response to a low-fiber diet may be adaptive by increasing salivary flow, thus reducing the acidity of the gastric tract (see review by Nicol, 1999).

Lack of stereotyped behavior in wild-caught animals in captivity

Callard *et al.* (2000) noted that wild-caught roof rats (*R. rattus*) maintained in laboratory cages were never observed backflipping even though it was predictably observed in their offspring. A similar result was reported by Sorensen and Randrup (1986), Cooper and Nicol (1996) and Schoenecker *et al.* (2000) working with voles (*Clethrionomys glareolus*). In the latter study, none of 92 wild-caught bank voles exhibited stereotyped behaviors when socially isolated in laboratory cages. However, 74 of 248 (30%) of their offspring (F1 generation) had developed stereotypies by 96 days of age. It is clear from these studies that being reared in a laboratory cage environment is responsible for the development of stereotyped behavior in these species. Wild-caught animals exhibit a different locomotor response to confinement than their cage-reared counterparts. Cooper and Nicol (1996) noted that wild-caught voles spent more time under cover in the laboratory than their offspring or voles from a stock several generations removed from the wild. This is a very appropriate response to aversive conditions in the wild but one that young rodents apparently do not acquire in laboratory cages. Does neurophysiological development in the wild in some way protect animals from developing atypical locomotor responses in captivity or atypical responses to external cues? Or, conversely, does the barrenness of the captive environment result in abnormal brain development?

Genetic predispositions

There is good evidence that animals can inherit a predisposition to develop stereotyped behaviors. Schoenecker and Heller (2000) reported a genetic pre-disposition for the development of stereotypies in laboratory-bred bank voles (*C. glareolus*) of the F2 generation. When one or both parents of the F2 generation exhibited stereotyped behavior, 53% of their socially isolated offspring also developed stereotypies. Only 8% of the offspring of nonstereotypic parents developed stereotypies. Callard *et al.* (2000) found litter differences in the rate of stereotyped backflipping behavior in roof rats (*R. rattus*). Schwaibold and Pillay (2001) reported that the development of stereotyped behavior in captive striped mice (*Rhabdomys pumilio*) was strongly related to its occurrence in their biological mother. In breeding experiments, females who had developed stereotyped behavior and nonstereotypic females were mated to nonstereotypic males. The incidence of stereotyped behavior was approximately four times greater in the offspring of stereotypic females than in young born to nonstereotypic females. Cross-fostering young between stereotypic and nonstereotypic females had no effect on patterns of transmission.

Effects of environmental enrichment

Environmental enrichment has been used to prevent or alleviate stereotyped behaviors in captive animals (see review by Carlstead and Shepherdson, 2000). Increasing the complexity of the living environment can reduce the frequency of injurious pecking in birds. Martrenchar *et al.* (2001) compared the incidence of wing, tail and head injuries inflicted by penmates in populations of domestic turkeys (*M. gallopavo*) housed either on wood shavings (5 cm thick) or on shavings with nonchopped straw and reflective galvanized iron sheets (15×20 cm) suspended at the birds' head level. Peck-induced injuries in the control populations were significantly greater than in the populations with the added straw and objects. For each peck at a neighboring bird in control pens, there were four to ten pecks at substrates in the experimental populations. It was hypothesized that the experimental treatment resulted in the redirection of injurious pecking behaviors from penmates to inanimate pecking stimuli. Alternatively, one could propose that the presence of inanimate pecking stimuli (a more natural situation) discouraged or precluded the redirection of pecks toward penmates (an unnatural response).

Grindrod and Cleaver (2001) reported that stereotyped circling behavior in captive seals (*Phoca vitulina*) was reduced by providing them with devices that increased feeding/foraging time. Callard *et al.* (2000) demonstrated that offering shelters to captive roof rats (*R. rattus*) reduced the frequency of stereotyped backflipping. Fourteen of 28 (50%) shelter-housed subjects performed 1000 or fewer backflips per day, while only 4 of 22 (18%) rats without shelters (controls) met this criterion. Nine of 22 (36%) control rats performed more than 4000 flips per day, while none of 28 shelter-housed subjects exhibited this level of backflipping. Successful enrichment programs should assist in alleviating the 'need' to perform stereotyped behaviors, if not preclude their development in the first place.

Ethical Issues Related to Domestication and the Release of Captive-reared Animals

There are no absolute answers to ethical questions related to breeding and maintaining animals in captivity. Philosophical differences in attitude are at the heart of this debate, with utilitarian viewpoints, on one hand, supporting the use of animals for food, fiber, companionship, research, work, species recovery and entertainment and, at the other extreme, idealistic notions giving equal rights to animals and humans (Hurnik and Lehman, 1988). Science may determine what constitutes good and bad welfare for captive animals from a biological point of view but in the end it is the collective decision of society that determines what is appropriate welfare from an ethical perspective (Bennett, 1996). Reasonable people agree that animals should be treated responsibly and efforts should be made to minimize the pain and suffering of animals under man's control. It is beyond the scope of this book to debate the pros and cons surrounding the ethics of animal use in our society, but several points should be made that relate specifically to the processes of domestication and the release of captive-reared animals in nature.

Captive Animals Have Value

While the intrinsic value of animals serves as an incentive to improve husbandry conditions, it is often their extrinsic value that determines the quality of care actually given (Webster, 2001). The decision to capture animals in the wild and bring them into captivity, to raise farm or companion animals, or to raise wild animals for release in nature is normally done with utilitarian purposes in mind.

In most cases, captive animals represent an investment and, accordingly, have extrinsic value whether it be to provide food and fiber for humans or to aid in the recovery efforts of an endangered species. Value is lost if animals are not managed to maintain good health and well-being. Yet, captive animals are not always treated with their best interests in mind.

Factors Influencing the Quality of Care Given to Captive Animals

A number of factors can either consciously or unconsciously influence the quality of care given to captive animals under man's control. First, respect for life and capacity for empathy toward animals often reflects the caretaker's philosophical attitudes towards animals and his/her basic personality traits, both of which can influence the quality of care provided (Hemsworth and Coleman, 1998; Podberscek and Gosling, 2000). Relative and absolute investment (in animals) in terms of money, time, and physical and emotional energy can encourage animal caretakers to provide a high level of health care to protect their investment. A cow or horse will usually receive more care than a laying hen. Pride of ownership also motivates people to protect the well-being of their animals. Animals can enhance the owner's feelings of self worth and serve as an extension of his or her personality or image. Familiarity with the species or individual animals strengthens personal involvement and attention to individuals. Social bonds with animals are often strengthened by familiarity and vice versa. Familiar animals frequently receive names and are more likely to be attributed with human-like characteristics. Relative abundance of the species can influence the perceived importance of individual animals. The worth of individual animals increases when the species is declared rare or endangered. Animals providing social feedback to their caretakers (e.g. dogs) encourage the formation of human–animal bonds, which promote caring attitudes toward specific individuals and, perhaps, animals in general. The perceived esthetic value of an animal or species may influence caretaker attitudes. Large attractive animals associated with esthetically pleasing places or experiences may encourage higher quality care of individual animals. The perceived degree of self-awareness of the species (e.g. nonhuman primates) can motivate humans to raise standards for animal care. The extent to which animals conflict with human interests (e.g. destroy property or inflict injury on humans) can influence the perceived worth of animals. Wild Norway rats (*R. norvegicus*) living freely on farms or in homes are viewed and treated very differently than domestic Norway rats in laboratories. Caretakers often make a special effort to provide enriched living environments for animals prone to exhibit stereotyped behaviors in captivity (King, 1993). Societal (peer) pressure and legal requirements for animal care encourage or coerce people to provide better care for animals under their control. For example, the European Union Directive 99/74/EC banned the use of battery cages for laying hens and requires a nest, a perch, litter for dust bathing and 750 cm^2 space per bird. Such directives must take into account the financial effects on producers, impacts on consumer prices and international trade as well as any costs to the government and the taxpayer (Bennett, 1996). Lastly, through the free-market system, consumers of animal products may 'demand' higher levels of animal care from producers than is commonly practiced. Many producers will comply if consumers are willing to pay the increased production costs either in the grocery store or through increased taxes. McInerney (1998) proposed that if animal welfare is perceived as a quality characteristic of livestock production, consumers will expect to pay the extra costs. He further proposed that the economic costs of reasonable improvements in animal welfare are likely to be relatively small. For example, McInerney examined the cost to farmers and consumers of implementing the changes described above for rearing caged laying hens and estimated that the improved husbandry would increase production costs by 28%, while the increased cost to the consumer for eggs would be about 18%. Since eggs account for only 1.3% of the food budget of the average British household, the increased cost to the consumer for this one commodity would average 0.22% of the total food budget. In spite of this 'bargain' for the welfare-sensitive consumer, the overall global demand for high-welfare foods is very weak (Webster, 2001). Either the average person considers the increased costs too high, which is unlikely, or the cheapest product is purchased without concern for animal welfare. It appears that although improved welfare for farm animals is commonly voiced, it is not a high-priority demand of the masses, even in developed countries.

Degree of Adaptation to the Captive Environment

Ethical issues related to the welfare of captive animals are rooted in concerns about the ability of captive animals to adapt, both physically and psychologically, to their environment. The subjective states of animals are obviously important to their well-being but the current inability to effectively monitor animal feelings demands a more functional approach to welfare studies, recognizing that, in general, ability to adapt to the captive environment and feelings of well-being should be positively correlated. Fraser *et al.* (1997) recognize three classes of problems which may arise when adaptation to the captive environment falls short, thus affecting the animal's quality of life. First, an animal may possess adaptations, based on its evolutionary history, which serve no useful function in captivity often because the original function is now achieved in some other way. Some of these long-standing adaptations are of little consequence in captivity. The camouflaging stripes of the zebra are nonfunctional in captivity and are of no consequence to the animal's quality of life. On the other hand, some former adaptations may lead to unpleasant subjective experiences, even though these experiences may not negatively impact the animal's biological functioning. A bucket-fed calf may experience frustration when it finishes its meal before its desire to suck has been satiated, even though it has obtained adequate nourishment. In other words, suffering may occur if animals experience a high and persistent motivation to perform actions that the environment does not permit. Secondly, if animals do not possess a full complement of adaptations to the captive environment, functional problems may arise which are not necessarily accompanied by subjective feelings. For example, an animal breathing polluted air may develop lung damage without being aware of the problem (i.e. be conscious of suffering), at least until the advanced stages of the pathology. Animals recently translocated to new environments often have not evolved subjective states to motivate adaptations to new environmental challenges. A third concern occurs when animals possess a full complement of adaptive responses but some responses prove inadequate when environmental challenges become extreme. In such cases, there is likely to be good correspondence between feelings and functioning. An animal may both feel and function poorly if it is exposed to climatic elements (e.g. temperature) that exceed

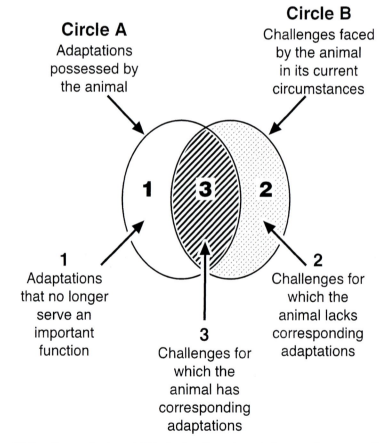

Fig. 20.6. Conceptual model illustrating three broad classes of concerns that may arise when the adaptations possessed by the animal (circle A) make an imperfect fit to the challenges it faces in its current environment (circle B). In areas 1 and 2, the animal's adaptations are not well matched to the challenges it faces, which may result in welfare problems (Fraser *et al.*, 1997).

its ability to adapt. Fraser *et al.* (1997) illustrate in Fig. 20.6 these three circumstances that can affect the quality of life of captive animals. This schematic is also useful in understanding the importance of pre-adaptations (to captivity) in animals undergoing domestication.

Effects of Artificial Selection on Welfare

Theoretically, animals in captivity experiencing the best welfare should leave more offspring for future generations than less-fit animals. Through natural selection in captivity, some genetically based adaptations to captivity will be attained over generations in captivity, a point discussed in earlier chapters. A highly stress-tolerant strain of red seabream (*Pagrus major*) has been produced in Japan after 30 years of captive rearing (Foscarini, 1988). In addition, there is evidence that artificial selection can improve the welfare of captive animals with respect to certain traits. Lines of animals have been selectively bred for divergence in a number of behavioral or physiological traits associated with response to stress (see review by Pottinger, 2000). Selection for the polled (hornless) condition in breeds of cattle (*B. taurus*) and sheep (*O. aries*) has possibly improved their welfare by reducing injury and fearfulness associated with agonistic interactions. Jones and Hocking (1999) discuss a number of selective breeding studies with Japanese quail (*C. japonica*) and chickens (*G. domesticus*), in which fearfulness, sociality, feather pecking, adrenocortical responsiveness and growth rate were manipulated, and point out how appropriate levels of selection for these traits could improve their welfare and productivity.

Adaptation to the captive environment may become complicated when artificial selection is intensely applied to very specific traits over many generations with little concern for its effect on the total phenotype. The pitfalls of overselection for a single trait are well known (see review by Jones and Hocking, 1999). For example, intensive selection of broiler chickens (*G. domesticus*) for rapid growth has repeatedly been associated with cardiac abnormalities, leg abnormalities and lameness (Lubritz *et al.*, 1995; Hughes and Curtis, 1997). More studies are urgently needed demonstrating the welfare implications of artificial selection for very specific traits and how correlated negative impacts on welfare can be avoided. The welfare problems associated with loss of fitness may be more serious than those associated with captive animals exhibiting abnormal behavior (Webster, 2001). Kangasniemi (1996, cited in Serenius *et al.*, 2001) reported that in Finland over 15% of the culled nonfarrowed gilts are culled due to leg weakness. Leg problems can result in decreased mobility and reduced food consumption.

Artificial selection has more than doubled the milk production of dairy cows (*B. taurus*) in the USA over the last 50 years (Rauw *et al.*, 1998). Fortunately, many improvements in dairy husbandry (e.g. nutrition, housing and veterinary practices) have accompanied this increased production but there is increasing concern about whether husbandry is keeping pace with demands on the animals. Fleischer *et al.* (2001) found positive correlations between milk yield and the incidence of

mastitis, milk fever, retained placenta, ketosis, ovarian cysts, foot problems and
displaced abomasum in a large sample of over 1000 dairy cows in Germany. As
milk production has increased, reproductive efficiency has declined (see review by
Rauw *et al.*, 1998; Lucy, 2001). While the increased biological demands of lactation may partly explain the decline in reproductive efficiency, Lucy has identified
many other factors contributing to the decline, including increasing herd sizes,
greater use of confinement housing, labor shortages and higher inbreeding
percentages. Future increases in milk production beyond the limits imposed
by the environment could compromise the animals' health if not accompanied
by a corresponding improvement in management techniques (Beilharz, 1998).
Hagen and Kephart (1980) reported that ovulation rate in a herd of primiparous
domestic (Yorkshire) swine (*S. scrofa*) was almost twice that of primiparous feral
gilts reared at the same facility (15.1 vs. 8.7 ova, respectively). Increased litter size
in domestic swine has been made possible by artificial selection, good management and health practices and excellent nutrition. At what point, though, will the
cost of improved husbandry or the risk to the animals' welfare be great enough
that selection for increased milk production or faster growth or increased litter size
is no longer justified? Is it ethical to continue selection until unacceptable levels of
health problems or reproductive performance are realized? Are scientists and/or
veterinarians prepared to set those limits and what ethical criteria will they use?
'Because the fundamental motivation for studying and promoting welfare is
ethical, and because the very concept of animal welfare forces one to make
and rationally defend value judgments about what ways of treating animals are
appropriate, ethics is among those disciplines that must comprise the field of
welfare' (Tannenbaum, 1991).

Production Efficiency, Profitability and Animal Welfare

To what extent are new domestications warranted to provide new products from
animals or other benefits to humans? How are the potential benefits to human
society weighted against the costs and benefits associated with life in captivity
for the animals involved, particularly in the early generations? For production
systems to be efficient and profitable, animal welfare may be less than optimal
(Sandøe *et al.*, 1996). How can an acceptable balance between production
efficiency, profitability and welfare be achieved? Considering the current rate of
growth of human populations and the importance of animal products for human
development, especially in underdeveloped countries, the search for ways to
provide animal products more efficiently and cheaply will undoubtedly continue.
Improved techniques for genetic modification of animals and improved animal
management practices must be developed to meet the increasing demand for
animal products, while protecting the environment and preventing further
encroachment on the world's undeveloped space to meet agricultural objectives.

New knowledge and tools are now available to genetically modify animals to
improve their well-being and usefulness to man (Mench, 1999; Gengler and

Druet, 2001). Many people oppose the use of genetic engineering even though it can benefit the animals involved or humankind (Spencer, 2001). Opposition to genetic change for its own sake, whether through gene manipulation/transfer or selective breeding, does not stand up to close scrutiny (Sandøe *et al.*, 1996). The gene pools of all animal populations, both wild and domestic, are already in a constant state of change through normal evolutionary processes. The existing gene pool of a population represents just one point in its evolutionary history. Secondly, genetic change by whatever means can be either beneficial or harmful. Yet, attempting to maintain the genetic status quo of animal populations does not necessarily benefit animals from a welfare standpoint and may be viewed as negative if opportunities to improve welfare are ignored. Advances in genetic engineering (e.g. animal transgenesis) must be accompanied by systematic risk assessment and investigations into possible harmful effects on animal well-being (Van Reenen *et al.*, 2001). To do otherwise is ethically irresponsible.

Ethical Concerns about Releasing Captive-reared Animals into the Wild

From a welfare standpoint, it is easier to ethically justify bringing animals into captivity from the wild than releasing captive-reared animals in nature. In both cases, individuals may not be totally prepared to meet the adaptations required for survival and reproduction in their new environment. However, in captivity, food, water, shelter, protection from predators, and health care are typically provided, whereas in the wild, captive-reared animals may not be prepared to meet challenges such as finding appropriate food items, adjusting to a new diet, avoiding predators, finding suitable shelter, navigation and orientation in an unfamiliar environment, integration into natural social groups and possessing the physical stamina to meet the demands of living in the wild. Failure to meet any one of these challenges could result in severe suffering and death. Persons responsible for release programs have a moral obligation to refrain from releasing animals that do not have a reasonably good chance of surviving in nature (Waples and Stagoll, 1997). It is ironic that a few radical animal rights groups will release captive-reared animals into environments that almost guarantee suffering and death in a relatively short period of time. As discussed in Chapter 19, preconditioning captive-reared animals to avoid predators and to search for natural foods improves survival following release in nature, but mortality is still decidedly greater than when animals are brought into captivity from the wild.

Does the Conservation Ethic Justify Captive-breeding Programs?

Does the conservation ethic justify bringing representatives of rare and endangered species into captivity solely to establish captive-breeding programs, knowing full well that some will not survive or reproduce because of the stress associated with life in captivity? To what extent can we justify sacrificing an animal's welfare to prevent extinction of its species?

The Role of Zoos in Conservation Programs

Zoos and wildlife parks have important roles in the conservation of rare and endangered species and in educating the public about animals and their habitats (Gippoliti and Carpaneto, 1997). Elimination of this role of zoos/wildlife parks will make them more vulnerable to anti-zoo sentiment, which considers it immoral to keep animals in captivity for any reason. The anti-zoo movement would like to replace live animals in zoos with 'electronic zoos' featuring videos, computers and other experiences of virtual zoos. While such developments should be encouraged for their intrinsic value, public support for the preservation and study of animals is best inspired by seeing and experiencing live animals first hand. It is interesting to note that broad support for the international drive to reduce whaling was not achieved until aquariums began to exhibit live whales and dolphins (Conway, 1978). Without public support (financial and otherwise), conservation efforts will be short-lived, and respect for all forms of animal life, wild and domestic, will suffer.

In addition to the ethical dilemma of whether to bring wild animals into captivity for display, various management practices are debated for their welfare implications. For example, techniques to prevent flight in captive birds incurs short-term handling stress and, perhaps, some pain but allows keepers to maintain them in larger, more esthetically pleasing natural areas (Hesterman *et al.*, 2001). Does the greater freedom allowed by the deflighting process justify the stress and pain incurred? What about more basic questions? Do birds suffer if they are not allowed to fly? Obviously, each management practice and associated handling event(s) must be weighed in the balance between benefit to the individual and species and any associated suffering or loss in productivity that may incur.

Conclusions

The welfare of captive animals is ultimately dependent on a different set of principles than for animals in nature. Domestication offers no guarantee that animals will experience acceptable levels of welfare in captivity. One can also not assume that captive wild animals will experience poorer welfare than highly domesticated species. Caretakers' attitudes can be key to good welfare. Stress may be minimized by genetic changes in captive populations, by instituting animal-sensitive management practices and housing designs, and by animals learning to adapt to the biological and physical features of their living environment. Efforts to minimize atypical behaviors in captive animals will benefit most from studies on early weaning (in mammals), environmental enrichment and genetic predispositions.

Further advances in genetic engineering represent an important opportunity to improve the productivity and welfare of captive animals while minimizing their impact on natural resources (i.e. the environment). Consumers of animal products and services (e.g. recreation) have a pivotal role in the quality of care provided to captive animals. Welfare standards can and will be raised if society is willing to pay the extra cost. The conservation of rare wild and domestic animal populations will ultimately depend not only on the preservation of genetic material but man's success in nurturing a compassion for animals in future generations. Exposure to captive animals will play an important role in this process.

References

Adams, D.B. (1976) The relation of scent-marking, olfactory investigation and specific postures in the isolation-induced fighting of rats. *Behaviour* 56, 286–297.

Adams, N. and Boice, R. (1989) Development of dominance in domestic rats in laboratory and seminatural environments. *Behavioural Processes* 19, 127–142.

Aengus, W.L. and Millam, J.R. (1999) Taming parent-reared orange-winged Amazon parrots by neonatal handling. *Zoo Biology* 18, 177–187.

Agnese, J.-F., Oteme, Z.J. and Gilles, S. (1995) Effects of domestication on genetic variability, fertility, survival and growth rate in a tropical siluriform: *Heterobranchus longifilis* Valenciennes 1840. *Aquaculture* 131, 197–204.

Ahola, L., Harri, M., Kasanen, S., Mononen, J. and Pyykönen, T. (2000) Effect of family housing of farmed silver foxes (*Vulpes vulpes*) in outdoor enclosures on some behavioural and physiological parameters. *Canadian Journal of Animal Science* 80, 427–434.

Ajayi, S.S. (1974) Giant rats for meat and some taboos. *Oryx* 12, 379–380.

Al-Abdul-Elah, K.M., Almatar, S., Abu-Rezq, T. and James, C.M. (2001) Development of hatchery technology for the silver pomfret *Pampus argenteus*: effect of microalgal species on larval survival. *Aquaculture Research* 32, 849–860.

Alibhai, S.K., Jewell, Z.C. and Towindo, S.S. (2001) Effects of immobilization on fertility in female black rhino (*Diceros bicornis*). *Journal of the Zoological Society of London* 253, 333–345.

Allendorf, F.W. (1983) Conservation biology of fishes. *Conservation Biology* 2, 145–148.

Allendorf, F.W. (1986) Genetic drift and the loss of alleles versus heterozygosity. *Zoo Biology* 5, 181–190.

Allendorf, F.W. (1993) Delay of adaptation to captive breeding by equalizing family size. *Conservation Biology* 7, 416–419.

Altbacker, V., Hudson, R. and Bilko, A. (1995) Rabbit-mothers' diet influences pups' later food choice. *Ethology* 99, 107–116.

Amalin, D.M., Peña, J.E., Reiskind, J. and McSorley, R. (2001) Comparison of the survival of three species of sac spiders on natural and artificial diets. *The Journal of Arachnology* 29, 253–262.

Amir, S.Z., Brown, Z.W. and Amit, Z. (1981) The role of endorphins in stress: evidence and speculations. *Neuroscience and Biobehavioral Reviews* 4, 77–86.

Anderson, J.R. and Chamove, A.S. (1984) Allowing captive primates to forage. In: *Standards in Laboratory Animals Management*. Universities Federation for Animal Welfare, Potters Bar, pp. 253–256.

Andersson, M., Nordin, E. and Jensen, P. (2001) Domestication effects on foraging strategies in fowl. *Applied Animal Behaviour Science* 72, 51–62.

Andrews, M.W. and Rosenblum, L.A. (1993) Live-social-video reward maintains joystick task performance in bonnet macaques. *Perceptual Motor Skills* 77, 755–763.

Anonymous (2001) Status of world aquaculture – 2000. *Aquaculture Magazine Buyer's Guide and Industry Directory* 2001, 6–38.

Anthony, D., Telegin, D.Y. and Brown, D. (1991) The origin of horseback riding. *Scientific American* 265, 94–100.

Appleby, M.C. and Lawrence, A.B. (1987) Food restriction as a cause of stereotypic behaviour in tethered gilts. *Animal Production* 46, 104–110.

Araújo, A., Arruda, M.F., Alencar, A.I., Albuquerque, F., Nascimento, M.C. and Yamamoto, M.E. (2000) Body weight of wild and captive common marmosets (*Callithrix jacchus*). *International Journal of Primatology* 21, 317–324.

Asa, C.S. and Valdespino, C. (1998) Canid reproductive biology: an integration of proximate mechanisms and ultimate causes. *American Zoologist* 38, 251–259.

Aspi, J. (2000) Inbreeding and outbreeding depression in male courtship song characters in *Drosophila montana*. *Heredity* 84, 273–282.

Atchley, W.R. and Fitch, W.M. (1991) Gene trees and the origins of inbred strains of mice. *Science* 254, 554–558.

Baccus, R., Ryman, N., Smith, M.H., Reuterwall, C. and Cameron, D. (1983) Genetic variability and differentiation of large mammals. *Journal of Mammalogy* 64, 109–120.

Baenninger, R. (1995) Some consequences of animal domestication for humans. *Anthrozoos* 8, 69–77.

Baker, A. (1994) Variation in the parental care systems of mammals and the impact on zoo breeding programs. *Zoo Biology* 13, 413–421.

Baker, C.M.A. and Manwell, C. (1981) "Fiercely feral": on the survival of domesticates without care from man. *Zeitschrift für Tierzüchtung und Züchtungsbiologie* 98, 241–257.

Baker, M.L., Gemmell, E. and Gemmell, R.T. (1998) Physiological changes in brushtail possums, *Trichosurus vulpecula*, transferred from the wild to captivity. *Journal of Experimental Zoology* 280, 203–212.

Baker, P.J., Harris, S., Robertson, C.P.J., Saunders, G. and White, P.C.L. (2001) Differences in the capture rate of cage-trapped red foxes *Vulpes vulpes* and an evaluation of rabies control measures in Britain. *Journal of Applied Ecology* 38, 823–835.

Bakken, M. (1993a) Reproduction in farmed silver fox vixens (*Vulpes vulpes*) in relation to own competition capacity and that of neighbouring vixens. *Journal of Animal Breeding and Genetics* 110, 305–311.

Bakken, M. (1993b) The relationship between competition capacity and reproduction in farmed silver fox vixens (*Vulpes vulpes*). *Journal of Animal Breeding and Genetics* 110, 147–155.

Ball, S.J., Adams, M., Possingham, H.P. and Keller, M.A. (2000) The genetic contribution of single male immigrants to small, inbred populations: a laboratory study using *Drosophila melanogaster*. *Heredity* 84, 677–684.

Ballou, J.D. (1984) Strategies for maintaining genetic diversity in captive populations through reproductive technology. *Zoo Biology* 3, 311–323.

Ballou, J.D. and Foose, T.J. (1996) Demographic and genetic management of captive populations. In: Kleiman, D.G., Allen, M.E., Thompson, K.V. and Lumpkin, S. (eds)

Wild Mammals in Captivity: Principles and Techniques. University of Chicago Press, Chicago, Illinois, pp. 363–383.

Balmford, A., Mace, G.M. and Leader-Williams, N. (1996) Designing the ark: setting priorities for captive breeding. *Conservation Biology* 10, 719–727.

Balon, E.K. (1995) Origin and domestication of the wild carp, *Cyprinus carpio*: from Roman gourmets to the swimming flowers. *Aquaculture* 129, 3–48.

Bancroft, D.R., Pemberton, J.M. and King, P. (1995) Extensive protein and microsatellite variability in an isolated cyclic ungulate population. *Heredity* 74, 326–336.

Banks, E.M. and Popham, T. (1975) Intraspecific agonistic behavior of captive brown lemmings, *Lemmus trimucronatus*. *Journal of Mammalogy* 56, 514–516.

Bardach, J.E., Ryther, J.H. and McLarney, W.O. (1972) *Aquaculture: the Farming and Husbandry of Freshwater and Marine Organisms.* John Wiley & Sons, New York.

Bardi, M., Petto, A.J. and Lee-Parritz, D.E. (2001) Parental failure in captive cotton-top tamarins (*Saguinus oedipus*). *American Journal of Primatology* 54, 159–169.

Barlow, R. (1981) Experimental evidence for interaction between heterosis and environment in animals. *Animal Breeding Abstracts* 49, 715–737.

Barnett, J.L., Hemsworth, P.H. and Newman, E.A. (1992) Fear of humans and its relationships with productivity in laying hens at commercial farms. *British Poultry Science* 33, 699–710.

Barnett, J.L., Hemsworth, P.H., Hennessy, D.P., McCallum, T.H. and Newman, E.A. (1994) The effects of modifying the amount of human contact on behavioural, physiological and production responses of laying hens. *Applied Animal Behaviour Science* 41, 87–100.

Barnett, J.L., Hemsworth, P.H., Cronin, G.M., Jongman, E.C. and Hutson, G.D. (2001) A review of the welfare issues for sows and piglets in relation to housing. *Australian Journal of Agricultural Research* 52, 1–28.

Barnett, S.A., Dickson, R.G. and Hocking, W.E. (1979) Genotype and environment in the social interactions of wild and domestic Norway rats. *Aggressive Behavior* 5, 105–119.

Barria, N. and Bradford, G.E. (1981a) Long-term selection for rapid gain in mice. I. Genetic analysis at the limit of response. *Journal of Animal Science* 52, 729–738.

Barria, N. and Bradford, G.E. (1981b) Long-term selection for rapid gain in mice. II. Correlated changes in reproduction. *Journal of Animal Science* 52, 739–747.

Bartlett, A.C. (1984) Genetic changes during insect domestication. In: King, E.G. and Leppla, N.C. (eds) *Advances and Challenges in Insect Rearing.* Agricultural Research Service, US Department of Agriculture, New Orleans, Louisiana, pp. 2–8.

Bartlett, A.C. (1985) Guidelines for genetic diversity in laboratory colony establishment and maintenance. In: Singh, P. and Moore, R.F. (eds) *Handbook of Insect Rearing*, Vol. 1. Elsevier, New York, pp. 7–17.

Bartlett, A.C. (1993) Maintaining genetic diversity in laboratory colonies of parasites and predators. In: Narang, S.K., Bartlett, A.C. and Faust, R.M. (eds) *Applications of Genetics to Arthropods of Biological Control Significance.* CRC Press, Boca Raton, Florida, pp. 133–145.

Bartley, M.M. (1992) Darwin and domestication: studies on inheritance. *Journal of the History of Biology* 25, 307–333.

Bateson, P. (ed.) (1983) *Mate Choice.* Cambridge University Press, New York.

Bauer, O.N., Egusa, S. and Hoffman, G.L. (1981) Parasitic infections of economic importance in fish. *Proceedings of the 4th International Congress of Parasitology, Review of Advances in Parasitology*, Warsaw, pp. 425–443.

Baxter, E. and Plowman, A.B. (2001) The effect of increasing dietary fibre on feeding, rumination and oral stereotypies in captive giraffes (*Giraffa camelopardalis*). *Animal Welfare* 10, 281–290.

Beattie, V.E., O'Connell, N.E. and Moss, B.W. (2000) Influence of environmental enrichment on the behaviour, performance and meat quality of domestic pigs. *Livestock Production Science* 65, 71–79.

Beck, A. (1973) *The Ecology of Stray Dogs: a Study of Free Ranging Urban Animals.* York Press, Baltimore, Maryland.

Beck, B.B. and Power, M.L. (1988) Correlates of sexual and maternal competence in captive gorillas. *Zoo Biology* 7, 339–350.

Beck, B.B., Rapaport, L.G., Stanley-Price, M.R. and Wilson, A.C. (1994) Reintroduction of captive-born animals. In: Olney, P.J.S., Mace, G.M. and Geistner, A.T.C. (eds) *Creative Conservation: Interactive Management of Wild and Captive Animals.* Chapman and Hall, London, pp. 265–286.

Beilharz, R.G. (1998) Environmental limit to genetic change. An alternative theorem of natural selection. *Journal of Animal Breeding and Genetics* 115, 433–437.

Beilharz, R.G. and Zeeb, K. (1982) Social dominance in dairy cattle. *Applied Animal Ethology* 8, 79–97.

Beilharz, R.G., Luxford, B.G. and Wilkinson, J.L. (1993) Quantitative genetics and evolution: is our understanding of genetics sufficient to explain evolution? *Journal of Animal Breeding and Genetics* 110, 161–170.

Belliveau, A.M., Farid, A., O'Connell, M. and Wright, J.M. (1999) Assessment of genetic variability in captive and wild American mink (*Mustela vison*) using microsatellite markers. *Canadian Journal of Animal Science* 79, 7–16.

Belyaev, D.K. (1969) Domestication of animals. *Science Journal* (London) 1, 47–52.

Belyaev, D.K. (1979) Destabilizing selection as a factor in domestication. *Journal of Heredity* 70, 301-308.

Belyaev, D.K., Plyusnina, I.Z. and Trut, L.N. (1984/1985) Domestication in the silver fox (*Vulpes fulvus* Desm): changes in physiological boundaries of the sensitive period of primary socialization. *Applied Animal Behaviour Science* 13, 359–370.

Bennett, R.M. (1996) People's willingness to pay for farm animal welfare. *Animal Welfare* 5, 3–11.

Berejikian, B.A. (1995) The effects of hatchery and wild ancestry and experience on the relative ability of steelhead trout fry (*Oncorhynchus mykiss*) to avoid a benthic predator. *Canadian Journal of Fisheries and Aquatic Sciences* 52, 2476–2482.

Berejikian, B.A., Tezak, E.P., Schroder, S.L., Knudsen, C.M. and Hard, J.J. (1997) Reproductive behavioral interactions between wild and captively reared coho salmon (*Oncorhynchus kisutch*). *ICES Journal of Marine Science* 54, 1040–1050.

Berejikian, B.A., Smith, R.J.F., Tezak, E.P., Schroder, S.L. and Knudsen, C.M. (1999) Chemical alarm signals and complex hatchery rearing habitats affect antipredator behavior and survival of chinook salmon (*Oncorhynchus tshawytscha*) juveniles. *Canadian Journal of Fisheries and Aquatic Sciences* 56, 830–838.

Berejikian, B.A., Tezak, E.P., Flagg, T.A., LaRae, A.L., Kummerow, E. and Mahnken, C.V.W. (2000) Social dominance, growth, and habitat use of age-O steelhead (*Oncorhynchus mykiss*) grown in enriched and conventional hatchery rearing environments. *Canadian Journal of Fisheries and Aquatic Sciences* 57, 628–636.

Berejikian, B.A., Tezak, E.P., Park, L., LaHood, E., Schroder, S.L. and Beall, E. (2001) Male competition and breeding success in captively reared and wild coho salmon (*Oncorhynchus kisutch*). *Canadian Journal of Fisheries and Aquatic Sciences* 58, 804–810.

Berger, J. (1986) *Wild Horses of the Great Basin: Social Competition and Population Size.* University of Chicago Press, Chicago, Illinois.

Berlocher, S.H. and Friedman, S. (1981) Loss of genetic variation in laboratory colonies of *Phormia regina. Entomologia Experimentalis et Applicata* 30, 205–208.

Berman, C.M. (1990) Intergenerational transmission of maternal rejection rates among free-ranging rhesus monkeys. *Animal Behaviour* 39, 329–337.

Bernon, D.E. and Siegel, P.B. (1983) Mating frequency in male chickens: crosses among selected and unselected lines. *Genetics, Selection and Evolution* 15, 445–454.

Berry, R.J. (1978) Genetic variation in wild house mice: where natural selection and history meet. *American Scientist* 66, 52–60.

Biggins, D.E., Godbey, J.L., Hanebury, L.R., Luce, B., Marinari, P.E., Matchett, M.R. and Vargas, A. (1998) The effect of rearing methods on survival of reintroduced black-footed ferrets. *Journal of Wildlife Management* 62, 643–653.

Biggins, D.E., Vargas, A., Godbey, J.L. and Anderson, S.H. (1999) Influence of prerelease experience on reintroduced black-footed ferrets (*Mustela nigripes*). *Biological Conservation* 89, 121–129.

Black, J.M. (1994) The nene, *Branta sandvicensis*, recovery initiative: research against extinction. *Ibis* 137, S153–S160.

Blanchard, R.J., Fukunaga, K., Blanchard, D.C. and Kelley, M.J. (1975) Conspecific aggression in the laboratory rat. *Journal of Comparative and Physiological Psychology* 89, 1204–1209.

Blanchard, D.C., Blanchard, R.J., Lee, E.M.C. and Williams, G. (1981a) Taming in the wild Norway rat following lesions of the basal ganglia. *Physiology and Behavior* 27, 995–1000.

Blanchard, D.C., Williams, G., Lee, E.M.C. and Blanchard, R.J. (1981b) Taming of wild *Rattus norvegicus* by lesions of the mesencephalic central gray. *Physiological Psychology* 9, 157–163.

Blanchard, R.J., Flannelly, K.J. and Blanchard, D.C. (1986) Defensive behaviors of laboratory and wild *Rattus norvegicus. Journal of Comparative Psychology* 100, 101–107.

Blanchard, D.C., Rodgers, R.J., Hendrie, C.A. and Hori, K. (1988) 'Taming' of wild rats (*Rattus rattus*) by $5HT_{1A}$ agonists buspirone and gepirone. *Pharmacology, Biochemistry and Behavior* 31, 269–278.

Blank, R.D., Campbell, G.R. and D'Eustachio, P. (1986) Possible derivation of the laboratory mouse genome from multiple wild *Mus* species. *Genetics* 114, 1257–1269.

Blass, E.M. and Teicher, M.H. (1980) Suckling. *Science* 210, 15–21.

Blaxter, K.L. (1974) Deer farming. *Mammal Review* 4, 119–122.

Blohowiak, C. (1987) Baffling black ducks. *Zoogoer* 16, 18–19.

Bloomsmith, M.A. and Lambeth, S.P. (1995) Effects of predictable versus unpredictable feeding schedules on chimpanzee behavior. *Applied Animal Behaviour Science* 44, 65–74.

Bloomsmith, M.A. and Lambeth, S.P. (2000) Videotapes as enrichment for captive chimpanzees (*Pan troglodytes*). *Zoo Biology* 19, 541–551.

Bloomsmith, M.A., Alford, P.L. and Maple, T.L. (1988) Successful feeding enrichment for captive chimpanzees. *American Journal of Primatology* 16, 155–164.

Blottner, S., Franz, C., Rohleder, M., Zinke, O. and Stuermer, I.W. (2000) Higher testicular activity in laboratory gerbils compared to wild Mongolian gerbils (*Meriones unguiculatus*). *Journal of Zoology, London* 250, 461–466.

Bluhm, C.K. (1988) Temporal patterns of pair formation and reproduction in annual cycles and associated endocrinology in waterfowl. In: Johnston, R.F. (ed.) *Current Ornithology*, Vol. 5. Plenum Press, New York, pp. 123–185.

Blus, L.J. (1971) Reproduction and survival of short-tailed shrews (*Blarina brevicauda*) in captivity. *Laboratory Animal Science* 21, 884–891.

Bodson, L. (1997) Motivations for pet-keeping in Ancient Greece and Rome: a preliminary survey. In: Podberscek, A.L., Paul, E.S. and Serpell, J.A. (eds) *Companion Animals and Us.* Cambridge University Press, Cambridge, pp. 27–41.

Boice, R. (1972) Some behavioral tests of domestication in Norway rats. *Behaviour* 42, 198–231.

Boice, R. (1973) Domestication. *Psychological Bulletin* 80, 215–230.

Boice, R. (1977) Burrows of wild and domestic rats: effects of domestication, outdoor raising, age, experience and maternal state. *Journal of Comparative and Physiological Psychology* 91, 649–661.

Boice, R. (1980) Domestication and degeneracy. In: Denney, M.R. (ed.) *Comparative Psychology: an Evolutionary Analysis of Animal Behavior.* John Wiley & Sons, New York, pp. 84–99.

Boice, R. (1981) Captivity and feralization. *Psychological Bulletin* 89, 407–421.

Boice, R. and Adams, N. (1983) Degrees of captivity and aggressive behavior in domestic Norway rats. *Bulletin of the Psychonomic Society* 21, 149–152.

Boice, R. and Williams, R.C. (1971) Delay in onset of tonic immobility in *Rana pipiens*. *Copeia* 4, 747–748.

Bökönyi, S. (1969) Archaeological problems of recognizing animal domestication. In: Ucko, P.J. and Dimbleby, G.W. (eds) *The Domestication and Exploitation of Plants and Animals.* Duckworth, London, pp. 219–229.

Bökönyi, S. (1974) *History of Domestic Mammals in Central and Eastern Europe.* Akademiai Kiado, Budapest.

Bökönyi, S. (1984) Horse. In: Mason, I.L. (ed.) *Evolution of Domesticated Animals.* Longman, London, pp. 162–173.

Boller, E.F. (1979) Behavioral aspects of quality in insectary production. In: Hoy, M.A. and McKelvey, J.J. Jr (eds) *Genetics in Relation to Insect Management.* The Rockefeller Foundation, New York, pp. 153–160.

Boreman, J. and Price, E.O. (1972) Social dominance in wild and domestic Norway rats (*Rattus norvegicus*). *Animal Behaviour* 20, 534–542.

Bouissou, M.F. (1970) Role du contact physique dans la manifestation des relations hierarchiques chez les bovins. *Annales de Zootechnie* 19, 279–285.

Box, H.O. (1973) *Organization in Animal Communities.* Butterworth, London.

Box, H.O. (1991) Training for life after release: simian primates as examples. In: Gipps, J.H.W. (ed.) *Beyond Captive Breeding: Reintroducing Endangered Mammals to the Wild.* Clarendon Press, Oxford, pp. 111–124.

Bradley, D.G. and Cunningham, E.P. (1999) Genetic aspects of domestication. In: Fries, R. and Ruvinsky, A. (eds) *The Genetics of Cattle.* CAB International, Wallingford, UK, pp. 15–31.

Brant, A.W. (1998) A brief history of the turkey. *World's Poultry Science Journal* 54, 365–373.

Bray, W.A., Lawrence, A.L. and Lester, L.J. (1990) Reproduction of eyestalk-ablated *Penaeus stylirostris* fed various levels of total dietary lipids. *Journal of the World Aquaculture Society* 20, 41–52.

Breuer, K., Hemsworth, P.H., Barnett, J.L., Matthews, L.R. and Coleman, G.J. (2000) Behavioural response to humans and productivity of commercial dairy cows. *Applied Animal Behaviour Science* 66, 273–288.

Brewer, B.A., Lacy, R.C., Foster, M.L. and Alaks, G. (1990) Inbreeding depression in insular and central populations of *Peromyscus* mice. *Journal of Heredity* 81, 257–266.

Brisbin, I.L. Jr (1974) The ecology of animal domestication: its relevance to man's environmental crises – past, present and future. *Association of Southeastern Biological Bulletin* 21, 3–8.

Brisbin, I.L. Jr (1977) The pariah. Its ecology and importance to the origin, development and study of pure-bred dogs. *Pure-Bred Dogs American Kennel Gazette* 94, 22–29.

Brisbin, I.L. Jr and Risch, T.S. (1997) Primitive dogs, their ecology and behavior: unique opportunities to study the early development of the human-canine bond. *Journal of the American Veterinary Medical Association* 210, 1122–1126.

Brittas, R., Marcström, V., Kenward, R.E. and Karlbom, M. (1992) Survival and breeding success of reared and wild ring-necked pheasants in Sweden. *Journal of Wildlife Management* 56, 368–376.

Bronson, F.H. (1998) Energy balance and ovulation: small cages versus natural habitats. *Reproduction, Fertility and Development* 10, 127–137.

Broom, D.M. (1988) The relationship between welfare and disease susceptibility in farm animals. In: Gibson, T.E. (ed.) *Animal Disease – a Welfare Problem*. British Veterinary Association Animal Welfare Foundation, London, pp. 22–29.

Browdy, C.L., Hadani, A., Samocha, T.M. and Loya, Y. (1986) The reproductive performance of wild and pond-reared *Penaeus semisulcatus* De Haan. *Aquaculture* 59, 251–258.

Brown, G.E. and Smith, R.J.F. (1998) Acquired predator recognition in juvenile rainbow trout (*Oncorhynchus mykiss*): conditioning hatchery-reared fish to recognize chemical cues of a predator. *Canadian Journal of Fisheries and Aquatic Sciences* 55, 611–617.

Brown, J.A., Wiseman, D. and Kean, P. (1997) The use of behavioural observations in the larvaculture of cold-water marine fish. *Aquaculture* 155, 297–306.

Bruell, J.H. (1967) Behavioral heterosis. In: Hirsch, J. (ed.) *Behavior-Genetic Analysis*. McGraw-Hill, New York, pp. 270–286.

Bryant, E.H., Backus, V.L., Clark, M.E. and Reed, D.H. (1999) Experimental tests of captive breeding for endangered species. *Conservation Biology* 13, 1487–1496.

Bubier, N.E., Lambert, M.S., Deeming, D.C., Ayres, L.L. and Sibly, R.M. (1996) Time budget and colour preferences (with special reference to feeding) of ostrich (*Struthio camelus*) chicks in captivity. *British Poultry Science* 37, 547–551.

Buchenauer, D. (1999) Genetics of behaviour in cattle. In: Fries, R. and Ruvinsky, A. (eds) *The Genetics of Cattle*. CAB International, Wallingford, UK, pp. 365–390.

Budiansky, S. (1992) *The Covenant of the Wild: Why Animals Chose Domestication*. William Morrow, New York.

Budiansky, S. (1994) A special relationship: the coevolution of human beings and domesticated animals. *Journal of the American Veterinary Medical Society* 204, 365–368.

Burda, H., Ballast, L. and Bruns, V. (1988) Cochlea in Old World mice and rats (Muridae). *Journal of Morphology* 198, 269–285.

Butler, R.G. (1980) Population size, social behaviour and dispersal in house mice: a quantitative investigation. *Animal Behaviour* 28, 78–85.

Büttner, D. (1992) Climbing on the cage lid, a regular component of locomotor activity in the mouse. *Journal of Experimental Animal Science* 34, 165–169.

Cabib, S. (1993) Neurobiological basis of stereotypies. In: Lawrence, A.B. and Rushen, J. (eds) *Stereotypic Animal Behaviour: Fundamentals and Applications to Welfare*. CAB International, Wallingford, UK, pp. 119–145.

Cade, T.J. and Fyfe, R.W. (1978) What makes peregrine falcons breed in captivity? In: Temple, S.A. (ed.) *Endangered Birds: Management Techniques for Preserving Threatened Species*. University of Wisconsin Press, Madison, Wisconsin, pp. 251–262.

Cagigas, M.E., Vázquez, E., Blanco, G. and Sánchez, J.A. (1999) Genetic effects of introduced hatchery stocks on indigenous brown trout (*Salmo trutta* L.) populations in Spain. *Ecology of Freshwater Fish* 8, 141–150.

Cairns, R.B. (1976) The ontogeny and phylogeny of social interactions. In: Hahn, M.E. and Simmel, E.C. (eds) *Communicative Behavior and Evolution*. Academic Press, New York, pp. 115–139.

Calhoun, J.B. (1962) Population density and social pathology. *Scientific American* 206, 139–148.

Callard, M.D., Bursten, S.N. and Price, E.O. (2000) Repetitive backflipping behaviour in captive roof rats (*Rattus rattus*) and the effects of cage enrichment. *Animal Welfare* 9, 139–152.

Cameron-Beaumont, C., Lowe, S.E. and Bradshaw, J.W.S. (2002) Evidence suggesting preadaptation to domestication throughout the small Felidae. *Biological Journal of the Linnean Society* 75, 361–366.

Carder, B. and Berkowitz, K. (1970) Rats' preference for earned in comparison with free food. *Science* 167, 1273–1274.

Carles, A.B., King, J.M. and Heath, B.R. (1981) Game domestication for animal production in Kenya – an analysis of growth in oryx, eland and zebu cattle. *Journal of Agricultural Science* 97, 453–463.

Carlstead, K. (1986) Predictability of feeding: its effect on agonistic behaviour and growth in grower pigs. *Applied Animal Behaviour Science* 16, 25–38.

Carlstead, K. (1996) Effects of captivity on the behavior of wild mammals. In: Kleiman, D.G., Allen, M.E., Thompson, K.V. and Lumpkin, S. (eds) *Wild Mammals in Captivity: Principles and Techniques*. University of Chicago Press, Chicago, Illinois, pp. 317–333.

Carlstead, K. and Shepherdson, D. (1994) Effects of environmental enrichment on reproduction. *Zoo Biology* 13, 447–458.

Carlstead, K. and Shepherdson, D. (2000) Alleviating stress in zoo animals with environmental enrichment. In: Moberg, G.P. and Mench, J.A. (eds) *The Biology of Animal Stress: Basic Principles and Implications for Animal Welfare*. CAB International, Wallingford, UK, pp. 337–354.

Carlstead, K., Seidensticker, J. and Baldwin, R. (1991) Environmental enrichment for zoo bears. *Zoo Biology* 10, 3–16.

Carlstead, K., Mellen, J. and Kleiman, D.G. (1999) Black rhinoceros (*Diceros bicornis*) in U.S. zoos: I. Individual behavior profiles and their relationship to breeding success. *Zoo Biology* 18, 17–34.

Carpenter, J.W., Gabel, R.R. and Goodwin, J.G. (1991) Captive breeding and reintroduction of the endangered masked bobwhite. *Zoo Biology* 10, 439–449.

Carson, H.L. (1990) Increased genetic variance after a population bottleneck. *Trends in Ecology and Evolution* 5, 328–330.

Case, L.P. (1999) *The Dog. Its Behavior, Nutrition, and Health*. Iowa State University Press, Ames, Iowa.

Cassinello, J., Gomendio, M. and Roldan, E.R.S. (2001) Relationship between coefficient of inbreeding and parasite burden in endangered gazelles. *Conservation Biology* 15, 1171–1174.

Castro, M.I., Beck, B.B., Kleiman, D.G., Ruiz-Miranda, C.R. and Rosenberger, A.L. (1998) In: Shepherdson, D.J., Mellen, J.D. and Hutchins, M. (eds) *Second Nature: Environmental Enrichment for Captive Animals*. Smithsonian Institution Press, Washington, DC, pp. 113–128.

Cavalli, R.O., Lavens, P. and Sorgeloos, P. (2001) Reproductive performance of *Macrobrachium rosenbergii* females in captivity. *Journal of the World Aquaculture Society* 32, 60–67.

Chamove, A.S. (1989) Enrichment in chimpanzees: unpredictable ropes and tools. *Journal of the Association of British Wild Animal Keepers* 16, 139–141.

Champoux, M., Byrne, E., DeLizio, R. and Suomi, S.J. (1992) Motherless mothers revisited: rhesus maternal behavior and rearing history. *Primates* 33, 251–255.

Chase, I.D., Bartolomeo, C. and Dugatkin, L.A. (1994) Aggressive interactions and inter-contest interval: how long do winners keep winning? *Animal Behaviour* 48, 393–400.

Cheng, K.M., Shoffner, R.N., Phillips, R.E. and Lee, F.B. (1979a) Mate preference in wild and domesticated (game farm) mallards: II. Pairing success. *Animal Behaviour* 27, 417–425.

Cheng, K.M., Shoffner, R.N., Phillips, R.E. and Shapiro, L.J. (1979b) Early imprinting in wild and game-farm mallards (*Anas platyrhynchos*): genotype and arousal. *Journal of Comparative and Physiological Psychology* 93, 929–938.

Chew, K. (1990) Global bivalve shellfish introductions. *World Aquaculture* 21, 9–22.

Chiyokubo, T., Shikano, T., Nakajima, M. and Fujio, Y. (1998) Genetic features of salinity tolerance in wild and domestic guppies (*Poecilia reticulata*). *Aquaculture* 167, 339–348.

Christiansen, J.S., Jorgensen, E.H. and Jobling, M. (1991) Oxygen consumption in relation to sustained exercise and social stress in Arctic charr (*Salvelinus alpinus* L.). *Journal of Experimental Zoology* 260, 149–156.

Christie, D.W. and Bell, E.T. (1971) Some observations on the seasonal incidence and frequency of oestrus in breeding bitches in Britain. *Journal of Small Animal Practice* 12, 159–167.

Clark, B.R. and Price, E.O. (1981) Domestication effects on sexual maturation and fecundity of Norway rats (*Rattus norvegicus*). *Journal of Reproduction and Fertility* 63, 215–220.

Clark, M.M. and Galef, B.G. Jr (1977) The role of the physical rearing environment in the domestication of the Mongolian gerbil (*Meriones unguiculatus*). *Animal Behaviour* 25, 298–316.

Clark, M.M. and Galef, B.G. Jr (1980) Effects of rearing environment on adrenal weights, sexual development and behavior in gerbils: an examination of Richter's domestication hypothesis. *Journal of Comparative and Physiological Psychology* 94, 857–863.

Clark, M.M. and Galef, B.G. Jr (1981) Environmental influence on development, behavior, and endocrine morphology of gerbils. *Physiology and Behavior* 27, 761–765.

Clark, M.M. and Galef, B.G. Jr (1982) Environmental effects on the ontogeny of exploratory and escape behaviors of Mongolian gerbils. *Developmental Psychobiology* 15, 121–129.

Clayton, D.L. and Paietta, J.V. (1972) Selection for circadian eclosion time in *Drosophila melanogaster*. *Science* 178, 994–995.

Cleary, J.J., Pankhurst, N.W. and Battaglene, S.C. (2000) The effect of capture and handling stress on plasma steroid levels and gonadal condition in wild and farmed snapper *Pagrus auratus* (Sparidae). *Journal of the World Aquaculture Society* 31, 558–569.

Clutton-Brock, J. (1977) Man-made dogs. *Science* 197, 1340–1342.

Clutton-Brock, J. (1981) *Domesticated Animals from Early Times*. British Museum/W. Heinemann, London.

Clutton-Brock, J. (1992) Domestication of animals. In: Jones, S., Martin, R. and Pilbeam, D. (eds) *The Cambridge Encyclopedia of Human Evolution*. Cambridge University Press, Cambridge, pp. 380–385.

Clutton-Brock, J. (1999) *A Natural History of Domesticated Mammals*, 2nd edn. Cambridge University Press, Cambridge.

Cohen, A.C. (2000) Feeding fitness and quality of domesticated and feral predators: effects of long-term rearing on artificial diet. *Biological Control* 17, 50–54.

Cohen, A.C. and Staten, R.T. (1993) Long-term culturing and quality assessment of predatory big-eyed bugs, *Geocoris punctipes*. In: Narang, S.K., Bartlett, A.C. and Faust, R.M. (eds) *Applications of Genetics to Arthropods of Biological Control Significance*. CRC Press, Boca Raton, Florida, pp. 121–132.

Cohen, J.A. and Fox, M.W. (1976) Vocalizations in wild canids and possible effects of domestication. *Behavioural Processes* 1, 77–92.

Coleman, G.J., Hemsworth, P.H., Hay, M. and Cox, M. (1998) Predicting stockperson behaviour towards pigs from attitudinal and job-related variables and empathy. *Applied Animal Behaviour Science* 58, 63–75.

Collette, J.C., Millam, J.R., Klasing, K.C. and Wakenell, P.S. (2000) Neonatal handling of Amazon parrots alters the stress response and immune function. *Applied Animal Behaviour Science* 66, 335–349.

Coman, B.J. and Brunner, H. (1972) Food habits of the feral house cat in Victoria. *Journal of Wildlife Management* 36, 848–853.

Connor, J.L. (1975) Genetic mechanisms controlling the domestication of a wild house mouse population (*Mus musculus* L.). *Journal of Comparative and Physiological Psychology* 89, 118–130.

Constable, P., Hinchcliff, K., Demma, N., Callahan, M., Dale, B., Fox, K., Adams, L., Wack, R. and Kramer, L. (1998) Serum biochemistry of captive and free-ranging gray wolves (*Canis lupus*). *Journal of Zoo and Wildlife Medicine* 29, 435–440.

Contreras-Sanchez, W.M., Schreck, C.B., Fitzpatrick, M.S. and Pereira, C.B. (1998) Effects of stress on the reproductive performance of rainbow trout (*Oncorhynchus mykiss*). *Biology of Reproduction* 58, 439–447.

Conway, W.G. (1978) Breeding endangered birds in captivity: the last resort. In: Temple, S.A. (ed.) *Endangered Birds: Management Techniques for Preserving Threatened Species*. University of Wisconsin Press, Madison, Wisconsin, pp. 225–230.

Cooke, T. (1978) *Exhibition and Pet Mice*. Saiga, Hindhead, UK.

Cooper, J.J. and Nicol, C.J. (1991) Stereotypic behaviour affects environmental preference in bank voles (*Clethrionomys glareolus*). *Animal Behaviour* 41, 971–977.

Cooper, J.J. and Nicol, C.J. (1996) Stereotypic behaviour in wild caught and laboratory-bred bank voles, *Clethrionomys glareolus*. *Animal Welfare* 5, 245–257.

Coppinger, R. and Coppinger, L. (2001) *Dogs*. Scribner, New York.

Coppinger, R. and Feinstein, M. (1991) 'Hark! Hark! The dogs do bark....' and bark and bark. *Smithsonian* 21, 119–129.

Coppinger, R. and Schneider, R. (1995) Evolution of working dogs. In: Serpell, J. (ed.) *The Domestic Dog. Its Evolution, Behaviour, and Interactions with People*. Cambridge University Press, Cambridge, pp. 21–47.

Coppinger, R. and Smith, C.K. (1983) The domestication of evolution. *Environmental Conservation* 10, 283–292.

Coppinger, R., Glendinning, J., Torop, E., Matthay, C., Sutherland, M. and Smith, C. (1987) Degree of behavioral neoteny differentiates canid polymorphs. *Ethology* 75, 89–108.

Coss, R.G. (1999) Effects of relaxed natural selection on the evolution of behavior. In: Foster, S.A. and Endler, J.A. (eds) *Geographic Variation in Behavior: Perspectives on Evolutionary Mechanisms*. Oxford University Press, New York, pp. 180–208.

Coss, R.G. and Biardi, J.E. (1997) Individual variation in the antisnake behavior of California ground squirrels (*Spermophilus beecheyi*). *Journal of Mammalogy* 73, 294–310.

Cottle, C.A. and Price, E.O. (1987) Effect of the nonagouti pelage-color allele on the behavior of captive wild Norway rats (*Rattus norvegicus*). *Journal of Comparative Psychology* 101, 390–394.

Coulson, T., Albon, S., Slate, J. and Pemberton, J. (1999) Microsatellite loci reveal sex-dependent responses to inbreeding and outbreeding in red deer calves. *Evolution* 53, 1951–1960.

Crabbe, J.C., Wahlsten, D. and Dudek, B.C. (1999) Genetics of mouse behavior: interactions with laboratory environment. *Science* 284, 1670–1672.

Craig, J.V. and Muir, W.M. (1989) Fearful and associated responses of caged White Leghorn hens: genetic parameter estimates. *Poultry Science* 68, 1040–1046.

Craig, J.V., Craig, T.P. and Dayton, A.D. (1983) Fearful behaviour by caged hens of two genetic stocks. *Applied Animal Ethology* 10, 263–273.

Crawford, M.A. (1974) The case for new domestic animals. *Oryx* 12, 351–360.

Cresswell, R.C. and Williams, R. (1983) Post-stocking movements and recapture of hatchery-reared trout released into flowing water – effect of prior acclimation to flow. *Journal of Fish Biology* 23, 265–276.

Crnokrak, P. and Roff, D.A. (1999) Inbreeding depression in the wild. *Heredity* 83, 260–270.

Cronin, G.M., Lefébure, B. and McClintock, S. (2000) A comparison of piglet production and survival in the Werribee Farrowing Pen and conventional farrowing crates at a commercial farm. *Australian Journal of Experimental Agriculture* 40, 17–23.

Cronin, M.A., Renecker, L., Pierson, B.J. and Patton, J.C. (1995) Genetic variation in domestic reindeer and wild caribou in Alaska. *Animal Genetics* 26, 427–434.

Crooijmans, R.P., Groen, A.F., Van-Kampen, A.J., Van-der-Beek, S., Van-der-Poel, J.J. and Groenen, M.A. (1996) Microsatellite polymorphism in commercial broiler and layer lines estimated using pooled blood samples. *Poultry Science* 73, 904–909.

Cross, T.F. and King, J. (1983) Genetic effects of hatchery rearing in Atlantic salmon. *Aquaculture* 33, 33–40.

Crozier, W.W. (1998) Genetic implications of hatchery rearing in Atlantic salmon: effects of rearing environment on genetic composition. *Journal of Fish Biology* 52, 1014–1025.

Crozier, W.W. and Moffett, I.J.J. (1989) Amount and distribution of biochemical genetic variation among wild populations and a hatchery stock of Atlantic salmon, *Salmo salar* L. from north-east Ireland. *Journal of Fish Biology* 35, 665–677.

Cseh, S. and Solti, L. (2000) Importance of assisted reproductive technologies in the conservation of wild, rare or indigenous ungulates: review article. *Acta Veterinaria Hungarica* 48, 313–323.

Csermely, D. (2000) Behaviour of hand-reared orphaned long-eared owls and tawny owls after release in the wild. *Italian Journal of Zoology* 67, 57–62.

Cummins, M.S. and Suomi, S.J. (1976) Long-term effects of social rehabilitation in rhesus monkeys. *Primates* 17, 43–52.

Cunningham, A.A. (1996) Disease risks of wildlife translocations. *Conservation Biology* 10, 349–353.

Cuomo-Benzo, M., Price, E.O. and Hartenstein, R. (1977) Catecholamine levels in whole brain of stressed and control domestic and wild rats (*Rattus norvegicus*). *Behavioural Processes* 2, 33–40.

Dahlgaard, J. and Hoffmann, A.A. (2000) Stress resistance and environmental dependency of inbreeding depression in *Drosophila melanogaster*. *Conservation Biology* 14, 1187–1192.

Daly, M. (1973) Early stimulation of rodents: a critical review of present interpretations. *British Journal of Psychology* 64, 435–460.

D'Andrea, P.S., Horta, C., Cerqueira, R. and Rey, L. (1996) Breeding of the water rat (*Nectomys squamipes*) in the laboratory. *Laboratory Animals* 30, 369–376.

Daniels, T. and Bekoff, M. (1989a) Feralization: the making of wild animals. *Behavioural Processes* 19, 79–94.

Daniels, T. and Bekoff, M. (1989b) Population and social biology of free-ranging dogs, *Canis familiaris*. *Journal of Mammalogy* 70, 754–762.

Daniels, T. and Bekoff, M. (1990) Domestication, exploitation, and rights. In: Bekoff, M. and Jamieson, D. (eds) *Interpretation and Explanation in the Study of Animal Behavior*, Vol. II: *Explanation, Evolution, and Adaptation*. Westview Press, Boulder, Colorado, pp. 345–377.

Danielsdottir, A.K., Marteinsdottir, G., Arnason, F. and Gudjonsson, S. (1997) Genetic structure of wild and reared Atlantic salmon (*Salmo salar* L.) populations in Iceland. *ICES Journal of Marine Science* 54, 986–997.

Darwin, C. (1859) *The Origin of Species*, 1958 edn. Mentor, New York.

Darwin, C. (1868) *The Variation of Animals and Plants under Domestication*, Vols 1 and 2. John Murray, London.

Davis, B.D. (1987) Bacterial domestication: underlying assumptions. *Science* 235, 1329, 1332–1335.

Davis, D.E. (1951) The relation between level of population and pregnancy of Norway rats. *Ecology* 32, 459–461.

Davis, H. and Balfour, D. (1992) *The Inevitable Bond: Examining Scientist–Animal Interactions*. Cambridge University Press, New York.

Davis, H. and Taylor, A. (2001) Discrimination between individual humans by domestic fowl (*Gallus gallus domesticus*). *British Poultry Science* 42, 276–279.

Dawkins, M.S. (1980) *Animal Suffering. The Science of Animal Welfare*. Chapman and Hall, New York.

Dawkins, M.S. (2000) Animal minds and animal emotions. *American Zoologist* 40, 883–888.

Dechambre, E. (1949) La theorie de la foetalisation et la formation des races de chiens et de porcs. *Mammalia* 13, 129–137.

Deeming, D.C. and Bubier, N.E. (1999) Behaviour in natural and captive environments. In: Deeming, D.C. (ed.) *The Ostrich. Biology, Production and Health*. CAB International, Wallingford, UK, pp. 83–104.

Delany, M.J. and Monro, R.H. (1985) Growth and development of wild and captive Nile rats, *Arvicanthis niloticus* (Rodentia: Muridae). *African Journal of Ecology* 23, 121–131.

Delaplane, K.S. and Mayer, D.F. (2000) *Crop Pollination by Bees*. CAB International, Wallingford, UK.

DeRose, M.A. and Roff, D.A. (1999) A comparison of inbreeding depression in life-history and morphological traits in animals. *Evolution* 53, 1288–1292.

Desforges, M.F. and Wood-Gush, D.G.M. (1976) Behavioural comparisons of Aylesbury and Mallard ducks: sexual behaviour. *Animal Behaviour* 24, 391–397.

Deverill, J.I., Adams, C.E. and Bean, C.W. (1999) Prior residence, aggression and territory acquisition in hatchery-reared and wild brown trout. *Journal of Fish Biology* 55, 868–875.

Devlin, R.H., Biagi, C.A., Yesaki, T.Y., Smailus, D.E. and Byatt, J.C. (2001) Growth of domesticated transgenic fish. *Nature* 409, 781–782.

de Waal, B.M. (1994) Chimpanzee's adaptive potential: a comparison of social life under captive and wild conditions. In: Wrangham, R.W., McGrew, W.C., de Waal, B.M. and Heltne, P.G. (eds) *Chimpanzee Cultures*. Harvard University Press, Cambridge, Massachusetts, pp. 243–260.

Dewsbury, D.A., Lanier, D.L. and Miglietta, A. (1980) A laboratory study of climbing behavior in 11 species of muroid rodents. *American Midland Naturalist* 103, 66–72.

Diefenbach, T.J. and Goldberg, J.I. (1990) Postembryonic expression of the serotonin phenotype in *Helisoma trivolvis*: comparison between laboratory-reared and wild-type strains. *Canadian Journal of Zoology* 68, 1382–1389.

Dieperink, C., Pedersen, S. and Pedersen, M.I. (2001) Estuarine predation on radiotagged wild and domesticated sea trout (*Salmo trutta* L.) smolts. *Ecology of Freshwater Fish* 10, 177–183.

DiLauro, M.N. (1998) Renewal of voluntary feeding by wild-caught Atlantic sturgeon juveniles in captivity. *Progressive Fish-Culturist* 60, 311–314.

Dobzhansky, T. and Pavlovsky, O. (1957) An experimental study of interaction between genetic drift and natural selection. *Evolution* 11, 311–319.

Dohm, M.R., Richardson, C.S. and Garland, T. Jr (1994) Exercise physiology of wild and random-bred laboratory house mice and their reciprocal hybrids. *American Journal of Physiology* 267, R1098–R1108.

Donaldson, E.M. (1996) Manipulation of reproduction in farmed fish. *Animal Reproduction Science* 42, 381–392.

Doroshov, S.I., Clark, W.H. Jr, Lutes, P.B., Swallow, R.L., Beer, K.E., McGuire, A.B. and Cochran, M.D. (1983) Artificial propagation of the white sturgeon, *Acipenser transmontanus* Richardson. *Aquaculture* 32, 93–104.

Dou, S., Seikai, T. and Tsukamoto, K. (2000) Feeding behaviour of Japanese flounder larvae under laboratory conditions. *Journal of Fish Biology* 56, 654–666.

Dowell, S.D. (1990) The development and anti-predator responses in grey partridges and common pheasants. In: Hill, D., Jenkins, D. and Garson, P. (eds) *Pheasants in Asia, 1989*. World Pheasant Association, Reading, UK, pp. 166–175.

Downs, J.F. (1960) Domestication: an examination of the changing social relationships between man and animals. *Kroeber Anthropological Society Papers, University of California, Berkeley* 22, 18–67 (reprinted by Johnson Reprint Corp., New York).

Doyle, R.W. and Talbot, A.J. (1986) Artificial selection on growth and correlated selection on competitive behaviour in fish. *Canadian Journal of Fisheries and Aquatic Sciences* 43, 1059–1064.

Duncan, I.J.H. and Hughes, B.O. (1972) Free and operant feeding in domestic fowl. *Animal Behaviour* 20, 775–777.

Duncan, I.J.H., Savory, C.J. and Wood-Gush, D.G.M. (1978) Observations on the reproductive behaviour of domestic fowl in the wild. *Applied Animal Ethology* 4, 29–42.

Duncan, I.J.H., Rushen, J. and Lawrence, A.B. (1993) Conclusions and implications for animal welfare. In: Lawrence, A.B. and Rushen, J. (eds) *Stereotypic Animal Behaviour: Fundamentals and Applications to Welfare*. CAB International, Wallingford, UK, pp. 193–206.

Dunnington, E.A., Stallard, L.C., Hillel, J. and Siegel, P.B. (1994) Genetic diversity among commercial chicken populations estimated from DNA fingerprints. *Poultry Science* 73, 1218–1225.

Dupont-Nivet, M., Mallard, J., Bonnet, J.C. and Blanc, J.M. (2001) Evolution of genetic variability in a population of the edible snail, *Helix aspersa* Müller, undergoing domestication and short-term selection. *Heredity* 87, 129–135.

Ebinger, P. (1975) Quantitative investigations of visual brain structures in wild and domestic sheep. *Zeitschrift für Anatomie und Entwicklungsgeschichte* 146, 313–323.

Ebinger, P. (1995) Domestication and plasticity of brain organization in mallards (*Anas platyrhynchos*). *Brain, Behavior and Evolution* 45, 286–300.

Ebinger, P. and Löhmer, R. (1984) Comparative quantitative investigations on brains of rock doves, domestic and urban pigeons (*Columba l. livia*). *Zeitschrift für zoologische Systematik und Evolutionsforschung* 22, 136–145.

Ebinger, P. and Löhmer, R. (1987) A volumetric comparison of brains between greylag geese (*Anser anser* L.) and domestic geese. *Journal für Hirnforschung* 28, 291–299.

Ebinger, P., Röhrs, M. and Pohlenz, J. (1989) Veränderungen von Hirn- und Augengrössen bei wilden und domestizierten Truthühnern (*Meleagris gallopavo* L., 1758). *Zeitschrift für zoologische Systematik und Evolutionsforschung* 27, 142–148.

Economopoulos, A.P. and Loukas, M. (1986) ADH allele frequency changes in olive fruit flies shift from olives to artificial larval food and vice versa, effect of temperature. *Entomologia Experimentalis et Applicata* 40, 215–221.

Edwards, S.A. and Fraser, D. (1997) Housing systems for farrowing and lactation. *The Pig Journal* 39, 77–89.

Eibl-Eibesfeldt, I. (1975) *Ethology: the Biology of Behavior*, 2nd edn. Holt, Rinehart and Winston, New York.

Eklund, A. (1996) The effects of inbreeding on aggression in wild male house mice (*Mus domesticus*). *Behaviour* 133, 883–901.

Ellis, D.H., Dobrott, S.J. and Goodwin, J.G. Jr (1978) Reintroduction techniques for masked bobwhites. In: Temple, S.A. (ed.) *Endangered Birds: Management Techniques for Preserving Threatened Species*. University of Wisconsin Press, Madison, Wisconsin, pp. 345–354.

Ellis, D.H., Gee, G.F., Hereford, S.G., Olsen, G.H., Chisolm, T.D., Nicolich, J.M., Sullivan, K.A., Thomas, N.J., Nagendran, M. and Hatfield, J.S. (2000) Post-release survival of hand-reared and parent-reared Mississippi sandhill cranes. *The Condor* 102, 104–112.

Elsey, R.M., Joanen, T., McNease, L. and Kinler, N. (1992) Growth rates and body condition factors of *Alligator mississippiensis* in coastal Louisiana wetlands: a comparison of wild and farm-released juveniles. *Comparative Biochemistry and Physiology* 103A, 667–672.

Endal, H.P., Taranger, G.L., Stefansson, S.O. and Hansen, T. (2000) Effects of continuous additional light on growth and sexual maturity in Atlantic salmon, *Salmo salar*, reared in sea cages. *Aquaculture* 191, 337–349.

Enders, R.K. (1945) Induced changes in the breeding habits of foxes. *Sociometry* 8, 53–55.

Enders, R.K. (1952) Reproduction in the mink (*Mustela vison*). *Proceedings American Philosophical Society* 96, 691–755.

English, L.J., Maguire, G.B. and Ward, R.D. (2000) Genetic variation of wild and hatchery populations of the Pacific oyster, *Crassostrea gigas* (Thunberg), in Australia. *Aquaculture* 187, 283–298.

Enz, C.A., Schäffer, E. and Müller, R. (2001) Importance of diet type, food particle size, and tank circulation for culture of Lake Hallwil whitefish larvae. *North American Journal of Aquaculture* 63, 321–327.

Estep, D.Q., Lanier, D.L. and Dewsbury, D.A. (1975) Copulatory behavior and nest building behavior of wild house mice (*Mus musculus*). *Animal Learning and Behavior* 3, 329–336.

Etches, R.J. (1996) *Reproduction in Poultry*. CAB International, Wallingford, UK.

Evans, D.O. and Willox, C.C. (1991) Loss of exploited, indigenous populations of lake trout, *Salvelinus namaycush*, by stocking of non-native stocks. *Canadian Journal of Fisheries and Aquatic Sciences* 48 (Suppl. 1), 134–147.

Evans, R.H. (1987) Rearing orphaned wild mammals. *Veterinary Clinics of North America: Small Animal Practice* 17, 755–783.

Eysenck, H.J. and Broadhurst, P.L. (1964) Experiments with animals: introduction. In: Eysenck, H.J. (ed.) *Experiments in Motivation*. MacMillan, New York, pp. 285–291.

Falconer, D.S. and Mackay, T.F.C. (1996) *Introduction to Quantitative Genetics*, 4th edn. Longman, London.

Falk-Petersen, I.-B., Hansen, T.K. and Sunde, L.M. (1999) Cultivation of the spotted wolffish *Anarhichas minor* (Olafsen) – a new candidate for cold-water fish farming. *Aquaculture Research* 30, 711–718.

FAO (1997) *Aquaculture Development. FAO Technical Guidelines for Responsible Fisheries*, No. 5. FAO, Rome.

Feddersen-Petersen, D.U. (2000) Vocalization of European wolves (*Canis lupus lupus*) and various dog breeds (*C. lupus* f. fam.). *Archives of Animal Breeding* (*Archiv für Tierzucht, Dummerstorf*) 43, 387–397.

Ferguson, M.M., Ihssen, P.E. and Hynes, J.D. (1991) Are cultured stocks of brown trout (*Salmo trutta*) and rainbow trout (*Oncorhynchus mykiss*) genetically similar to their source populations? *Canadian Journal of Fisheries and Aquatic Sciences* 48 (Suppl. 1), 118–123.

Festing, M.F.W. (1969) Inbred mice in research. *Nature, London* 221, 716.

Festing, M.F.W. (1979a) *Inbred Strains in Biomedical Research*. Oxford University Press, New York.

Festing, M.F.W. (1979b) Inbred strains. In: Baker, H.J., Lindsey, J.R. and Weisbroth, S.H. (eds) *The Laboratory Rat:* Vol. 1. *Biology and Diseases*. Academic Press, New York, pp. 55–72.

Festing, M.F.W. (1989) Inbred strains of mice. In: Lyon, M.F. and Searle, A.G. (eds) *Genetic Variants and Strains of the Laboratory Mouse*, 2nd edn. Oxford University Press, New York, pp. 636–648.

Festing, M.F.W. and Lovell, D.P. (1981) Domestication and development of the mouse as a laboratory animal. *Symposium of the Zoological Society of London* 47, 43–62.

Fiumera, A.C., Parker, P.G. and Fuerst, P.A. (2000) Effective population size and maintenance of genetic diversity in captive-bred populations of a Lake Victoria cichlid. *Conservation Biology* 14, 886–892.

Fix, J.E. (1975) Selective predation by a ferret (*Mustela putorius* Linnaeus) on wild, domestic and hybrid Norway rats (*Rattus norvegicus* Berkenhout). MS thesis, State University of New York College of Environmental Science and Forestry, Syracuse, New York.

Fleischer, P., Metzner, M., Beyerbach, M., Hoedemaker, M. and Klee, W. (2001) The relationship between milk yield and the incidence of some diseases in dairy cows. *Journal of Dairy Science* 84, 2025–2035.

Fleming, I.A. and Einum, S. (1997) Experimental tests of genetic divergence of farmed from wild Atlantic salmon due to domestication. *ICES Journal of Marine Science* 54, 1051–1063.

Fleming, I.A. and Gross, M.R. (1992) Reproductive behavior of hatchery and wild coho salmon (*Oncorhynchus kisutch*): does it differ? *Aquaculture* 103, 101–121.

Fleming, I.A. and Gross, M.R. (1993) Breeding success of hatchery and wild coho salmon (*Oncorhynchus kisutch*) in competition. *Ecological Applications* 3, 230–245.

Fleming, I.A. and Gross, M.R. (1994) Breeding competition in a Pacific salmon (coho: *Oncorhynchus kisutch*): measures of natural and sexual selection. *Evolution* 48, 637–657.

Fleming, I.A., Jonnson, B., Gross, M.R. and Lamberg, A. (1996) An experimental study of the reproductive behaviour and success of farmed and wild Atlantic salmon (*Salmo salar*). *Journal of Applied Ecology* 33, 893–905.

Fleming, I.A., Lamberg, A. and Jonsson, B. (1997) Effects of early experience on the reproductive performance of Atlantic salmon. *Behavioral Ecology* 8, 470–480.

Foley, C.W., Seerly, R.W., Hansen, W.J. and Curtis, W.E. (1971) Thermoregulatory response to cold environment by neonatal wild and domestic piglets. *Journal of Animal Science* 32, 926–929.

Forbes, J.M. (1995) *Voluntary Food Intake and Diet Selection in Farm Animals*. CAB International, Wallingford, UK.

Foscarini, R. (1988) A review: intensive farming procedure for red sea bream (*Pagrus major*) in Japan. *Aquaculture* 72, 191–246.

Fox, M.W. (1968) The influence of domestication upon behaviour of animals. In: Fox, M.W. (ed.) *Abnormal Behavior in Animals*. W.B. Saunders, Philadelphia, Pennsylvania, pp. 64–76.

Fox, M.W. (1971) *Behavior of Wolves and Related Canids*. Krieger Publishing Co., Malabar.

Fox, M.W. (1978) *The Dog: Its Domestication and Behavior*. Garland Press, New York.

Frank, H. (1980) Evolution of canine information processing under conditions of natural and artificial selection. *Zeitschrift für Tierpsychologie* 53, 389–399.

Frank, H. and Frank, M.G. (1982a) On the effects of domestication on canine social development and behavior. *Applied Animal Ethology* 8, 507–525.

Frank, H. and Frank, M.G. (1982b) Comparison of problem-solving performance in six-week-old wolves and dogs. *Animal Behaviour* 30, 95–98.

Frank, H. and Frank, M.G. (1985) Comparative manipulation-test performance in ten-week-old wolves (*Canis lupus*) and Alaskan malamutes (*Canis familiaris*): a piagetian interpretation. *Journal of Comparative Psychology* 99, 266–274.

Frank, H., Frank, M.G., Hasselbach, L.M. and Littleton, D.M. (1989) Motivation and insight in wolf (*Canis lupus*) and Alaskan malamute (*Canis familiaris*): visual discrimination learning. *Bulletin of the Psychonomic Society* 27, 455–458.

Frankel, O.H. (1959) Variation under domestication. *Australian Journal of Science* 22, 27–32.

Frankham, R. and Loebel, D.A. (1992) Modeling problems in conservation genetics using captive *Drosophila* populations: rapid genetic adaptation to captivity. *Zoo Biology* 11, 333–342.

Frankham, R., Hemmer, H., Ryder, O.A., Cothran, E.G., Soule, M.E., Murray, N.D. and Snyder, M. (1986) Selection in captive populations. *Zoo Biology* 5, 127–138.

Fraser, A.F. (1968) Behavioral disorders in domestic animals. In: Fox, M.W. (ed.) *Abnormal Behavior in Animals*. W.B. Saunders, Philadelphia, Pennsylvania, pp. 179–187.

Fraser, D. (1978) Observations on the behavioural development of suckling and early-weaned piglets during the first six weeks after birth. *Animal Behaviour* 26, 22–30.

Fraser, D., Weary, D.M., Pajor, E.A. and Milligan, B.N. (1997) A scientific conception of animal welfare that reflects ethical concerns. *Animal Welfare* 6, 187–205.

Frölich, K. and Flach, E.J. (1998) Long-term viral serology of semi-free-living and captive ungulates. *Journal of Zoo and Wildlife Medicine* 29, 165–170.

Fuerst, P.A. and Maruyama, T. (1986) Considerations on the conservation of alleles and of genic heterozygosity in small managed populations. *Zoo Biology* 5, 171–180.

Fuller, J.L. and Clark, L.D. (1968) Genotype and behavioral vulnerability to isolation in dogs. *Journal of Comparative and Physiological Psychology* 66, 151–156.

Fuller, J.L. and Thompson, W.R. (1960) *Behavior Genetics*. John Wiley & Sons, New York.

Fumihito, A., Miyake, T., Sumi, S., Takada, M., Ohno, S. and Kondo, N. (1994) One subspecies of the red junglefowl (*Gallus gallus gallus*) suffices as the matriarchic ancestor of all domestic breeds. *Proceedings of the National Academy of Sciences USA* 91, 12505–12509.

Fumihito, A., Miyake, T., Takada, M., Shingu, R., Endo, T., Gojobori, T., Kondo, N. and Ohno, S. (1996) Monophyletic origin and unique dispersal patterns of domestic fowls. *Proceedings of the National Academy of Sciences USA* 93, 6792–6795.

Galef, B.G. Jr (1970) Aggression and timidity: responses to novelty in feral Norway rats. *Journal of Comparative and Physiological Psychology* 70, 370–381.

Galef, B.G. Jr and Allen, C. (1995) A new model system for studying behavioural traditions in animals. *Animal Behaviour* 50, 705–717.

Galef, B.G. Jr and Clark, M.M. (1971) Social factors in the poison avoidance and feeding behavior of wild and domesticated rat pups. *Journal of Comparative and Physiological Psychology* 78, 341–357.

Gallo, L., Carnier, P., Cassandro, M., Mantovani, R., Bailoni, L., Contiero, B. and Bittante, G. (1996) Change in body condition score of Holstein cows as affected by parity and mature equivalent milk yield. *Journal of Dairy Science* 79, 1009–1015.

García-Marin, J.L., Jorde, P.J., Ryman, N., Utter, F. and Pla, C. (1991) Management implications of genetic differentiation between native and hatchery populations of brown trout (*Salmo trutta*) in Spain. *Aquaculture* 95, 235–249.

García-Marin, J.L., Sanz, N. and Pla, C. (1999) Erosion of the native genetic resources of brown trout in Spain. *Ecology of Freshwater Fish* 8, 151–158.

Geiser, F., Sink, H.S., Stahl, B., Mansergh, I.M. and Broome, L.S. (1990) Differences in the physiological response to cold in wild and laboratory-bred mountain pygmy possums, *Burramys parvus* (Marsupialia). *Australian Wildlife Research* 17, 535–539.

Geist, V. (1971) A behavioral approach to the management of wild ungulates. In: Duffey, E. and Watt, A.S. (eds) *The Scientific Management of Animal and Plant Communities for Conservation*. Blackwell Scientific Publications, Oxford, pp. 413–424.

Gengler, N. and Druet, T. (2001) Impact of biotechnology on animal breeding and genetic progress. In: Renaville, R. and Burny, A. (eds) *Biotechnology in Animal Husbandry*. Kluwer Academic Publishers, Boston, Massachusetts, pp. 33–45.

Gibbons, E.F., Durrant, B.S. and Demarest, J. (1995) *Conservation of Endangered Species in Captivity: An Interdisciplinary Approach*. State University of New York Press, Albany.

Gibert, P., Moreteau, B., Moreteau, J.-C. and David, J.R. (1998) Genetic variability of quantitative traits in *Drosophila melanogaster* (fruit fly) natural populations: analysis of wild-living flies and of several laboratory generations. *Heredity* 80, 326–335.

Gille, U. and Salomon, F.-V. (1994) Heart and body growth in ducks. *Growth, Development and Aging* 58, 75–81.

Gille, U. and Salomon, F.-V. (2000) Brain growth in mallards, Pekin and Muscovy ducks (Anatidae). *Journal Zoological Society of London* 252, 399–404.

Gillespie, D., Frye, F.L., Stockham, S.L. and Fredeking, T. (2000) Blood values in wild and captive Komodo dragons (*Varanus komodoensis*). *Zoo Biology* 19, 495–509.

Ginsburg, B.E. and Hiestand, L. (1992) Humanity's 'best friend': the origins of our inevitable bond with dogs. In: Davis, H. and Balfour, D. (eds) *The Inevitable Bond: Examining Scientist–Animal Interactions*. Cambridge University Press, New York, pp. 93–108.

Giovambattista, G., Ripoli, M.V., Peral-Garcia, P. and Bouzat, J.L. (2001) Indigenous domestic breeds as reservoirs of genetic diversity: the Argentinean Creole cattle. *Animal Genetics* 32, 240–247.

Gippoliti, S. and Carpaneto, G.M. (1997) Captive breeding, zoos, and good sense. *Conservation Biology* 11, 806–807.

Giuffra, E., Kijas, J.M.H., Amarger, V., Carlborg, O., Jeon, J.-T. and Andersson, L. (2000) The origin of the domestic pig: independent domestication and subsequent introgression. *Genetics* 154, 1785–1791.

Gloyd, J. (1992) Wolf hybrids – a biological time bomb? *Journal of the American Veterinary Medical Association* 201, 381–382.

Goldfoot, D.A. (1977) Rearing conditions which support or inhibit later sexual potential of laboratory-born rhesus monkeys: hypotheses and diagnostic behaviors. *Laboratory Animal Science* 27, 548–556.

Gon, S.M. and Price, E.O. (1984) Invertebrate domestication: behavioral considerations. *BioScience* 34, 575–579.

Goodwin, D., Bradshaw, J.W.S. and Wickens, S.M. (1997) Paedomorphosis affects agonistic visual signals of domestic dogs. *Animal Behaviour* 53, 297–304.

Gosling, E.M. (1982) Genetic variability in hatchery-produced Pacific oysters (*Crassostrea gigas* Thunberg). *Aquaculture* 26, 273–287.

Gould, S.J. (1977) *Ontogeny and Phylogeny*. Belknap Press, Cambridge.

Gould, S.J. (1986) The egg-a-day barrier. *Natural History* 95, 16–24.

Grandin, T. (ed.) (1993) *Livestock Handling and Transport*. CAB International, Wallingford, UK.

Grandin, T. (1997) Assessment of stress during handling and transport. *Journal of Animal Science* 75, 249–257.

Grandin, T. (ed.) (1998) *Genetics and the Behavior of Domestic Animals*. Academic Press, San Diego, California.

Graves, H.B. (1984) Behavior and ecology of wild and feral swine (*Sus scrofa*). *Journal of Animal Science* 58, 482–492.

Green, E.L. (1966) Breeding systems. In: Green, E.L. (ed.) *Biology of the Laboratory Mouse*. McGraw-Hill, New York, pp. 11–22.

Griffin, A.S., Blumstein, D.T. and Evans, C.S. (2000) Training captive-bred or translocated animals to avoid predators. *Conservation Biology* 14, 1317–1326.

Griffith, B., Scott, J.M., Carpenter, J.W. and Reed, C. (1989) Translocation as a species conservation tool: status and strategy. *Science* 245, 477–480.

Grindrod, J.A.E. and Cleaver, J.A. (2001) Environmental enrichment reduces the performance of stereotypic circling behaviour in captive common seals (*Phoca vitulina*). *Animal Welfare* 10, 53–63.

Gross, M.R. (1998) One species with two biologies: Atlantic salmon (*Salmo salar*) in the wild and in aquaculture. *Canadian Journal of Fisheries and Aquatic Sciences* 55, 131–144.

Gross, W.B. and Siegel, P.B. (1979) Adaptation of chickens to their handler, and experimental results. *Avian Diseases* 23, 708–714.

Gross, W.B. and Siegel, P.B. (1988) Environment–genetic influences on immunocompetence. *Journal of Animal Science* 66, 2091–2094.

Grutter, A.S. and Pankhurst, N.W. (2000) The effects of capture, handling, confinement and ectoparasite load on plasma levels of cortisol, glucose and lactate in the coral reef fish *Hemigymnus melapterus*. *Journal of Fish Biology* 57, 391–401.

Gustafsson, M., Jensen, P., de Jonge, F.H. and Schuurman, T. (1999a) Domestication effects on foraging strategies in pigs (*Sus scrofa*). *Applied Animal Behaviour Science* 62, 305–317.

Gustafsson, M., Jensen, P., de Jonge, F.H., Illmann, G. and Spinka, M. (1999b) Maternal behaviour of domestic sows and crosses between domestic sows and wild boar. *Applied Animal Behaviour Science* 65, 29–42.

Guttinger, H.R. (1985) Consequences of domestication on the song structures in the canary. *Behaviour* 94, 254–278.

Ha, J.C., Robinette, R.L. and Davis, A. (2000) Survival and reproduction in the first two years following a large-scale primate colony move and social reorganization. *American Journal of Primatology* 50, 131–138.

Haase, E. (2000) Comparison of reproductive biological parameters in male wolves and domestic dogs. *Zeitschrift für Säugetierkunde* 65, 257–270.

Haase, E. and Donham, R.S. (1980) Hormones and domestication. In: Epple, A. and Stetson, M.H. (eds) *Avian Endocrinology*. Academic Press, New York, pp. 549–565.

Haenlein, G.F.W., Holdren, R.D. and Yoon, Y.M. (1966) Comparative response of horses and sheep to different physical forms of alfalfa hay. *Journal of Animal Science* 25, 740–743.

Haensly, T.F., Meyers, S.M., Crawford, J.A. and Castillo, W.J. (1985) Treatments affecting post-release survival and productivity of pen-reared ring-necked pheasants. *Wildlife Society Bulletin* 13, 521–528.

Hagen, D.R. and Kephart, K.B. (1980) Reproduction in domestic and feral swine. 1. Comparison of ovulatory rate and litter size. *Biology of Reproduction* 22, 550–552.

Haldane, J.B.S. (1949) Suggestions as to the quantitative measurement of rates of evolution. *Evolution* 3, 51–56.

Haldane, J.B.S. (1954) The statics of evolution. In: Huxley, J., Hardy, A.C. and Ford, E.B. (eds) *Evolution as a Process*. George Allen & Unwin, London, pp. 109–121.

Hale, E.B. (1969) Domestication and the evolution of behavior. In: Hafez, E.S.E. (ed.) *The Behaviour of Domestic Animals*, 2nd edn. Bailliere, Tindall & Cassell, London, pp. 22–42.

Hale, E.B. and Schein, M.W. (1962) The behaviour of turkeys. In: Hafez, E.S.E. (ed.) *The Behaviour of Domestic Animals*. Bailliére, Tindall and Cox, London, pp. 531–564.

Hall, A.L. (2001) The effect of stocking density on the welfare and behaviour of broiler chickens reared commercially. *Animal Welfare* 10, 23–40.

Hamamura, Y. (ed.) (2001) *Silkworm Rearing on Artificial Diet*. Science Publishers, Enfield, New Hampshire.

Hameed, A.S.S. (1997) Quality of eggs produced from wild and captive spawners of *Penaeus indicus* H. Milne Edwards and their bacterial load. *Aquaculture Research* 28, 301–303.

Hamlin, H.J. and Kling, L.J. (2001) The culture and early weaning of larval haddock (*Melanogrammus aeglefinus*) using a microparticulate diet. *Aquaculture* 201, 61–72.

Hammer, R.P. Jr, Hori, K.M., Blanchard, R.J. and Blanchard, D.C. (1992) Domestication alters 5-HT$_{1A}$ receptor binding in rat brain. *Pharmacology, Biochemistry and Behavior* 42, 25–28.

Hammond, J. (1962) Some changes in the form of sheep and pigs under domestication. *Journal of Animal Breeding and Genetics* 77, 156–158.

Hanlon, A.J. (1997) The welfare of farmed red deer. *Deer Farming* 53, 14–16.

Hanlon, A.J. (1999) Red deer. In: Ewbank, R., Kim-Madslien, F. and Hart, C.B. (eds) *Management and Welfare of Farm Animals*, 4th edn. Universities Federation for Animal Welfare, Wheathampstead, UK, pp. 179–192.

Hanlon, A.J., Rhind, S.M., Reid, H.W., Burrells, C., Lawrence, A.B., Milne, J.A. and McMillen, S.R. (1994) Relationship between immune response, liveweight gain, behaviour and adrenal function in red deer (*Cervus elaphus*) calves derived from wild and farmed stock, maintained at two housing densities. *Applied Animal Behaviour Science* 41, 243–255.

Hannah, A.C. and Brotman, B. (1990) Procedures for improving maternal behavior in captive chimpanzees. *Zoo Biology* 9, 233–240.

Hansen, C.P.B. and Jeppesen, L.L. (2001a) Use of water for swimming and its relationship to temperature and other factors in farm mink (*Mustela vison*). *Acta Agriculture Scandinavica*, Section A, Animal Science 51, 89–93.

Hansen, C.P.B. and Jeppesen, L.L. (2001b) Swimming activity of farm mink (*Mustela vison*) and its relation to stereotypies. *Acta Agriculture Scandinavica*, Section A, Animal Science 51, 71–76.

Hansen, L.P., Jacobsen, J.A. and Lund, R.A. (1993) High numbers of farmed Atlantic salmon, *Salmo salar* L., observed in oceanic waters north of the Faroe Islands. *Aquaculture and Fisheries Management* 24, 777–781.

Hansen, L.T. and Berthelsen, H. (2000) The effect of environmental enrichment on the behaviour of caged rabbits (*Oryctolagus cuniculus*). *Applied Animal Behaviour Science* 68, 163–178.

Hansen, S.W. and Damgaard, B. (1991) Effect of environmental stress and immobilization on stress physiological variables in farmed mink. *Behavioural Processes* 25, 191–204.

Hanson, J.P., Larson, M.E. and Snowdon, C.T. (1976) The effects of control over high intensity noise on plasma cortisol levels in rhesus monkeys. *Behavioral Biology* 16, 333–340.

Hard, J.J., Berejikian, B.A., Tezak, E.P., Schroder, S.L., Knudsen, C.M. and Parker, L.T. (2000) Evidence for morphometric differentiation of wild and captively reared adult coho salmon: a geometric analysis. *Environmental Biology of Fishes* 58, 61–73.

Harlow, H.F., Dodsworth, R.O. and Harlow, M.K. (1965) Total social isolation in monkeys. *Proceedings of the National Academy of Sciences USA* 54, 90–96.

Harri, M., Mononen, J., Rekilä, T., Korhonen, H. and Niemelä, P. (1998) Effects of top nest boxes on growth, fur quality and behaviour of blue foxes during their growing season. *Acta Agriculturae Scandinavica, Section A, Animal Science* 48, 184–191.

Harri, M., Lindblom, J., Malinen, H., Hyttinen, M., Lapveteläinen, T., Eskola, S. and Helminen, H.J. (1999) Effect of access to a running wheel on behavior of C57BL/6J mice. *Laboratory Animal Science* 49, 401–405.

Harrington, F.H. and Mech, L.D. (1978) Wolf vocalization. In: Hall, R.L. and Sharp, H.S. (eds) *Wolf and Man: Evolution in Parallel*. Academic Press, New York, pp. 109–132.

Harrison, R.O., Ford, S.P., Young, J.W., Conley, A.J. and Freeman, A.E. (1990) Increased production versus reproductive and energy status of high producing dairy cows. *Journal of Dairy Science* 73, 2749–2758.

Hart, P.R. and Purser, G.J. (1996) Weaning of hatchery-reared greenback flounder (*Rhombosolea tapirina* Günther) from live to artificial diets: effects of age and duration of the changeover period. *Aquaculture* 145, 171–181.

Hartl, G.B. (1987) Biochemical differentiation between the wild rabbit (*Oryctolagus cuniculus* L.), the domestic rabbit and the brown hare (*Lepus europaeus* Pallus). *Zeitschrift für zoologische Systematik und Evolutionsforschung* 24, 309–316.

Hartley, F.G.L., Follett, B.K., Harris, S., Hirst, D. and McNeilly, A.S. (1994) The endocrinology of gestation failure in foxes (*Vulpes vulpes*). *Journal of Reproduction and Fertility* 100, 341–346.

Hartt, E.W., Harvey, N.C., Leete, A.J. and Preston, K. (1994) Effects of age at pairing on reproduction in captive California condors (*Gymnogyps californianus*). *Zoo Biology* 13, 3–11.

Hasegawa, E. and Tanaka, S. (1996) Sexual maturation in *Locusta migratoria* females: laboratory vs. field conditions. *Applied Entomology and Zoology* 31, 279–290.

Hassin, S., de Monbrison, D., Hanin, Y., Elizur, A., Zohar, Y. and Popper, D.M. (1997) Domestication of the white grouper (*Epinephelus aeneus*): 1. Growth and reproduction. *Aquaculture* 156, 305–316.

Håstein, T. and Lindstad, T. (1991) Diseases in wild and cultured salmon: possible interactions. *Aquaculture* 98, 277–288.

Hattori, S., Noboru, Y. and Yamanouchi, K. (1986) Domestication of the Watase's shrew, *Crocidura horsfieldi watasei*, for a laboratory animal. *Japanese Journal of Experimental Medicine* 56, 75–79.

Hauber, W. (1998) Involvement of basal ganglia transmitter systems in motor initiation. *Progress in Neurobiology* 56, 507–540.

Hayssen, V. (1997) Effects of the nonagouti coat-color allele on behavior of deer mice (*Peromyscus maniculatus*): a comparison with Norway rats (*Rattus norvegicus*). *Journal of Comparative Psychology* 111, 419–423.

Hayssen, V., Gunawardhana, S. and Meyer, J. (1994) The agouti coat-color locus may influence brain catecholamines: regional differences in norepinephrine and dopamine in the brains of two color-morphs of deer mice (*Peromyscus maniculatus*). *Comparative Biochemistry and Physiology* 107C, 51–55.

Hazary, R.C., Staines, H.J. and Wishart, G.J. (2001) Assessing the effect of mating ratio on broiler breeder performance by quantifying sperm:egg interaction. *Journal of Applied Poultry Research* 10, 1–4.

Heaton, M.B. (1976) Developing visual function in the red junglefowl embryo. *Journal of Comparative and Physiological Psychology* 90, 53–56.

Hebert, P.L. and Bard, K. (2000) Orangutan use of vertical space in an innovative habitat. *Zoo Biology* 19, 239–251.

Hedgecock, D. and Sly, F. (1990) Genetic drift and effective population sizes of hatchery-propagated stocks of the Pacific oyster, *Crassostrea gigas. Aquaculture* 88, 21–38.

Hedgecock, D., Shleser, R.A. and Nelson, K. (1976) Applications of biochemical genetics to aquaculture. *Journal of the Fisheries Research Board of Canada* 33, 1108–1119.

Hediger, H. (1938) Tierpsychologie und Haustierforschung. *Zeitschrift für Tierpsychologie* 2, 29–46.

Hediger, H. (1964) *Wild Animals in Captivity*. Dover, New York.

Hedrick, P.W. and Kalinowski, S.T. (2000) Inbreeding depression in conservation biology. *Annual Review of Ecology and Systematics* 31, 139–162.

Hedrick, P.W., Dowling, T.E., Minckley, W.L., Tibbets, C.A., Demarais, B.D. and Marsh, P.C. (2000) Establishing a captive broodstock for the endangered bonytail chub (*Gila elegans*). *Journal of Heredity* 91, 35–39.

Hegmann, J.P. (1975) The response to selection for altered conduction velocity in mice. *Behavioral Biology* 13, 413–423.

Heinrich, B. (1988) Why do ravens fear their food? *Condor* 90, 950–952.

Heller, K.E., Houbak, B. and Jeppesen, L.L. (1988) Stress during mother–infant separation in ranch mink. *Behavioural Processes* 17, 217–227.

Hellwing, S. (1971) Maintenance and reproduction in the white toothed shrew, *Crocidura russula monacha*, in captivity. *Zeitschrift für Säugetierkunde* 36, 103–113.

Hemmer, H. (1988) Ethological aspects of deer farming. In: Reid, H.W. (ed.) *The Management and Health of Farmed Deer*. Kluwer Academic Publishers, Dordrecht, pp. 129–140.

Hemmer, H. (1990) *Domestication: the Decline of Environmental Appreciation*. Cambridge University Press, Cambridge.

Hemmingsen, W., Lysne, D.A., Eidnes, T. and Skorping, A. (1993) The occurrence of larval ascaridoid nematodes in wild-caught and in caged and artificially fed Atlantic cod, *Gadus morhua* L., in Norwegian waters. *Fisheries Research* 15, 379–386.

Hemsworth, P.H. and Barnett, J.L. (1987) Human–animal interactions. In: Price, E.O. (ed.) *Farm Animal Behavior, Veterinary Clinics of North America – Food Animal Practice*, Vol. 3. W.B. Saunders, Philadelphia, Pennsylvania, pp. 339–356.

Hemsworth, P.H. and Barnett, J.L. (2000) Human–animal interactions and animal stress. In: Moberg, G.P. and Mench, J.A. (eds) *The Biology of Animal Stress: Basic Principles and Implications for Animal Welfare*. CAB International, Wallingford, UK, pp. 309–335.

Hemsworth, P.H. and Coleman, G.J. (1998) *Human–Livestock Interactions*. CAB International, Wallingford, UK.

Hemsworth, P.H., Brand, A. and Willems, P.J. (1981) The behavioural response of sows to the presence of human beings and their productivity. *Livestock Production Science* 8, 67–74.

Hemsworth, P.H., Gonyou, H.W. and Dziuk, P.J. (1986) Human communication with pigs: the behavioural response of pigs to specific human signals. *Applied Animal Behaviour Science* 15, 45–54.

Hemsworth, P.H., Barnett, J.L., Coleman, G.J. and Hansen, C. (1989) A study of the relationships between the attitudinal and behavioural profiles of stockpeople and the level of fear of humans and the reproductive performance of commercial pigs. *Applied Animal Behaviour Science* 23, 301–314.

Hemsworth, P.H., Barnett, J.L., Treacy, D. and Madgwick, P. (1990) The heritability of the trait fear of humans and the association between this trait and subsequent reproductive performance of gilts. *Applied Animal Behaviour Science* 25, 85–95.

Hemsworth, P.H., Coleman, G.J. and Barnett, J.L. (1994a) Improving the attitude and behaviour of stockpersons toward pigs and the consequences on the behaviour and reproductive performance of commercial pigs. *Applied Animal Behaviour Science* 39, 349–362.

Hemsworth, P.H., Coleman, G.J., Barnett, J.L. and Jones, R.B. (1994b) Behavioural responses of humans and the productivity of commercial broiler chickens. *Applied Animal Behaviour Science* 41, 101–114.

Hemsworth, P.H., Coleman, G.J., Barnett, J.L. and Borg, S. (2000) Relationships between human–animal interactions and productivity of commercial dairy cows. *Journal of Animal Science* 78, 2821–2831.

Henderson, J.V. and Waran, N.K. (2001) Reducing equine stereotypies using an Equiball. *Animal Welfare* 10, 73–80.

Henderson, N. (1970) Genetic influences on the behavior of mice can be obscured by laboratory rearing. *Journal of Comparative and Physiological Psychology* 72, 505–511.

Hennig, C.W. and Dunlap, W.P. (1978) Tonic immobility in *Anolis carolinensis*: effects of time and conditions of captivity. *Behavioral Biology* 23, 75–86.

Herre, W. and Röhrs, M. (1977) Zoological considerations on the origins of farming and domestication. In: Reed, C. (ed.) *Origin of Agriculture*. Mouton, The Hague, pp. 245–279.

Herre, W. and Röhrs, M. (1990) *Haustiere – zoologisch gesehen*, 2nd edn. G. Fischer, Stuttgart.

Hertig, A.T., King, N.E. and MacKey, J. (1971) Spontaneous abortion in wild caught rhesus monkeys *Macaca mulatta*. *Laboratory Animal Science* 21, 510–519.

Heslinga, G.A. and Fitt, W.K. (1987) The domestication of reef-dwelling clams. *BioScience* 37, 332–339.

Hessler, E., Tester, J.R., Siniff, D.B. and Nelson, M.M. (1970) A biotelemetry study of survival in pen-reared pheasants released in selected habitats. *Journal of Wildlife Management* 34, 267–274.

Hesterman, H., Gregory, N.G. and Boardman, W.S.J. (2001) Deflighting procedures and their welfare implications in captive birds. *Animal Welfare* 10, 405–419.

Hesthagen, T., Fløystad, L., Hegge, O., Staurnes, M. and Skurdal, J. (1999) Comparative life-history characteristics of native and hatchery-reared brown trout, *Salmo trutta* L., in a sub-alpine reservoir. *Fisheries Management and Ecology* 6, 47–61.

Hetzer, H.O. and Miller, R.H. (1972) Rate of growth as influenced by selection for high and low fatness in swine. *Journal of Animal Science* 35, 730–742.

Hickman, R.W. and Tait, M.J. (2001) Experimental rearing of juvenile New Zealand turbot *Colistium nudipinnis* (Waite 1910): a potential new aquaculture species. *Aquaculture Research* 32, 727–737.

Hilborn, R. and Eggers, D. (2000) A review of the hatchery programs for pink salmon in Prince William Sound and Kodiak Island, Alaska. *Transactions of the American Fisheries Society* 129, 333–350.

Hill, D. and Robertson, P. (1988) *The Pheasant: Ecology, Management and Conservation*. BSP Professional Books, London.

Hindar, K., Ryman, N. and Utter, F. (1991) Genetic effects of cultured fish on natural fish populations. *Canadian Journal of Fisheries and Aquatic Sciences* 48, 945–957.

Hoefs, M. and Nowlan, U. (1997) Comparison of horn growth in captive and free-ranging Dall's rams. *Journal of Wildlife Management* 61, 1154–1160.

Hoffmann, A.A., Hallas, R., Sinclair, C. and Partridge, L. (2001) Rapid loss of stress resistance in *Drosophila melanogaster* under adaptation to laboratory culture. *Evolution* 55, 436–438.

Hohenboken, W.D. (1999) Applications of sexed semen in cattle production. *Theriogenology* 52, 1421–1433.

Höhn, M., Kronschnabl, M. and Gansloßer, U. (2000) Similarities and differences in activities and agonistic behavior of male eastern grey kangaroos (*Macropus giganteus*) in captivity and the wild. *Zoo Biology* 19, 529–539.

Hoikkala, A., Aspi, J. and Suvanto, L. (1998) Male courtship song frequency as an indicator of male genetic quality in an insect species, *Drosophila montana*. *Proceedings of the Royal Society B* 265, 503–508.

Holmes, L.N., Song, G.K. and Price, E.O. (1987) Head partitions facilitate feeding by subordinate horses in the presence of dominant pen-mates. *Applied Animal Behaviour Science* 19, 179–182.

Honacki, J.H., Kinman, K.E. and Koeppl, J.W. (eds) (1982) *Mammal Species of the World: a Taxonomic and Geographic Reference*. Allen Press and Association of Systematic Collections, Lawrence, Kansas.

Horwich, R.H. (1989) Use of surrogate parent models and age periods in a successful release of hand-reared sandhill cranes. *Zoo Biology* 8, 379–390.

Howard, W.E. (1949) Dispersal, amount of inbreeding and longevity in a local population of prairie deermice on the George Reserve, Southern Michigan. *Contributions of the Laboratory of Vertebrate Biology*, No. 43. University of Michigan, Ann Arbor, Michigan.

Howell, P., Jones, K., Scarnecchia, D., Lavoy, L., Kendra, W. and Ortmann, D. (1985) Stock assessment of Columbia River anadromous salmonids. Vol. I: Chinook, coho, chum, and sockeye salmon stock summaries. Final Report, Project No. 83–335, Bonneville Power Administration, Portland, Oregon. (Cited in Waples, 1991.)

Hubrecht, R.C., Serpell, J.A. and Poole, T.B. (1992) Correlates of pen size and housing conditions on the behaviour of kenneled dogs. *Applied Animal Behaviour Science* 34, 365–383.

Huck, U.W. and Price, E.O. (1975) Differential effects of environmental enrichment on the open-field behavior of wild and domestic Norway rats. *Journal of Comparative and Physiological Psychology* 89, 892–898.

Huck, U.W. and Price, E.O. (1976) Effect of the post-weaning environment on the climbing behaviour of wild and domestic Norway rats. *Animal Behaviour* 24, 364–371.

Hughes, B.O. and Curtis, P.E. (1997) Health and disease. In: Appleby, M.C. and Hughes, B.O. (eds) *Animal Welfare*. CAB International, Wallingford, UK, pp. 109–125.

Hughes, B.O. and Duncan, I.J.H. (1988) The notion of ethological 'need', models of motivation and animal welfare. *Animal Behaviour* 36, 1696–1707.

Hughes, C.W., Pottinger, H.J. and Safron, J. (1981) On the origin of domesticity: a test of Keeler's 'black-gene' hypothesis. *Bulletin of the Psychonomic Society* 17, 289–292.

Hughes, K. and Sokolowski, M.B. (1996) Natural selection in the laboratory for a change in resistance by *Drosophila melanogaster* to the parasitoid wasp *Asobara tabida*. *Journal of Insect Behavior* 9, 477–491.

Huon, F., Meunier-Salaün, M.-C. and Faure, J.M. (1986) Feeder design and available feeding space influence the feeding behaviour of hens. *Applied Animal Behaviour Science* 15, 65–70.

Hurnik, J.F. and Lehman, H. (1988) Ethics and farm animal welfare. *Journal of Agricultural Ethics* 1, 305–318.

Hyams, E. (1972) *Animals in the Service of Man: 10,000 Years of Domestication*. J.M. Dent & Sons, London.

Inglis, I.R. and Ferguson, N.J.K. (1986) Starlings search for food rather than eat freely available identical food. *Animal Behaviour* 34, 614–617.

Inglis, I.R., Forkman, B. and Lazarus, J. (1997) Free food or earned food? A review and fuzzy model of contrafreeloading. *Animal Behaviour* 53, 1171–1191.

Isaac, E. (1970) *Geography of Domestication*. Prentice-Hall, Englewood Cliffs, New Jersey.

Iverson, J.B. (1978) The impact of feral cats and dogs on populations of the West Indian rock iguana, *Cyclura carinata*. *Biological Conservation* 14, 63–73.

Jackson, S. and Diamond, J. (1996) Metabolic and digestive responses to artificial selection in chickens. *Evolution* 50, 1638–1650.

Jenkins, D. (1961) Social behavior in the partridge (*Perdix perdix*). *Ibis* 103, 155–188.

Jeppesen, L.L., Heller, K.E. and Dalsgaard, T. (2000) Effects of early weaning and housing conditions on the development of stereotypies in farmed mink. *Applied Animal Behaviour Science* 68, 85–92.

Jewell, P. (1969) Wild mammals and their potential for new domestication. In: Ucko, P. and Dimbleby, G. (eds) *The Domestication and Exploitation of Plants and Animals*. Duckworth, London, pp. 101–109.

Jimenez, J.A., Hughes, K., Alaks, G., Graham, L. and Lacy, R.C. (1994) An experimental study of inbreeding depression in a natural habitat. *Science* 266, 271–273.

Johnsen, B.O. and Jensen, A.J. (1991) The *Gyrodactylus* story in Norway. *Aquaculture* 98, 289–302.

Johnson, G.D. and Boyce, M.S. (1991) Survival, growth, and reproduction of captive-reared sage grouse. *Wildlife Society Bulletin* 19, 88–93

Johnson, K.G., Tyrrell, J., Rowe, J.B. and Pethick, D.W. (1998) Behavioural changes in stabled horses given non-therapeutic levels of virginiamycin as Founderguard. *Equine Veterinary Journal* 30, 139–143.

Johnsson, J.I. and Abrahams, M.V. (1991) Interbreeding with domestic strain increases foraging under threat of predation in juvenile steelhead trout (*Oncorhynchus mykiss*): an experimental study. *Canadian Journal of Fisheries and Aquatic Sciences* 48, 243–247.

Johnsson, J.I., Petersson, E., Jönsson, E., Björnsson, B.T. and Järvi, T. (1996) Domestication and growth hormone alter antipredation behaviour and growth pattern in juvenile brown trout, *Salmo trutta*. *Canadian Journal of Fisheries and Aquatic Sciences* 53, 1546–1554.

Johnsson, J.I., Höjesjö, J. and Fleming, I.A. (2001) Behavioural and heart rate responses to predation risk in wild and domesticated Atlantic salmon. *Canadian Journal of Fisheries and Aquatic Sciences* 58, 788–794.

Jones, G.F. (1998) Genetic aspects of domestication, common breeds and their origin. In: Rothschild, M.F. and Ruvinsky, A. (eds) *The Genetics of the Pig.* CAB International, Wallingford, UK, pp.17–50.

Jones, R.B. (1993) Reduction of the domestic chick's fear of human beings by regular handling and related treatments. *Animal Behaviour* 46, 991–998.

Jones, R.B. and Hocking, P.M. (1999) Genetic selection for poultry behaviour: big bad wolf or friend in need? *Animal Welfare* 8, 343–359.

Jones, R.B., Satterlee, D.G. and Ryder, F.H. (1994) Fear of humans in Japanese quail selected for low or high adrenocortical response. *Physiology and Behavior* 56, 379–383.

Jones, R.D. Jr (1966) Raising caribou for an Aleutian introduction. *Journal of Wildlife Management* 30, 453–460.

Jonsson, B. (1997) A review of ecological and behavioural interactions between cultured and wild Atlantic salmon. *ICES Journal of Marine Science* 54, 1031–1039.

Jonsson, B. and Fleming, I.A. (1993) Enhancement of wild populations. In: Sundnes, G. (ed.) *Human Impact of Self-recruiting Populations.* Tapir Publishers, Trondheim, pp. 209–238.

Jonsson, B., Jonsson, N. and Hansen, L.P. (1991) Differences in life history and migratory behaviour between wild and hatchery reared Atlantic salmon in nature. *Aquaculture* 98, 69–78.

Jonssonn, S., Brønnøs, E. and Lundqvist, H. (1999) Stocking of brown trout, *Salmo trutta* L.: effects of acclimatization. *Fisheries Management and Ecology* 6, 459–473.

Jørgensen, B. and Vestergaard, T. (1990) Genetics of leg weakness in boars at the Danish pig breeding stations. *Acta Agriculturae Scandinavica, Section A, Animal Science* 40, 59–69.

Justines, G. and Johnson, K.M. (1970) Observations on the laboratory breeding of the cricetine rodent *Calomys callosus. Laboratory Animal Care* 20, 57–60.

Kalinowski, S.T., Hedrick, P.W. and Miller, P.S. (1999) No inbreeding depression observed in Mexican and red wolf captive breeding programs. *Conservation Biology* 13, 1371–1377.

Kalinowski, S.T., Hedrick, P.W. and Miller, P.S. (2000) Inbreeding depression in the Speke's gazelle captive breeding program. *Conservation Biology* 14, 1375–1384.

Kanid'hev, A.N., Kostyunin, G.M. and Salmin, S.A. (1970) Hatchery propagation of the pink and chum salmons as a means of increasing the salmon stocks of Sakhalin. *Journal of Ichthyology* 10, 249–259.

Kantanen, J., Olsaker, I., Holm, L.-E., Lien, S., Vilkki, J., Brusgaard, K., Eythorsdottir, E., Danell, B. and Adalsteinsson, S. (2000) Genetic diversity and population structure of 20 North European cattle breeds. *The American Genetic Association* 91, 446–457.

Kardong, K. (1993) The predatory behavior of the Northern Pacific rattlesnake (*Crotalus viridis oreganus*): laboratory versus wild mice as prey. *Herpetologica* 49, 457–463.

Kavanau, J.L. (1964) Behavior: confinement, adaptation and compulsory regimens in laboratory studies. *Science* 143, 490.

Kawahara, T. (1972) Genetic changes occurring in wild quails due to 'natural selection' under domestication. *Annual Report of the National Institute of Genetics Japan* 22, 111–112.

Kay, R.N.B. and Staines, B.W. (1981) The nutrition of the red deer (*Cervus elaphus*). *Nutrition Abstracts and Reviews (B)* 51, 601–622.

Keeler, C.E. (1931) *The Laboratory Mouse: Its Origin, Heredity and Culture.* Harvard University Press, Cambridge, Massachusetts.

Keeler, C.E. (1942) The association of the black (non-agouti) gene with behavior in the Norway rat. *Journal of Heredity* 33, 371–384.

Keeler, C.E. (1975) Genetics of behavior variations in color phases of the red fox. In: Fox, M.W. (ed.) *The Wild Canids: Their Systematics, Behavioral Ecology and Evolution*. Van Nostrand-Reinhold, New York, pp. 399–413.

Keeler, C.E., Ridgway, S., Lipscomb, L. and Fromm, E. (1968) The genetics of adrenal size and tameness in colorphase foxes. *Journal of Heredity* 59, 82–84.

Keller, L.F., Arcese, P., Smith, J.N.M., Hochachka, W.M. and Stearns, S.C. (1994) Selection against inbred song sparrows during a natural population bottleneck. *Nature* 372, 356–357.

Kellison, G.T., Eggleston, D.B. and Burke, J.S. (2000) Comparative behaviour and survival of hatchery-reared versus wild summer flounder (*Paralichthys dentatus*). *Canadian Journal of Fisheries and Aquatic Sciences* 57, 1870–1877.

Kells, A., Dawkins, M.S. and Cortina Borja, M. (2001) The effect of a 'Freedom Food' enrichment on the behaviour of broilers on commercial farms. *Animal Welfare* 10, 347–356.

Kemble, E.D., Blanchard, D.C., Blanchard, R.J. and Takushi, R. (1984) Taming in wild rats following medial amygdaloid lesions. *Physiology and Behavior* 32, 131–134.

Kempinger, J.J. (1996) Habitat, growth and food of young lake sturgeons in the Lake Winnebago system, Wisconsin. *North American Journal of Fisheries Management* 16, 102–114.

Kenagy, G.J. and Trombulak, S.C. (1986) Size and function of mammalian testes in relation to body size. *Journal of Mammalogy* 67, 1–22.

Kenward, R.E., Hall, D.G., Walls, S.S. and Hodder, K.H. (2001) Factors affecting predation by buzzards *Buteo buteo* on released pheasants *Phasianus colchicus*. *Journal of Applied Ecology* 38, 813–822.

Kepler, C.B. (1978) Captive propagation of whooping cranes: a behavioral approach. In: Temple, S.A. (ed.) *Endangered Birds: Management Techniques for Preserving Threatened Species*. University of Wisconsin Press, Madison, Wisconsin, pp. 231–241.

Kerr, S.G.C., Wood-Gush, D.G.M., Moser, H. and Whittemore, C.T. (1988) Enrichment of the production environment and the enhancement of welfare through the use of the Edinburgh Family Pen System of Pig Production. *Research and Development in Agriculture* 5, 171–186.

Kessel, A. and Brent, L. (2001) The rehabilitation of captive baboons. *Journal of Medical Primatology* 30, 71–80.

King, C.E. (1993) Environmental enrichment: is it for the birds? *Zoo Biology* 12, 509–512.

King, E.G. (1984) *Advances and Challenges in Insect Rearing*. Agricultural Research Service, US Department of Agriculture, New Orleans, Louisiana.

King, E.G. and Leppla, N.C. (eds) (1984) *Advances and Challenges in Insect Rearing*. Agricultural Research Service (Southern Region), US Department of Agriculture, New Orleans, Louisiana.

King, H.D. (1929) Life processes and size of the body organs of the gray Norway rat during ten generations in captivity. Part 1. Life processes. *American Anatomical Memoirs* 14, 5–72.

King, H.D. (1939) Life processes in gray Norway rats during fourteen years in captivity. *American Anatomical Memoirs* 17, 1–72.

King, H.D. and Donaldson, H.H. (1929) Life processes and size of the body and organs of the gray Norway rat during ten generations in captivity. *American Anatomical Memoirs* 14, 1–106.

Kirkwood, J.K., Gaskin, C.D. and Markham, J. (1987) Perinatal mortality and season of birth in captive wild ungulates. *The Veterinary Record* 120, 386–390.

Kisling, V.N. Jr (ed.) (2001) *Zoo and Aquarium History. Ancient Animal Collections to Zoological Gardens.* CRC Press, Boca Raton, Florida.

Kleiman, D.G. (1989) Reintroduction of captive mammals for conservation. *BioScience* 39, 152–161.

Kleiman, D.G. (1996) Reintroduction programs. In: Kleiman, D.G., Allen, M.E., Thompson, K.V. and Lumpkin, S. (eds) *Wild Mammals in Captivity: Principles and Techniques.* University of Chicago Press, Chicago, Illinois, pp. 297–305.

Kleiman, D.G., Stanley-Price, M.R. and Beck, B.B. (1994) Criteria for reintroductions. In: Olney, P.J.S., Mace, G.M. and Feistner, A.T.C. (eds) *Creative Conservation: Interactive Management of Wild and Captive Animals.* Chapman and Hall, London, pp. 287–303.

Klotchkov, D.V., Trapezov, O.V. and Kharlamova, A.V. (1998) Folliculogenesis, onset of puberty and fecundity of mink (*Mustela vison*) selectively bred for docility or aggressiveness. *Theriogenology* 49, 1545–1553.

Knipling, E.F. (1984) What colonization of insects means to research and pest management. In: King, E.G. and Leppla, N.C. (eds) *Advances and Challenges in Insect Rearing.* Agricultural Research Service (Southern Region), US Department of Agriculture, New Orleans, Louisiana, pp. ix–xi.

Koebele, B.P. (1985) Growth and the size hierarchy effect: an experimental assessment of three proposed mechanisms; activity differences, disproportional food acquisition, physiological stress. *Environmental Biology of Fishes* 12, 181–188.

Korhonen, H., Jauhiainen, L., Niemelä, P., Harri, M. and Sauna-aho, R. (2001) Physiological and behavioural responses in blue foxes (*Alopex lagopus*): comparisons between space quantity and floor material. *Animal Science* 72, 375–387.

Koteja, P., Garland, T. Jr, Sax, J.K., Swallow, J.G. and Carter, P.A. (1999) Behaviour of house mice artificially selected for high levels of voluntary wheel running. *Animal Behaviour* 58, 1307–1318.

Kovach, J.K. (1974) The behavior of Japanese quail: review of literature from a bioethological perspective. *Applied Animal Ethology* 1, 77–102.

Koyama, J., Nakamori, H. and Kuba, H. (1986) Mating behavior of wild and mass-reared strains of the melon fly, *Dacus cucurbitae* (Diptera: Tephritidae), in a field cage. *Applied Entomology and Zoology* 21, 203–209.

Kraus, G.D., Graves, H.B. and Zervanos, S.M. (1987) Survival of wild and game-farm cock pheasants released in Pennsylvania. *Journal of Wildlife Management* 51, 555–559.

Kretchmer, K. and Fox, M.W. (1975) Effects of domestication on animal behavior. *Veterinary Record* 96, 102–108.

Kristiansen, T.S., Otterå, H. and Svåsand, T. (2000) Size-dependent mortality of juvenile Atlantic cod, estimated from recaptures of released reared cod and tagged wild cod. *Journal of Fish Biology* 56, 687–712.

Krohn, C.C., Jago, J.G. and Boivin, X. (2001) The effect of early handling on the socialisation of young calves to humans. *Applied Animal Behaviour Science* 74, 121–133.

Kronenberger, J.-P. and Medioni, J. (1985) Food neophobia in wild and laboratory mice (*Mus musculus domesticus*). *Behavioural Processes* 11, 53–59.

Kruska, D. (1980) Domestikationsbedingte Hirngrössenänderungen bei Säugetieren. *Zeitschrift für zoologische Systematik und Evolutionsforschung* 18, 161–195.

Kruska, D. (1982) Hirngrößenänderungen bei Tylopoden während der Stammesgeschichte und in der Domestikation. *Verhandlungen Deutschen Zoologischen Gesellschaft* 1982, 173–183.

Kruska, D. (1987) How fast can total brain size change in mammals? *Journal für Hirnforschung* 28, 59–70.

Kruska, D. (1988) Mammalian domestication and its effect on brain structure and behavior. In: Jerison, H.J. and Jerison, I. (eds) *Intelligence and Evolutionary Biology*, NATO ASI Series, Vol. G17. Springer-Verlag, Berlin, pp. 211–250.

Kruska, D. (1996) The effect of domestication on brain size and composition in the mink (*Mustela vison*). *Journal of Zoology, London* 239, 645–661.

Kruska, D. and Röhrs, M. (1974) Comparative-quantitative investigations on brains of feral pigs from the Galapagos Islands and of European domestic pigs. *Zeitschrift für Anatomie und Entwicklungsgeschichte* 144, 61–73.

Kruska, D. and Schreiber, A. (1999) Comparative morphometrical and biochemical-genetic investigations in wild and ranch mink (*Mustela vison*: Carnivora: Mammalia). *Acta Theriologica* 44, 377–392.

Kujawa, R., Kucharczyk, D. and Mamcarz, A. (1999) A model system for keeping spawners of wild and domestic fish before artificial spawning. *Aquacultural Engineering* 20, 85–89.

Künzl, C. and Sachser, N. (2000) Behavioural and endocrinological changes during the process of domestication in guinea pigs. *Archives of Animal Breeding (Archiv für Tierzucht, Dummerstorf)* 43, 153–158.

Kyle, R. (1987) Rodents under the carving knife. *New Scientist* 114 (1556), 58–62.

Labisky, R.F. (1959) Report of attempts to establish Japanese quail in Illinois. *Journal of Wildlife Management* 25, 290–295.

Lacy, R.C. (1987) Loss of genetic diversity from managed populations: interacting effects of drift, mutation, immigration, selection and population subdivision. *Conservation Biology* 1, 143–158.

Lacy, R.C. and Ballou, J.D. (1998) Effectiveness of selection in reducing the genetic load in populations of *Peromyscus polionotus* during generations of inbreeding. *Evolution* 52, 900–909.

Lacy, R.C., Petric, A. and Warneke, M. (1993) Inbreeding and outbreeding in captive populations of wild animal species. In: Thornhill, N.W. (ed.) *The Natural History of Inbreeding and Outbreeding: Theoretical and Empirical Perspectives*. University of Chicago Press, Chicago, Illinois, pp. 352–374.

Lagadic, H. and Faure, J.-M. (1987) Preferences of domestic hens for cage size and floor types as measured by operant conditioning. *Applied Animal Behaviour Science* 19, 147–155.

Laikre, L. and Ryman, N. (1991) Inbreeding depression in a captive wolf (*Canis lupus*) population. *Conservation Biology* 5, 33–40.

Lam, K., Rupniak, N.M.J. and Iversen, S.D. (1991) Use of a grooming and foraging substrate to reduce cage stereotypies in macaques. *Journal of Medical Primatology* 20, 104–109.

Lankin, V.S. (1997) Factors of diversity of domestic behaviour in sheep. *Genetics, Selection and Evolution* 29, 73–92.

Lankin, V.S. (1999) Domestic behavior in sheep: the role of behavioral polymorphism in the regulation of stress reactions. *Russian Journal of Genetics* 35, 952–959.

Lapveteläinen, T., Tiihonen, A., Koskela, P., Nevalainen, T., Lindblom, J., Király, K., Halonen, P. and Helminen, H.J. (1997) Training a large number of laboratory mice using running wheel and analyzing running behavior by use of a computer-assisted system. *Laboratory Animal Science* 47, 172–179.

LaRue, C. and Hoffman, K. (1981) Artificial incubation and hand-rearing of cranes. *International Zoo Yearbook* 21, 215–217.

Launey, S. and Hedgecock, D. (2001) High genetic load in the Pacific oyster *Crassostrea gigas. Genetics* 159, 255–265.

Laurent, E.L. (1997) Children, 'insects' and play in Japan. In: Podberscek, A.L., Paul, E.S. and Serpell, J.A. (eds) *Companion Animals and Us.* Cambridge University Press, Cambridge, pp. 61–89.

Lawrence, A.B. and Rushen, J. (eds) (1993) *Stereotypic Animal Behaviour: Fundamentals and Applications to Welfare.* CAB International, Wallingford, UK.

Leider, S.A., Hulett, P.L., Loch, J.J. and Chilcote, M.W. (1990) Electrophoretic comparison of the reproductive success of naturally spawning transplanted and wild steelhead trout through the returning adult stage. *Aquaculture* 88, 239–252.

Le Neindre, P., Poindron, P., Trillat, G. and Orgeur, P. (1993) Influence of breed on reactivity of sheep to humans. *Genetics, Selection and Evolution* 25, 447–458.

Le Neindre, P., Trillat, G., Sapa, J., Menissier, F., Bonnet, J.N. and Chupin, J.M. (1995) Individual differences in docility in Limousine cattle. *Journal of Animal Science* 73, 2249–2253.

Lensink, B.J., Fernandez, X., Cozzi, G., Florand, L. and Veissier, I. (2001) The influence of farmers' behavior on calves' reactions to transport and quality of veal meat. *Journal of Animal Science* 79, 642–652.

Leopold, A.S. (1944) The nature of heritable wildness in turkeys. *Condor* 46, 133–197.

Leppla, N.C., Huettel, M.D., Chambers, D.L. and Turner, W.K. (1976) Comparative life history and respiratory activity of 'wild' and colonized Caribbean fruit flies (Diptera: Tephritidae). *Entomophaga* 21, 353–357.

Leppla, N.C., Guy, R.H., Chambers, D.L. and Workman, R.B. (1980) Laboratory colonization of a noctuid moth. *Annals of the Entomological Society of America* 73, 568–571.

Leppla, N.C., Huettel, M.D., Chambers, D.L., Ashley, T.R., Miyashita, D.H., Wong, T.T.Y. and Harris, E.J. (1983) Strategies for colonization and maintenance of the Mediterranean fruit fly. *Entomologia Experimentalis et Applicata* 33, 89–96.

Lerner, I.M. (1954) *Genetic Homeostasis.* Oliver and Boyd, Edinburgh.

Letts, G.A. (1964) Feral animals in the Northern Territory. *Australian Veterinary Journal* 40, 84–88.

Leyhausen, P. (1988) The tame and the wild – another just-so-story? In: Turner, D.C. and Bateson, P. (eds) *The Domestic Cat: the Biology of its Behaviour.* Cambridge University Press, Cambridge, pp. 57–66.

Lickliter, R. and Ness, J.W. (1990) Domestication and comparative psychology: status and strategy. *Journal of Comparative Psychology* 104, 211–218.

Liimatainen, J., Hoikkala, A. and Shelly, T. (1997) Courtship behavior in *Ceratitis capitata* (Diptera: Tephritidae): comparison of wild and mass-reared males. *Annals of the Entomological Society of America* 90, 836–843.

Line, S.W., Morgan, K.N. and Markowitz, H. (1991) Simple toys do not alter the behavior of aged rhesus monkeys. *Zoo Biology* 10, 473–484.

Linley, T.J. (2001) Influence of short-term estuarine rearing on the ocean survival and size at return of coho salmon in Southeastern Alaska. *North American Journal of Aquaculture* 63, 306–311.

Linnaeus, C. (1758) *Systema Naturae,* facsimile edition, 1958. Trustees of the British Museum (Natural History), London.

Liu, D., Diorio, J., Tannenbaum, B., Caldji, C., Francis, D., Freedman, A., Sharma, S., Pearson, D., Plotsky, P.M. and Meaney, M.J. (1997) Maternal care, hippocampal glucocorticoid receptors, and hypothalamic–pituitary–adrenal responses to stress. *Science* 277, 1659–1662.

Liukkonen-Anttila, T., Putaala, A. and Hissa, R. (1999) Does shifting from a commercial to a natural diet affect the nutritional status of hand-reared grey partridges, *Perdix perdix?* *Wildlife Biology* 5, 147–156.

Lockard, R.B. (1968) The albino rat: a defensible choice or a bad habit? *American Psychologist* 23, 734–742.

Loebel, D.A., Nurthen, R.K., Frankham, R., Briscoe, D.A. and Craven, D. (1992) Modeling problems in conservation genetics using captive *Drosophila* populations: consequences of equalizing founder representation. *Zoo Biology* 11, 319–332.

Loftin, R. (1995) Captive breeding of endangered species. In: Norton, B.G., Hutchins, M., Stevens, E.F. and Maple, T.L. (eds) *Ethics on the Ark: Zoos, Animal Welfare and Wildlife Conservation*. Smithsonian Institution Press, Washington, DC, pp. 164–180.

Logan, C.A. (2001) '[A]re Norway rats ..... things?': diversity versus generality in the use of albino rats in experiments on development and sexuality. *Journal of the History of Biology* 34, 287–314.

Lore, R.K. and Flannelly, K.J. (1981) Comparative studies of wild and domestic rats: some difficulties in isolating the effects of genotype and environment. *Aggressive Behavior* 7, 253–257.

Lorenz, K. (1965) *Evolution and Modification of Behavior*. University of Chicago Press, Chicago, Illinois.

Lott, D.F. (1991) *Intraspecific Variation in the Social Systems of Wild Vertebrates*. Cambridge University Press, New York.

Lu, D., Willard, D., Patel, I.R., Kadwell, S., Overton, L., Kost, T., Luther, M., Chen, W., Woychik, R.P., Wilkison, W.O. and Cone, R.D. (1994) Agouti protein is an antagonist of the melanocyte-stimulating hormone receptor. *Nature* 371, 799–802.

Lubritz, D.L., Smith, J.L. and McPherson, B.N. (1995) Heritability of ascites and the ratio of right to total ventral weight in broiler breeder male lines. *Poultry Science* 74, 1237–1241.

Lucy, M.C. (2001) Reproductive loss in high-producing dairy cattle: where will it end? *Journal of Dairy Science* 84, 1277–1293.

Lush, J.L. (1945) *Animal Breeding Plans*. Iowa State College Press, Ames, Iowa.

Lynch, J.J. and Bell, A.K. (1987) The transmission from generation to generation in sheep of the learned behaviour of eating grain supplements. *Australian Veterinary Journal* 64, 291–292.

Lynch, M. (1991) The genetic interpretation of inbreeding depression and outbreeding depression. *Evolution* 45, 622–629.

Lynch, M. and Walsh, B. (1998) *Genetics and Analysis of Quantitative Traits*. Sinauer Associates, Sunderland, Massachusetts.

Lyons, D.M. (1989) Individual differences in temperament of dairy goats and the inhibition of milk ejection. *Applied Animal Behaviour Science* 22, 269–282.

Lyons, D.M., Price, E.O. and Moberg, G.P. (1988a) Individual differences in temperament of dairy goats: constancy and change. *Animal Behaviour* 36, 1323–1333.

Lyons, D.M., Price, E.O. and Moberg, G.P. (1988b) Social modulation of pituitary-adrenal responsiveness and individual differences in behavior of young domestic goats. *Physiology and Behavior* 43, 451–458.

MacDonald, K.B. and Ginsberg, B.E. (1981) Induction of normal prepubertal behavior in wolves with restricted rearing. *Behavioral and Neural Biology* 33, 133–162.

Macfarlane, W.V. (1968) Adaptation of ruminants to tropics and deserts. In: Hafez, E.S.E. (ed.) *Adaptation of Domestic Animals*. Lea & Febiger, Philadelphia, Pennsylvania, pp. 164–182.

Mäki, K., Groen, A.F., Liinamo, A.-E. and Ojala, M. (2001) Population structure, inbreeding trend and their association with hip and elbow dysplasia in dogs. *Animal Science* 73, 217–228.

Malecha, S.R. (1980) Development and general characterization of genetic stocks of *Macrobrachium rosenbergii* and their hybrids for domestication. *University of Hawaii Sea Grant Quarterly* 2, 1–6.

Malmkvist, J. and Hansen, S.W. (2001) The welfare of farmed mink (*Mustela vison*) in relation to behavioural selection: a review. *Animal Welfare* 10, 41–52.

Manwell, C. and Baker, C.M.A. (1984) Domestication of the dog: hunter, food, bed-warmer, or emotional object? *Zeitschrift für Tierzüchtung Züchtungsbiologie* 101, 241–256.

Margan, S.H., Nurthen, R.K., Montgomery, M.E., Woodworth, L.M., Lowe, E.H., Briscoe, D.A. and Frankham, R. (1998) Single large or several small? Population fragmentation in the captive management of endangered species. *Zoo Biology* 17, 467–480.

Markowitz, H. (1982) *Behavioral Enrichment in the Zoo*. Van Nostrand Reinhold, New York.

Markowitz, H. and LaForse, S. (1987) Artificial prey as behavioral enrichment devices for felines. *Applied Animal Behaviour Science* 18, 31–44.

Markowitz, H. and Line, S.W. (1989) Primate research models and environmental enrichment. In: Segal, E.F. (ed.) *Housing, Care and Psychological Well-being of Captive and Laboratory Primates*. Noyes Publications, Park Ridge, New Jersey, pp. 203–212.

Markowitz, H. and Woodworth, G. (1978) Experimental analysis and control of group behavior. In: Markowitz, H. and Stevens, V. (eds) *Behavior of Captive Wild Animals*. Nelson-Hall, Chicago, Illinois, pp. 107–131.

Markowitz, T.M., Dally, M.R., Gursky, K. and Price, E.O. (1998) Early handling increases lamb affinity for humans. *Animal Behaviour* 55, 573–587.

Marliave, J.B., Gergits, W.F. and Aota, S. (1993) F_{10} pandalid shrimp: sex determination; DNA and dopamine as indicators of domestication; and outcrossing for wild pigment pattern. *Zoo Biology* 12, 435–451.

Marsden, M.D. (1993) Feeding practices have a greater effect than housing practices on the behaviour and welfare of the horse. In: Collins, E. and Boon, C. (eds) *Proceedings of the 4th International Symposium on Livestock Environment*. American Society of Agricultural Engineers, St Joseph, Missouri, pp. 314–318.

Marshall, J.T. (1981) Taxonomy. In: Foster, H.L., Small, J.D. and Fox, J.G. (eds) *The Mouse in Biomedical Research*, Vol. 1. Academic Press, New York, pp. 17–27.

Marston, J.H. (1972) The Mongolian gerbil. In: *The UFAW Handbook on the Care and Management of Laboratory Animals*. Longman, Harlow, UK, pp. 257–268.

Martin, J.T. (1975) Hormonal influences in the evolution and ontogeny of imprinting behavior in the duck. *Progress in Brain Research* 42, 357–366.

Martrenchar, A., Huonnic, D. and Cotte, J.P. (2001) Influence of environmental enrichment on injurious pecking and perching behaviour in young turkeys. *British Poultry Science* 42, 161–170.

Mason, G.J. (1991a) Stereotypies: a critical review. *Animal Behaviour* 41, 1015–1037.

Mason, G.J. (1991b) Stereotypies and suffering. *Behavioural Processes* 25, 103–115.

Mason, G.J. (1996) Early weaning enhances later development of stereotypy in mink. In: Duncan, I.J.H., Widowski, T. and Haley, D.B. (eds) *Proceedings of the 30th International Congress of the International Society for Applied Ethology*, p. 16.

Mason, G.J., Cooper, J. and Clarebrough, C. (2001) Frustrations of fur-farmed mink. *Nature* 410, 35–36.

Mason, I.L. (1984) *Evolution of Domesticated Animals*. Longman, New York.

Mason, L.J., Pashley, D.P. and Johnson, S.J. (1987) The laboratory as an altered habitat: phenotypic and genetic consequences of colonization. *Florida Entomologist* 70, 49–58.

Mason, W.A. (1978) Social experience and primate cognitive development. In: Burghardt, G.M. and Bekoff, M. (eds) *The Development of Behavior*. Garland Press, New York, pp. 233–253.

Mateo, J.M., Estep, D.Q. and McCann, J.S. (1991) Effects of differential handling on the behaviour of domestic ewes (*Ovis aries*). *Applied Animal Behaviour Science* 32, 45–54.

Mathis, A., Chivers, D.P. and Smith, R.J.F. (1996) Cultural transmission of predator recognition in fishes: intraspecific and interspecific learning. *Animal Behaviour* 51, 185–201.

Matos, M., Rose, M.R., Rocha Pite, M.T., Rego, C. and Avelar, T. (2000) Adaptation to the laboratory environment in *Drosophila subobscura*. *Journal of Evolutionary Biology* 13, 9–19.

Maynard, D.J., Flagg, T.A. and Mahnken, C.V.W. (1995) A review of semi-natural culture strategies for enhancing the post-release survival of anadromous salmonids. *American Fisheries Society Symposium* 15, 307–314.

McBride, G., Parer, I.P. and Foenander, F. (1969) The social organization and behaviour of the feral domestic fowl. *Animal Behaviour Monograph* 2, 127–181.

McCabe, B.J., Cipolla-Neto, J., Horn, G. and Bateson, P. (1982) Amnesic effects of bilateral lesions placed in the hyperstriatum ventrale of the chick after imprinting. *Experimental Brain Research* 45, 13–21.

McCall, T.C., Brown, R.D. and DeYoung, C.A. (1988) Mortality of pen-raised and wild whitetailed deer bucks. *Wildlife Society Bulletin* 16, 380–384.

McCort, W.D. (1984) Behavior of feral horses and ponies. *Journal of Animal Science* 493–499.

McDonald, D.G., Milligan, C.L., McFarlane, W.J., Croke, S., Currie, S., Hooke, B., Angus, R.B., Tufts, B.L. and Davidson, K. (1998) Condition and performance of juvenile Atlantic salmon (*Salmo salar*): effects of rearing practices on hatchery fish and comparison with wild fish. *Canadian Journal of Fisheries and Aquatic Sciences* 55, 1208–1219.

McGregor, S.E. (1976) *Insect Pollination of Cultivated Crop Plants*. Agriculture Handbook 496, US Department of Agriculture, Washington, DC.

McInerney, J.P. (1998) The economics of welfare. In: Michell, A.R. and Ewbank, R. (eds) *Ethics, Welfare, Law and Market Forces: the Veterinary Interface*. UFAW, Wheathampstead, UK, pp. 115–134.

McKnight, T. (1964) *Feral livestock in Anglo-America*. University of California Publications in Geography, Vol. 16. University of California Press, Berkeley, California.

McKnight, T. (1976) *Friendly Vermin: a Survey of Feral Livestock in Australia*. University of California Publications in Geography, Vol. 21. University of California Press, Berkeley, California.

McLean, E., Devlin, R.H., Byatt, J.C., Clark, W.C. and Donaldson, E.M. (1997) Impact of a conrolled release formulation of recombinant bovine growth hormone upon growth and seawater adaptation in coho (*Oncorhynchus kisutch*) and chinook (*Oncorhynchus tshawytscha*) salmon. *Aquaculture* 156, 113–128.

McLean, I.G., Hölzer, C. and Studholme, B.J.S. (1999) Teaching predator-recognition to a naive bird: implications for management. *Biological Conservation* 87, 123–130.

McNeil, W.J. (1991) Expansion of cultured Pacific salmon into marine ecosystems. *Aquaculture* 98, 173–183.

McVicar, A.H. (1997) Disease and parasite implications of the coexistence of wild and cultured Atlantic salmon populations. *ICES Journal of Marine Science* 54, 1093–1103.

Meadow, R.H. (1984) Animal domestication in the Middle East: a view from the eastern margin. In: Clutton-Brock, J. and Grigson, C. (eds) *Animals and Archaeology*, Vol. 3. *Early Herders and their Flocks*. International Series 202, British Archaeological Reports, Oxford, pp. 309–337.

Meadow, R.H. (1989) Osteological evidence for the process of domestication. In: Clutton-Brock, J. (ed.) *The Walking Larder: Patterns of Domestication, Pastoralism, and Predation*. Unwin Hyman, London, pp. 80–90.

Mech, L.D. (1970) *The Wolf*. Natural History Press, Garden City, New York.

Medina, A., Vila, Y., Mourente, G. and Rodriguez, A. (1996) A comparative study of the ovarian development in wild and pond-reared shrimp, *Penaeus kerathurus* (Forskal, 1775). *Aquaculture* 148, 63–75.

Meffert, L.M. (1999) How speciation experiments relate to conservation biology. *BioScience* 49, 701–711.

Meffert, L.M. and Bryant, E.H. (1991) Mating propensity and courtship behavior in serially bottlenecked lines of the housefly. *Evolution* 45, 293–306.

Mellen, J.D. (1991) Factors influencing reproductive success in small captive exotic felids (*Felis* spp.): a multiple regression analysis. *Zoo Biology* 10, 95–110.

Mellen, J.D. (1992) Effects of early rearing experience on subsequent adult sexual behavior using domestic cats (*Felis catus*) as a model for exotic small felids. *Zoo Biology* 11, 17–32.

Mellen, J. and Sevenich MacPhee, M. (2001) Philosophy of environmental enrichment: past, present and future. *Zoo Biology* 20, 211–226.

Mello, D.A. (1981) Studies on reproduction and longevity of *Calomys callosus* under laboratory conditions (Rodentia-Cricetidae). *Revista Brasileira de Biologia* 41, 841–843.

Mello, D.A. (1986) Breeding of wild-caught rodent cricetidae *Holochilus brasiliensis* under laboratory conditions. *Laboratory Animals* 20, 195–196.

Melzack, R. (1968) Early experience: a neuropsychological approach to heredity–environment interactions. In: Newton, G. and Levine, S. (eds) *Early Experience and Behavior*. Thomas, Springfield, Massachusetts, pp. 65–82.

Menache, S. (1997) Hunting and attachment to dogs in the Pre-Modern Period. In: Podberscek, A.L., Paul, E.S. and Serpell, J.A. (eds) *Companion Animals and Us*. Cambridge University Press, Cambridge, pp. 42–60.

Mench, J.A. (1999) Ethics, animal welfare and transgenic farm animals. In: Murray, J.D., Anderson, G.B., Oberbauer, A.M. and McGloughlin, M.M. (eds) *Transgenic Animals in Agriculture*. CAB International, Wallingford, UK, pp. 251–268.

Mendl, M., Burman, O., Laughlin, K. and Paul, E. (2001) Animal memory and animal welfare. *Animal Welfare* 10, S141-S159.

Meretsky, V., Snyder, N.F.R., Beissinger, S.R., Clendenen, D.A. and Wiley, J.W. (2000) Demography of the California condor: implications for reestablishment. *Conservation Biology* 14, 957–967.

Meretsky, V., Snyder, N.F.R., Beissinger, S.R., Clendenen, D.A. and Wiley, J.W. (2001) Quantity versus quality in California condor reintroduction: reply to Beres and Starfield. *Conservation Biology* 15, 1449–1451.

Mesa, M.G. (1991) Variation in feeding, aggression, and position choice between hatchery and wild cutthroat trout in an artificial stream. *Transactions of the American Fisheries Society* 120, 723–727.

Messent, P.R. and Serpell, J.A. (1981) An historical and biological view of the pet-owner bond. In: Fogle, B. (ed.) *Interrelations Between People and Pets*. Charles Thomas, Springfield, Illinois, pp. 5–22.

Meyer, W., Schnapper, A., Hulmann, G. and Seger, H. (2000) Domestication-related variations of the hair cuticula pattern in mammals. *Journal of Animal Breeding and Genetics* 117, 281–283.

Meyers, J.M. and Miller, D.L. (1992) Post-release activity of captive- and wild-reared bald eagles. *Journal of Wildlife Management* 56, 744–749.

Millam, J.R., Roudybush, T.E. and Grau, C.R. (1988) Influence of environmental manipulation and nest-box availability on reproductive success of captive cockatiels (*Nymphicus hollandicus*). *Zoo Biology* 7, 25–34.

Millán, J., Gortázar, C. and Villafuerte, R. (2001) Marked differences in the splanchnometry of farm-bred and wild red-legged partridges (*Alectoris rufa*). *Poultry Science* 80, 972–975.

Miller, D.B. (1977) Social displays of mallard ducks (*Anas platyrhynchos*): effects of domestication. *Journal of Comparative and Physiological Psychology* 91, 221–232.

Miller, D.B. and Gottlieb, G. (1981) Effects of domestication on production and perception of mallard maternal alarm calls: developmental lag in behavioral arousal. *Journal of Comparative and Physiological Psychology* 95, 205–219.

Miller, N. (1911) Reproduction in the brown rat (*Mus norvegicus*). *American Naturalist* 45, 623–635.

Miller, P.S. (1994) Is inbreeding depression more severe in a stressful environment? *Zoo Biology* 13, 195–208.

Miller, P.S. and Hedrick, P.W. (1993) Inbreeding and fitness in captive populations: lessons from *Drosophila*. *Zoo Biology* 12, 333–351.

Miller, P.S. and Hedrick, P.W. (2001) Purging of inbreeding depression and fitness decline in bottlenecked populations of *Drosophila melanogaster*. *Journal of Evolutionary Biology* 14, 595–601.

Milligan, C.L., Hooke, G.B. and Johnson, C. (2000) Sustained swimming at low velocity following a bout of exhaustive exercise enhances metabolic recovery in rainbow trout. *Journal of Experimental Biology* 203, 921–926.

Milonas, P.G. and Savopoulou-Soultani, M. (2000) Temperature dependent development of the parasitoid *Colpoclypeus florus* (Hymenoptera: Eulophidae) in the laboratory. *Journal of Economic Entomology* 93, 1627–1632.

Minckler, J. and Pease, F.D. (1938) A colony of albino rats existing under feral conditions. *Science* 87, 460–461.

Mineka, S., Gunnar, M. and Champoux, M. (1986) Control and early socioemotional development: infant rhesus monkeys reared in controllable versus uncontrollable environments. *Child Development* 57, 1241–1256.

Mirza, R.S. and Chivers, D.P. (2000) Predator-recognition training enhances survival of brook trout: evidence from laboratory and field-enclosure studies. *Canadian Journal of Zoology* 78, 2198–2208.

Mischke, C.C. and Morris, J.E. (1998) Growth and survival of larval bluegill in the laboratory under different feeding regimes. *Progressive Fish Culturist* 60, 206–213.

Mischke, C.C., Dvorak, G.D. and Morris, J.E. (2001) Growth and survival of hybrid sunfish larvae in the laboratory under different feeding and temperature regimes. *North American Journal of Aquaculture* 63, 265–271.

Mitchell, D., Beatty, E.T. and Cox, P.K. (1977) Behavioral differences between two populations of wild rats: implications for domestication research. *Behavioral Biology* 19, 206–216.

Mjolnerod, I.B., Refseth, U.H., Karlsen, E., Balstad, T., Jakobsen, K.S. and Hindar, K. (1997) Genetic differences between two wild and one farmed population of Atlantic salmon (*Salmo salar*) revealed by three classes of genetic markers. *Hereditas* 127, 239–248.

Moav, R., Brody, T. and Hulata, G. (1978) Genetic improvement of wild fish populations. *Science* 201, 1090–1094.

Mononen, J., Kasanen, S., Harri, M., Sepponen, J. and Rekilä, T. (2001) The effects of elevated platforms and concealment screens on the welfare of blue foxes. *Animal Welfare* 10, 373–385.

Moore, C.L., Whittlestone, W.G., Mullord, M., Priest, P.N., Kilgour, R. and Albright, J.L. (1975) Behavior responses of dairy cows trained to activate a feeding device. *Journal of Dairy Science* 58, 1531–1535.

Moore, I.T., Greene, M.J. and Mason, R.T. (2001) Environmental and seasonal adaptations of the adrenocortical and gonadal responses to capture stress in two populations of the male garter snake, *Thamnophis sirtalis*. *Journal of Experimental Zoology* 289, 99–108.

Moran, G. (1987) Applied dimensions of comparative psychology. *Journal of Comparative Psychology* 101, 277–281.

Moreno, G. (1994) Genetic architecture, genetic behavior, and character evolution. *Annual Review of Ecology and Systematics* 25, 31–44.

Morey, D.F. (1994) The early evolution of the domestic dog. *Scientific American* 82, 336–347.

Moring, J.R. (2000) The creation of the first public salmon hatchery in the United States. *Fisheries* 25, 6–12.

Mork, O.I., Bjerkeng, B. and Rye, M. (1999) Aggressive interactions in pure and mixed groups of juvenile farmed and hatchery-reared wild Atlantic salmon, *Salmo salar* L., in relation to tank substrate. *Aquaculture Research* 30, 571–578.

Moscarella, R.A. and Aguilera, M. (1999) Growth and reproduction of *Oryzomys albigularis* (Rodentia: Sigmodontinae) under laboratory conditions. *Mammalia* 63, 349–362.

Moulton, M.P., Sanderson, J.G. and Labisky, R.F. (2001) Patterns of success in game bird (Aves: Galliformes) introductions to the Hawaiian islands and New Zealand. *Evolutionary Ecology Research* 3, 507–519.

Moyle, P.B. (1969) Comparative behavior of young brook trout of domestic and wild origin. *Progressive Fish Culturist* 31, 51–59.

Munakata, A., Björnsson, B.T., Jönsson, E., Amano, M., Ikuta, K., Kitamura, S., Kurokawa, T. and Aida, K. (2000) Post-release adaptation processes of hatchery-reared honmasu salmon parr. *Journal of Fish Biology* 56, 163–172.

Muntzing, A. (1959) Darwin's views on variation under domestication in the light of present-day knowledge. *Proceedings of the American Philosophical Society* 103, 190–220.

Murphy, M.R. (1985) History of the capture and domestication of the Syrian golden hamster (*Mesocricetus auratus* Waterhouse). In: Siegel, H. (ed.) *The Hamster: Reproduction and Behavior*. Plenum, New York, pp. 3–20.

Mydans, C. (1973) Orangutans can return to the wild with some help. *Smithsonian* 4, 26–33.

Mylonas, C.C. and Zohar, Y. (2001) Use of GnRHa-delivery systems for the control of reproduction in fish. *Reviews in Fish Biology and Fisheries* 10, 463–491.

Nachtsheim, H. and Stengel, H. (1977) *Vom Wildtier zum Haustier*. Paul Parey, Berlin.

Naumenko, E.V., Popova, N.K., Nikulina, E.M., Dygalo, N.N., Shishkina, G.T., Boradin, P.M. and Markel, A.L. (1989) Behavior, adrenocortical activity, and brain monoamines in Norway rats selected for reduced aggressiveness towards man. *Pharmacology, Biochemistry and Behavior* 33, 85–91.

Nelson, K., Hedgecock, D., Borgeson, W., Johnson, E., Daggett, R. and Aronstein, D. (1980) Density-dependent growth inhibition in lobsters, *Homarus* (Decapoda, Nephropidae). *Biological Bulletin* 159, 162–176.

Nesbitt, W.H. (1975) Ecology of a feral dog pack on a wildlife refuge. In: Fox, M.W. (ed.) *The Wild Canids: Their Systematics, Behavioral Ecology and Evolution*. Van Nostrand-Reinhold, New York, pp. 391–396.

Neumann, K., Maak, S., Stuermer, I.W., von Lengerken, G. and Gattermann, R. (2001) Low microsatellite variation in laboratory gerbils. *Journal of Heredity* 92, 71–74.

Nevison, C.M., Hurst, J.L. and Barnard, C.J. (1999) Strain-specific effects of cage enrichment in male laboratory mice (*Mus musculus*). *Animal Welfare* 8, 361–379.

New, M.B. (1995) Status of freshwater prawn farming: a review. *Aquaculture Research* 26, 1–54.

Newberry, R.C. (1995) Environmental enrichment: increasing the biological relevance of captive environments. *Applied Animal Behaviour Science* 44, 229–243.

Newberry, R.C. and Wood-Gush, D.G.M. (1988) Development of some behaviour patterns in piglets under semi-natural conditions. *Animal Production* 46, 103–109.

Nicol, C. (1999) Understanding equine stereotypies. *Equine Veterinary Journal* 28 (Suppl.), 20–25.

Nielson, W.T.A. (1971) Dispersal studies of a natural population of apple maggot adults. *Journal of Economic Entomology* 64, 648–653.

Nikoletseas, M. and Lore, R.K. (1981) Aggression in domesticated rats reared in a burrow-digging environment. *Aggressive Behavior* 7, 245–252.

Nikulina, E.M. (1990) Brain catecholamines during domestication of the silver fox (*Vulpes vulpes*). *Journal of Evolution, Biochemistry and Physiology* 26, 118–121.

Nikulina, E.M., Trapezov, O.V. and Popova, N.K. (1985a) Content of monoamines in the brain of minks differing by reaction to humans. *Zhurn Vysshey Nervn Deyatelnosti* 35, 1142–1146.

Nikulina, E.M., Borodin, P.M., and Popova, N.K. (1985b). Changes in several forms of aggressive behavior and in the content of monoamines in the brain during selection for domestication in wild rats. *Zhurn Vysshey Nervn Deyatelnosti* 35, 703–709.

Nimon, A.J. and Broom, D.M. (2001) The welfare of farmed foxes *Vulpes vulpes* and *Alopex lagopus* in relation to housing and management: a review. *Animal Welfare* 10, 223–248.

Nogueira, S.S.C., Nogueira-Filho, S.L.G., Otta, E., Dias, C.T.S. and Carvalho, A. (1999) Determination of the causes of infanticide in capybara (*Hydrochaeris hydrochaeris*) groups in captivity. *Applied Animal Behaviour Science* 62, 351–357.

Noon, C. (1991) Resocialization of a group of ex-laboratory chimpanzees, *Pan troglodytes*. *Journal of Medical Primatology* 20, 375–381.

Nordeide, J.T. and Salvanes, A.G.V. (1991) Observations on reared newly released and wild cod (*Gadus morhua*) and their potential predators. *ICES Journal of Marine Science* 192, 139–146.

Norris, A.T., Bradley, D.G. and Cunningham, E.P. (1999) Microsatellite genetic variation between and within farmed and wild Atlantic salmon (*Salmo salar*) populations. *Aquaculture* 180, 247–264.

Norris, D.E., Shurtleff, A.C., Touré, Y.T. and Lanzaro, G.C. (2001) Microsatellite DNA polymorphism and heterozygosity among field and laboratory populations of *Anopheles gambiae* s.s. (Diptera: Culicidae). *Journal of Medical Entomology* 38, 336–340.

O'Brien, P.H. (1984) Feral goat home range: influence of social class and environmental variables. *Applied Animal Behaviour Science* 12, 373–385.

Ochieng'-Odero, J.P.R. (1994) Does adaptation occur in insect rearing systems, or is it a case of selection, acclimatization and domestication? *Insect Science and its Applications* 15, 1–7.

Ödberg, F.O. (1986) The jumping stereotypy in the bank vole (*Clethrionomys glareolus*). *Biology of Behaviour* 11, 130–143.

O'Donohue, T.L. and Dorsa, D.M. (1982) The opiomelanotropinergic neuronal and endocrine systems. *Peptides* 3, 353–395.

Odum, R.A. (1994) Assimilation of new founders into existing captive populations. *Zoo Biology* 13, 187–190.

Ohno, K., Niwa, Y., Kato, S., Koyasu, K., Oda, S. and Kondo, K. (1992) The domestication of *Crocidura dsinezumi* as a new laboratory animal. *Experimental Animals* 41, 449–454.

Olla, B.L. and Davis, M.W. (1989) The role of learning and stress in predator avoidance of hatchery and wild steelhead trout (*Oncorhynchus kisutch*) juveniles. *Aquaculture* 76, 209–214.

Olla, B.L., Davis, M.W. and Ryer, C.H. (1994) Behavioural deficits in hatchery-reared fish: potential effects on survival following release. *Aquaculture Fisheries Management* 25 (Suppl. 1), 19–34.

Olson-Rutz, K.M., Urness, P.J. and Urness, L.A. (1986) Dam-raised fawns, an alternative to bottle feeding. *Great Basin Naturalist* 46, 217–219.

O'Neill, P.L., Novak, M.A. and Suomi, S.J. (1991) Normalizing laboratory-reared rhesus macaque (*Macaca mulatta*) behavior with exposure to complex outdoor enclosures. *Zoo Biology* 10, 237–245.

Osadchuk, L.V., Braastad, B.O., Hovland, A.L. and Bakken, M. (2001) Morphometric and hormonal changes following persistent handling in pregnant blue fox vixens (*Alopex lagopus*). *Animal Science* 72, 407–414.

Özer, A. and Erdem, O. (1999) The relationship between occurrence of ectoparasites, temperature and culture conditions: a comparison of farmed and wild common carp (*Cyprinus cario* L., 1758) in the Sinop region of northern Turkey. *Journal of Natural History* 33, 483–491.

Pakandl, M. (1994) The prevalence of intestinal protozoa in wild and domestic pigs. *Veterinary Medicine, Czechoslovakia* 39, 377–380.

Palma, J., Alarcon, J.A., Alvarez, C., Zouros, E., Magoulas, A. and Andrade, J.P. (2001) Developmental stability and genetic heterozygosity in wild and cultured stocks of gilthead sea bream (*Sparus aurata*). *Journal of the Marine Biological Association of the United Kingdom* 81, 283–288.

Pankhurst, N.W. (2001) Stress inhibition of reproductive endocrine processes in a natural population of the spiny damselfish *Acanthochromis polyacanthus*. *Marine Freshwater Research* 52, 753–761.

Pankhurst, N.W. and Van Der Kraak, G. (1997) Effects of stress on reproduction and growth of fish. In: Iwama, G.K., Pickering, A.D., Sumpter, J.P. and Schreck, C.B. (eds) *Fish Stress and Health in Aquaculture*. Cambridge University Press, Cambridge, pp. 73–93.

Pascual, M., Sagarra, E. and Serra, L. (2000) Interspecific competition in the laboratory between *Drosophila subobscura* and *D. azteca*. *American Midland Naturalist* 144,19–27.

Paterson, A.M. and Pearce, G.P. (1992) Growth, response to humans and corticosteroids in male pigs housed individually and subjected to pleasant, unpleasant or minimal handling during rearing. *Applied Animal Behaviour Science* 34, 315–328.

Patterson-Kane, E.G., Harper, D.N. and Hunt, M. (2001) The cage preferences of laboratory rats. *Laboratory Animals* 35, 74–79.

Peake, S. (1999) Substrate preferences of juvenile hatchery-reared lake sturgeon, *Acipenser fulvescens*. *Environmental Biology of Fishes* 56, 367–374.

Pearce-Kelly, P., Mace, G.M. and Clarke, D. (1995) The release of captive bred snails (*Partula taeniata*) into a semi-natural environment. *Biodiversity and Conservation* 4, 645–663.

Pedersen, V. and Jeppesen, L.L. (1990) Effects of early handling on later behaviour and stress responses in the silver fox (*Vulpes vulpes*). *Applied Animal Behaviour Science* 26, 383–393.

Peebles, J.B. (1979) The influence of shelter utilization on growth and survival in *Macrobrachium rosenbergii*. Paper presented at the Tenth Annual Meeting of the World Mariculture Society, Honolulu, Hawaii, 22–26 January.

Perez-Enriquez, R., Takagi, M. and Taniguchi, N. (1999) Genetic variability and pedigree tracing of a hatchery-reared stock of red sea bream (*Pagrus major*) used for stock enhancement, based on microsatellite DNA markers. *Aquaculture* 173, 411–421.

Peterka, M. and Hartl, G.B. (1992) Biochemical-genetic variation and differentiation in wild and domestic rabbits. *Zeitschrift für zoologische Systematik und Evolutionsforschung* 30, 129–141.

Philippart, J.C. (1995) Is captive breeding an effective solution for the preservation of endemic species? *Biological Conservation* 72, 281–295.

Phillips, R.E. (1964) 'Wildness' in the mallard duck: effects of brain lesions and stimulation on 'escape behavior' and reproduction. *Journal of Comparative Neurology* 122, 139–155.

Plogmann, D. and Kruska, D. (1990) Volumetric comparison of auditory structures in the brains of European wild boars (*Sus scrofa*) and domestic pigs (*Sus scrofa* f. dom.). *Brain, Behavior and Evolution* 35, 146–155.

Plomin, R., DeFries, J.C. and McClearn, G.E. (1980) *Behavioral Genetics*. W.H. Freeman, San Francisco, California.

Podberscek, A.L. and Gosling, S.D. (2000) Personality research on pets and their owners: conceptual issues and review. In: Podberscek, A.L., Paul, E.S. and Serpell, J.A. (eds) *Companion Animals and Us*. Cambridge University Press, Cambridge, pp. 143–167.

Polsky, R.H. (1995) Wolf hybrids: are they suitable as pets? *Veterinary Medicine* 90, 1122–1124.

Popova, N.K., Voitenko, N.N., Kulikov, A.V. and Avgustinovich, D.F. (1991a) Evidence for the involvement of central serotonin in mechanism of domestication of silver foxes. *Pharmacology, Biochemistry and Behavior* 40, 751–756.

Popova, N.K., Kulikov, A.V., Nikulina, E.M., Kozlachkova, E.Y. and Maslova, G.B. (1991b) Serotonin metabolism and serotonergic receptors in Norway rats selected for low aggressiveness to man. *Aggressive Behavior* 17, 207–213.

Popovic, S. (1988) Influence of aerodynamic properties of the nasal cavity on the function of the olfactory sense in wild and domestic pigs. *Acta Veterinaria* (Beograd) 38, 287–292.

Popovic, S. (1992) Some morphological features of the nasal airways in domestic and wild pigs especially important in regard to aerodynamics. *Acta Veterinaria* (Beograd) 42, 167–174.

Pottinger, T.G. (2000) Genetic selection to reduce stress in animals. In: Moberg, G.P. and Mench, J.A. (eds) *The Biology of Animal Stress: Basic Principles and Implications for Animal Welfare*. CAB International, Wallingford, UK, pp. 291–308.

Pottinger, T.G. and Carrick, T.R. (1999) Modification of the plasma cortisol response to stress in rainbow trout by selective breeding. *General and Comparative Endocrinology* 116, 122–132.

Pottinger, T.G. and Pickering, A.D. (1997) Genetic basis to the stress response: selective breeding for stress-tolerant fish. In: Iwama, G.K., Pickering, A.D., Sumpter, J.P. and

Schreck, C.B. (eds) *Fish Stress and Health in Aquaculture*. Cambridge University Press, Cambridge, pp. 171–193.

Pray, L.A., Schwartz, J.M., Goodnight, C.J. and Stevens, L. (1994) Environmental dependency of inbreeding depression: implications for conservation biology. *Conservation Biology* 8, 562–568.

Price, E.O. (1966) The influence of light on reproduction in *Peromyscus maniculatus gracilis*. *Journal of Mammalogy* 47, 343–344.

Price, E.O. (1967) The effect of reproductive performance on the domestication of the prairie deermouse, *Peromyscus maniculatus bairdii*. *Evolution* 21, 762–770.

Price, E.O. (1969) The effect of early outdoor experience on the activity of wild and semi-domestic deermice. *Developmental Psychobiology* 2, 60–67.

Price, E.O. (1972) Novelty-induced self-food deprivation in wild and semi-domestic deermice (*Peromyscus maniculatus bairdii*). *Behaviour* 41, 91–104.

Price, E.O. (1973) Some behavioral differences between wild and domestic Norway rats: gnawing and platform jumping. *Animal Learning and Behavior* 1, 312–316.

Price, E.O. (1976a) The laboratory animal and its environment. In: McScheehy, T. (ed.) *Control of the Animal House Environment*. Laboratory Animal Handbook no. 7, Laboratory Animals, London, pp. 7–23.

Price, E.O. (1976b) Food deprivation effects on the running-wheel activity of wild and domestic Norway rats. *Behavioural Processes* 1, 191–196.

Price, E.O. (1978) Genotype versus experience effects on aggression in wild and domestic Norway rats. *Behaviour* 64, 340–353.

Price, E.O. (1980) Sexual behaviour and reproductive competition in male wild and domestic Norway rats. *Animal Behaviour* 28, 657–667.

Price, E.O. (1984) Behavioral aspects of animal domestication. *Quarterly Review of Biology* 59, 1–32.

Price, E.O. (1987) Male sexual behavior. In: Price, E.O. (ed.) *Farm Animal Behavior. Veterinary Clinics of North America: Food Animal Practice*, Vol. 3. W.B. Saunders, Philadelphia, Pennsylvania, pp. 405–422.

Price, E.O. and Belanger, P.L. (1977) Maternal behavior of wild and domestic stocks of Norway rats. *Behavioral Biology* 20, 60–69.

Price, E.O. and Huck, U.W. (1976) Open-field behavior of wild and domestic Norway rats. *Animal Learning and Behavior* 4, 125–130.

Price, E.O. and King, J.A. (1968) Domestication and adaptation. In: Hafez, E.S.E. (ed.) *Adaptation of Domestic Animals*. Lea & Febiger, Philadelphia, Pennsylvania, pp. 34–45.

Price, E.O. and Loomis, S. (1971) Maternal influence on the response of wild and domestic Norway rats to a novel environment. *Developmental Psychobiology* 6, 203–208.

Price, E.O. and Stehn, R. (1977) Self-imposed food deprivation and wheel-running activity in field-trapped Norway rats as a function of environmental change, time in captivity and age at capture. *Behavioural Processes* 2, 255–264.

Price, E.O. and Wallach, S.J.R. (1990) Physical isolation of hand-reared Hereford bulls increases their aggressiveness toward humans. *Applied Animal Behaviour Science* 27, 263–267.

Price, E.O., Belanger, P.L. and Duncan, R.A. (1976) Competitive dominance of wild and domestic Norway rats (*Rattus norvegicus*). *Animal Behaviour* 24, 589–599.

Price, E.O., Borgwardt, R., Blackshaw, J.K., Blackshaw, A., Dally, M.R. and Erhard, H. (1994) Effect of early experience on the sexual performance of yearling rams. *Applied Animal Behaviour Science* 42, 41–48.

Price, T., Turelli, M. and Slatkin, M. (1993) Peak shifts produced by correlated response to selection. *Evolution* 47, 280–290.

Primavera, J.H. (1985) A review of maturation and reproduction in closed thelycum penaeids. In: Taki, Y., Primavera, J.H. and Llobrera, J.A. (eds) *Proceedings of the First International Conference on the Culture of Penaeid Prawns/Shrimps, Philippines, 1984.* Aquaculture Department, Southeast Asian Fisheries Development Center, Tigbauan, Iloilo, Philippines, pp. 47–64.

Prince, H.H., Siegel, P.B. and Cornwell, G.W. (1970) Inheritance of egg production and juvenile growth in mallards. *Auk* 87, 342–352.

Proshold, G.I. and Bartell, J.A. (1972) Difference in radiosensitivity of two colonies of tobacco budworms, *Heliothis virescens* (Lepidoptera: Noctuidae). *Canadian Entomologist* 104, 995–1002.

Provenza, F.D. (1995) Postingestive feedback as an elementary determinant of food preference and intake in ruminants. *Journal of Range Management* 48, 2–17.

Pullar, E.M. (1950) The wild (feral) pigs of Australia and their role in the spread of infectious diseases. *Australian Veterinary Journal* 26, 99–110.

Putaala, A. and Hissa, R. (1995) Effects of hand-rearing on physiology and anatomy in the grey partridge. *Wildlife Biology* 1, 27–31.

Putaala, A., Oksa, J., Rintamäki, H. and Hissa, R. (1997) Effects of hand-rearing and radiotransmitters on flight of gray partridge. *Journal of Wildlife Management* 61, 1345–1351.

Pyörnilä, A.E.I., Putaala, A.P. and Hissa, R.K. (1998) Fibre types in breast and leg muscles of hand-reared and wild grey partridge (*Perdix perdix*). *Canadian Journal of Zoology* 76, 236–242.

Raina, A.K., Stadelbacher, E.A. and Ridgway, R.L. (1989) Comparison of sex pheromone composition and pheromone-mediated male behavior of laboratory-reared and wild *Heliothis zea* (Lepidoptera: Noctuidae). *Journal of Chemical Ecology* 15, 1259–1265.

Ralls, K., Ballou, J.D. and Templeton, A. (1988) Estimates of lethal equivalents and the cost of inbreeding in mammals. *Conservation Biology* 2, 185–193.

Randolph, S.E., Rogers, D.J. and Kiilu, J. (1990) Rapid changes in the reproductive cycle of wild-caught tsetse, *Glossina pallidipes* Austen, when brought into the laboratory. *Insect Science and its Application* 11, 347–354.

Rao, K.J. (1991) Reproductive biology of the giant freshwater prawn *Macrobrachium rosenbergii* (de Man) from Lake Kolleru (Andhra Pradesh). *Indian Journal of Animal Sciences* 61, 780–787.

Ratner, S.C. and Boice, R. (1975) Effects of domestication on behaviour. In: Hafez, E.S.E. (ed.) *The Behaviour of Domestic Animals*, 3rd edn. Williams and Wilkins, Baltimore, Maryland, pp. 3–19.

Rauw, W.M., Kanis, E., Noordhuizen-Stassen, E.N. and Grommers, F.J. (1998) Undesirable side effects of selection for high production efficiency in farm animals: a review. *Livestock Production Science* 56, 15–33.

Reed, C.A. (1980) The beginnings of animal domestication. In: Cole, H.H. and Garrett, W.N. (eds) *Animal Agriculture: the Biology, Husbandry and Use of Domestic Animals.* W.H. Freeman, San Francisco, California, pp. 3–20.

Reed, D.H. and Bryant, E.H. (2000) The evolution of senescence under curtailed life span in laboratory populations of *Musca domestica* (the housefly). *Heredity* 85, 115–121.

Reinhardt, U.G. (2001) Selection for surface feeding in farmed and sea-ranched masu salmon juveniles. *Transactions of the American Fisheries Society* 130, 155–158.

Reinhardt, V. (1989) Behavioral responses of unrelated adult male rhesus monkeys familiarized and paired for the purpose of environmental enrichment. *American Journal of Primatology* 17, 243–248.

Reinhardt, V. (1991) Training adult rhesus monkeys to actively co-operate during in-homecage venipuncture. *Animal Technology* 42, 11–17.

Reinhardt, V. (1994) Caged rhesus macaques voluntarily work for ordinary food. *Primates* 35, 95–98.

Reinhardt, V. (1996) Refining the blood collection procedure for macaques. *Laboratory Animals* 25, 32–35.

Reinhardt, V., Houser, W.D., Eisele, S.G. and Champoux, M. (1987) Social enrichment with infants of the environment of singly caged adult rhesus monkeys. *Zoo Biology* 6, 365–371.

Rekilä, T., Mononen, J. and Harri, M. (1996) Effect of inside-cage and outside-cage environment on behaviour test-performance of blue foxes (*Alopex lagopus*). *Acta Agriculturae Scandinavica, Section A, Animal Science* 46, 247–252.

Renner, M.J. and Rosenzweig, M.R. (1987) *Enriched and Impoverished Environments: Effects on Behavior*. Springer-Verlag, New York.

Rhodes, J.S. and Quinn, T.P. (1998) Factors affecting the outcome of territorial contests between hatchery and naturally reared coho salmon parr in the laboratory. *Journal of Fish Biology* 53, 1220–1230.

Rich, S.T. (1968) The Mongolian gerbil (*Meriones unguiculatus*) in research. *Laboratory Animal Care* 18, 235–243.

Richards, C. and Leberg, P.L. (1996) Temporal changes in allele frequencies and a population's history of severe bottlenecks. *Conservation Biology* 10, 832–839.

Richards, R.A. (1998) Darwin, domestic breeding and artificial selection. *Endeavour* 22, 106–109.

Richardson, C.S., Dohm, M.R. and Garland, T. Jr (1994) Metabolism and thermo-regulation in crosses between wild and random-bred laboratory house mice (*Mus domesticus*). *Physiological Zoology* 67, 944–975.

Richardson, M.K. (1995) Heterochrony and the phylotypic period. *Developmental Biology* 172, 412–421.

Richter, C.P. (1949) Domestication of the Norway rat and its implication for the problems of stress. *Proceedings of the Association for Research on Nervous and Mental Diseases* 29, 19–47.

Ricker, J.P. and Hirsch, J. (1985) Evolution of an instinct under long-term divergent selection for geotaxis in domesticated populations of *Drosophila melanogaster*. *Journal of Comparative Psychology* 99, 380–390.

Ridgeway, W. (1905) *The History and Influence of the Thoroughbred Horse*. Cambridge Biological Series, Cambridge.

Rindos, D. (1980) Symbiosis, instability and the origins and spread of agriculture: a new model. *Current Anthropology* 21, 751–772.

Ringo, J., Barton, K. and Dowse, H. (1986) The effect of genetic drift on mating propensity, courtship behaviour, and postmating fitness in *Drosophila simulans*. *Behaviour* 97, 226–233.

Robinson, B.W. and Doyle, R.W. (1990) Phenotypic correlations among behavior and growth variables in tilapia: implications for domestication selection. *Aquaculture* 85, 177–186.

Robinson, R. (1965) *Genetics of the Norway Rat*. Pergamon Press, Oxford.

Robinson, W.S., Nowogrodzki, R. and Morse, R.A. (1989) The value of honey bees as pollinators of U.S. crops. *American Bee Journal* 129, 411–423, 477–487.

Roest, A.I. (1991) Captive reproduction in Heermann's kangaroo rat, *Dipodomys heermanni*. *Zoo Biology* 10, 127–137.

Röhrs, V.M. (1985) Cephalization, neocorticalization and the effects of domestication on brains of mammals. In: Duncker, H.-R. and Fleischer, G. (eds) *Functional Morphology in Vertebrates*. G. Fischer, Stuttgart, pp. 544–547.

Röhrs, V.M. (1986) Domestikationsbedingte Hirnänderungen bei Musteliden. *Zeitschrift zoologische Systematik und Evolutionsforschung* 24, 231–239.

Röhrs, V.M. and Kruska, D. (1969) Der Einfluß der Domestikation auf das Zentralnervensystem und Verhalten von Schweinen. *Deutsche Tierärztliche Wochenschrift* 76, 497–536.

Röhrs, M. and Ebinger, P. (1998) Sind Zooprzewalskipferde Hauspferde? *Berliner und Münchener Tierärztliche Wochenschrift* 111, 273–280.

Romanov, M.N. and Weigend, S. (2001) Analysis of genetic relationships between various populations of domestic and jungle fowl using microsatellite markers. *Poultry Science* 80, 1057–1063.

Rood, J.P. (1972) Ecological and behavioural comparisons of three genera of Argentine cavies. *Animal Behaviour Monographs* 5, 1–83.

Roseberry, J.L., Ellsworth, D.L. and Klimstra, W.D. (1987) Comparative post-release behavior and survival of wild, semi-wild, and game-farm bobwhites. *Wildlife Society Bulletin* 15, 449–455.

Rosenzweig, M.R. and Bennett, E.L. (1996) Psychobiology of plasticity: effects of training and experience on brain and behavior. *Behavior and Brain Research* 78, 57–65.

Rössler, Y. (1975a) The ability to inseminate: a comparison between laboratory-reared and field populations of the Mediterranean fruit fly (*Ceratitis capitata*). *Entomologia Experimentalis et Applicata* 18, 255–260.

Rössler, Y. (1975b) Reproductive differences between laboratory-reared and field-collected populations of the Mediterranean fruit fly, *Ceratitis capitata*. *Annals of the Entomological Society of America* 68, 987–991.

Roumeyer, A. and Bouissou, M.F. (1992) Assessment of fear reaction in domestic sheep and influence of breed and rearing conditions. *Applied Animal Behaviour Science* 34, 93–119.

Rowe, D.K., Thorpe, J.E. and Shanks, A.M. (1991) The role of fat stores in the maturation of male Atlantic salmon (*Salmo salar*) parr. *Canadian Journal of Fisheries and Aquatic Sciences* 48, 405–413.

Rushen, J. and de Passillé, A.M. (1992) The scientific assessment of the impact of housing on animal welfare: a critical review. *Canadian Journal of Animal Science* 72, 721–743.

Rushen, J., Taylor, A.A. and de Passillé, A.M. (1999) Domestic animals' fear of humans and its effect on their welfare. *Applied Animal Behaviour Science* 65, 285–303.

Ruzzante, D.E. (1994) Domestication effects on aggressive and schooling behavior in fish. *Aquaculture* 120, 1–24.

Ruzzante, D.E. and Doyle, R.W. (1993) Evolution of behavior in a resource-rich, structured environment: selection experiments with medaka (*Oryzias latipes*). *Evolution* 47, 456–470.

Rybarczyk, P., Koba, Y., Rushen, J., Tanida, H. and de Passillé, A.M. (2001) Can cows discriminate people by their faces? *Applied Animal Behaviour Science* 74, 175–189.

Ryman, N. and Stahl, G. (1980) Genetic changes in hatchery stocks of brown trout (*Salmo trutta*). *Canadian Journal of Fisheries and Aquatic Sciences* 37, 82–87.

Salonius, K. and Iwama, G.K. (1993) Effects of early rearing environment on stress response, immune function and disease resistance in juvenile coho (*Oncorhynchus kisutch*) and chinook salmon (*O. tschawytscha*). *Canadian Journal of Fisheries and Aquatic Sciences* 50, 759–766.

Sambrook, T.D. and Buchanan-Smith, H.M. (1997) Control and complexity in novel object enrichment. *Animal Welfare* 6, 207–216.

Sandnabba, N.K. (1997) Territorial behaviour and social organization as a function of the level of aggressiveness of male mice. *Ethology* 103, 566–577.

Sandøe, P., Holtug, N. and Simonsen, H.B. (1996) Ethical limits to domestication. *Journal of Agricultural and Environmental Ethics* 9, 114–122.

Sarrazin, F. and Legendre, S. (2000) Demographic approach to releasing adults versus young in reintroductions. *Conservation Biology* 14, 488–500.

Sato, Y., Fenerich-Verani, N., Verani, J.R., Vieira, L.J.S. and Godinho, H.P. (2000) Induced reproductive responses of the neotropical anostomid fish *Leporinus elongatus* Val. under captive breeding. *Aquaculture Research* 31, 189–193.

Savory, C.J. (1974) Growth and behaviour of chicks fed on pellets or mash. *British Poultry Science* 15, 281–286.

Schassburger, R.M. (1987) Wolf vocalization: An integrated model of structure, motivation, and ontogeny. In: Frank, H. (ed.) *Man and Wolf.* Dr W. Junk Publishers, Dordrecht, pp. 313–347.

Schladweiler, J.L. and Tester, J.R. (1972) Survival and behavior of hand-reared mallards released in the wild. *Journal of Wildlife Management* 36, 1118–1127.

Schoenecker, B. and Heller, K.E. (2000) Indication of a genetic basis of stereotypies in laboratory-bred bank voles (*Clethrionomys glareolus*). *Applied Animal Behaviour Science* 68, 339–347.

Schoenecker, B., Heller, K.E. and Freimanis, T. (2000) Development of stereotypies and polydipsia in wild caught bank voles (*Clethrionomys glareolus*) and their laboratory-bred offspring. *Applied Animal Behaviour Science* 68, 349–357.

Schoonhoven, L.M. (1967) Loss of hostplant specificity by *Manduca sexta* after rearing on an artificial diet. *Entomologia Experimentalis et Applicata* 10, 270–272.

Schreck, C.B., Olla, B.L. and Davis, M.W. (1997) Behavioral responses to stress. In: Iwama, G.K., Pickering, A.D., Sumpter, J.P. and Schreck, C.B. (eds) *Fish Stress and Health in Aquaculture.* Cambridge University Press, Cambridge, pp. 145–170.

Schreck, C.B., Contreras-Sanchez, W. and Fitzpatrick, M.S. (2001) Effects of stress on fish reproduction, gamete quality, and progeny. *Aquaculture* 197, 3–24.

Schroeder, W.J., Miyabara, R.Y. and Chambers, D.L. (1972) Protein products for rearing three species of larval Tephritidae. *Journal of Economic Entomology* 65, 969–972.

Schroth, K.-E. (1991) Survival, movements, and habitat selection of released capercaillie in the north-east Black Forest in 1984–1989. *Ornis Scandanavica* 22, 249–254.

Schudnagis, R. (1975) Vergleichend quantitative Untersuchungen an Organen, insbesondere am Gehirn von Wild- und Hausform der Graugans (*Anser anser* Linnaeus, 1758). *Zeitschrift für Tierzüchtung und Züchtungsbiologie* 92, 73–105.

Schutz, J.R. and Harrell, R.M. (1999) Use of sex reversal in striped bass to create an all-male population. *North American Journal of Aquaculture* 61, 97–106.

Schütz, K.E. and Jensen, P. (2001) Effects of resource allocation on behavioural strategies: a comparison of red junglefowl (*Gallus gallus*) and two domesticated breeds of poultry. *Ethology* 107, 753–765.

Schütz, K.E., Forkman, B. and Jensen, P. (2001) Domestication effects on foraging strategy, social behaviour and different fear responses: a comparison between the red junglefowl (*Gallus gallus*) and a modern layer strain. *Applied Animal Behaviour Science* 74, 1–14.

Schwaibold, U. and Pillay, N. (2001) Stereotypic behaviour is genetically transmitted in the African striped mouse *Rhabdomys pumilio*. *Applied Animal Behaviour Science* 74, 273–280.

Schwarz, E. and Schwarz, H. (1943) The wild and commensal stocks of the house mouse, *Mus musculus* Linnaeus. *Journal of Mammalogy* 24, 59–72.

Schwentker, V. (1963) The gerbil, a new laboratory animal. *Illinois Veterinarian* 6, 5–9.

Scott, J.P. and Fuller, J.L. (1965) *Genetics and the Social Behavior of the Dog*. University of Chicago Press, Chicago, Illinois.

Scott, M.D. and Causey, K. (1973) Ecology of feral dogs in Alabama. *Journal of Wildlife Management* 37, 253–265.

Seabrook, M.F. (1972) A study to determine the influence of the herdsman's personality on milk yield. *Journal of Agricultural Labour Science* 1, 44–59.

Searle, J.B. (1984) Breeding the common shrew (*Sorex araneus*) in captivity. *Laboratory Animals* 18, 359–363.

Seidel, G.E. Jr and Johnson, L.A. (1999) Sexing mammalian sperm – overview. *Theriogenology* 52, 1267–1272.

Selman, I.E., McEwan, E.D. and Fisher, E.W. (1970) Studies on natural suckling in cattle during the first eight hours post-partum. I. Behavioural studies. *Animal Behaviour* 18, 276–283.

Serenius, T., Sevón-Aimonen, M.-L. and Mäntysaari, E.A. (2001) The genetics of leg weakness in Finnish Large White and Landrace populations. *Livestock Production Science* 69, 101–111.

Serpell, J.A. (1996) Evidence for an association between pet behavior and owner attachment levels. *Applied Animal Behaviour Science* 47, 49–60.

Setchell, B.P. (1992) Domestication and reproduction. *Animal Reproduction Science* 28, 195–202.

Sgrò, C.M. and Partridge, L. (2000) Evolutionary responses of the life history of wild-caught *Drosophila melanogaster* to two standard methods of laboratory culture. *American Naturalist* 156, 341–353.

Shackelford, R.M. (1949) Origin of the American ranch-bred mink. *American Fur Breeder* 22, 12–14.

Shackelford, R.M. (1984) American mink. In: Mason, I.L. (ed.) *Evolution of Domesticated Animals*. Longman, New York, pp. 229–233.

Shank, C.C. (1972) Some aspects of social behavior in a population of feral goats (*Capra hircus*). *Zeitschrift für Tierpsychologie* 30, 488–528.

Sharma, J.G. and Chakrabarti, R. (1999) Larval rearing of common carp *Cyprinus carpio*: a comparison between natural and artificial diets under three stocking densities. *Journal of the World Aquaculture Society* 30, 490–495.

Sharp, P.M. (1984) The effect of inbreeding on competitive male-mating ability in *Drosophila melanogaster*. *Genetics* 106, 601–612.

Sharpe, P.B., Woolf, A. and Roby, D.D. (1998) Raising and monitoring tame ruffed grouse (*Bonasa umbellus*) for field studies. *American Midland Naturalist* 139, 39–48.

Sheldon, P. (1993) The evolution of form. In: Skelton, P. (ed.) *Evolution: a Biological and Palaeontological Approach*. Addison-Wesley, Wokingham, UK, pp. 723–732.

Shelton, M. and Wade, D. (1979) Predatory losses: a serious livestock problem. *Animal Industry Today* 2, 4–9.

Shepherdson, D.J., Brownback, T. and Tinkler, D. (1990) Putting the wild back into zoos: enriching the zoo environment. *Applied Animal Behaviour Science* 28, 300 (Abstr.).

Shepherdson, D.J., Carlstead, K., Mellen, J.D. and Seidensticker, J. (1993) The influence of food presentation on the behavior of small cats in confined environments. *Zoo Biology* 12, 203–216.

Shepherdson, D.J., Mellen, J. and Hutchins, M. (eds) (1997) *Second Nature: Environmental Enrichment for Captive Animals*. Smithsonian Institute Press, Washington, DC.

Sherwin, C.M. (1998) The use and perceived importance of three resources which provide caged laboratory mice the opportunity for extended locomotion. *Applied Animal Behaviour Science* 55, 353–367.

Shields, W.M. (1982) *Philopatry, Inbreeding and the Evolution of Sex*. State University of New York Press, Albany, New York.

Shikano, T., Chiyokubo, T., Nakadate, M. and Fujio, Y. (2000) The relationship between allozyme heterozygosity and salinity tolerance in wild and domestic populations of the guppy (*Poecilia reticulata*). *Aquaculture* 184, 233–245.

Shikano, T., Chiyokubo, T. and Taniguchi, N. (2001) Effect of inbreeding on salinity tolerance in the guppy (*Poecilia reticulata*). *Aquaculture* 202, 45–55.

Shrimpton, J.M., Bernier, N.J., Iwama, G.K. and Randall, D.J. (1994) Differences in measurements of smolt development between wild and hatchery-reared juvenile coho salmon (*Oncorhynchus kisutch*) before and after saltwater exposure. *Canadian Journal of Fisheries and Aquatic Sciences* 51, 2170–2178.

Siegel, P.B. (1975) Behavioural genetics. In: Hafez, E.S.E. (ed.) *The Behaviour of Domestic Animals*, 3rd edn. Williams and Wilkins, Baltimore, Maryland, pp. 20–42.

Skaala, O., Jorstad, K.E. and Borgstrom, R. (1996) Genetic impact on two wild brown trout (*Salmo trutta*) populations after release of non-indigenous hatchery spawners. *Canadian Journal of Fisheries and Aquatic Sciences* 53, 2027–2035.

Slate, J., Kruuk, L.E.B., Marshall, T.C., Pemberton, J.M. and Clutton-Brock, T.H. (2000) Inbreeding depression influences lifetime breeding success in a wild population of red deer (*Cervus elaphus*). *Proceedings of the Royal Society of London, Series B* 267, 1657–1662.

Slater, P.J.B. and Clayton, N.S. (1991) Domestication and song learning in zebra finches *Taeniopygia guttata*. *Emu* 91, 126–128.

Slaugh, B.T., Flinders, J.T., Roberson, J.A. and Johnston, N.P. (1992) Effect of rearing method on chukar survival. *Great Basin Naturalist* 52, 25–28.

Sloan, R.J. (1973) Relationships between behavior and reproduction in captive wild Norway rats (*Rattus norvegicus*). PhD thesis, State University of New York College of Environmental Science and Forestry, Syracuse, New York.

Smith, J., Hurst, J.L. and Barnard, C.J. (1994) Comparing behaviour in wild and laboratory strains of the house mouse: levels of comparison and functional inference. *Behavioural Processes* 32, 79–86.

Smith, R.H. (1972) Wildness and domestication in *Mus musculus*: a behavioral analysis. *Journal of Comparative and Physiological Psychology* 79, 22–29.

Smith, R.H. (1978) Open-field freezing as a stable parameter of wildness in *Mus musculus*. *Behavioral Biology* 23, 67–74.

Smith, R.H. and Connor, J.L. (1978) Behavioral effects of laboratory rearing in wild *Mus musculus*. *Behavioral Biology* 24, 387–392.

Smits, J. (1990) Comments on game farming. *Canadian Veterinary Journal* 31, 676–677.

Smythe, N. (1987) The paca (*Cuniculus paca*) as a domestic source of protein for the neotropical, humid lowlands. *Applied Animal Behaviour Science* 17, 155–170.

Snyder, N.F.R., Derrickson, S.R., Beissinger, S.R., Wiley, J.W., Smith, T.B., Toone, W.D. and Miller, B. (1997) Limitations of captive breeding: reply to Gippoliti and Carpeneto. *Conservation Biology* 11, 808–810.

Sorensen, G. and Randrup, A. (1986) Possible protective value of severe psychopathology against lethal effects of an unfavourable millieu. *Stress Medicine* 2, 103–105.

Sørensen, P. (1992) Genetics of leg disorders. In: Whitehead, C.C. (ed.) *Bone Biology and Skeletal Disorders in Poultry*. Carfax Publishing, Abingdon, UK, pp. 213–229.

Soulé, M.E., Gilpin, M., Conway, W. and Foose, T. (1986) The millennium ark: how long a voyage, how many staterooms, how many passengers? *Zoo Biology* 5, 101–113.

Spencer, S. (2001) Consumer attitudes to animal biotechnology: broader ethical considerations. In: Renaville, R. and Burny, A. (eds) *Biotechnology in Animal Husbandry*. Kluwer Academic Publishers, Boston, Massachusetts, pp. 345–357.

Spiegel, R., Price, E.O. and Huck, U.W. (1974) Differential vulnerability of wild, domestic and hybrid Norway rats to predation by great-horned owls. *Journal of Mammalogy* 55, 386–392.

Špinka, M., Illmann, G., de Jonge, F., Andersson, M., Schuurman, T. and Jensen, P. (2000) Dimensions of maternal behaviour characteristics in domestic and wild × domestic crossbred sows. *Applied Animal Behaviour Science* 70, 99–114.

Spivey, M. (1973) A school opens for hunting birds to learn to live. *Smithsonian* 4, 32–38.

Spurway, H. (1955) The causes of domestication: an attempt to integrate some ideas of Konrad Lorenz with evolution theory. *Journal of Genetics* 53, 325–362.

Staats, J. (1981) List of inbred strains. In: Green, M.C. (ed.) *Genetic Variants and Strains of the Laboratory Mouse*. G. Fischer Verlag, New York, pp. 373–376.

Stahl, G. (1983) Differences in the amount and distribution of genetic variation between natural populations and hatchery stocks of Atlantic salmon. *Aquaculture* 33, 23–32.

Stahnke, A. (1987) Verhaltensunterschiede zwischen Wild- und Hausmeerschweinchen. *Zeitschrift für Säugetierkunde* 52, 294–307.

Stanley-Price, M.R. (1985) Game domestication for animal production in Kenya: feeding trials with oryx, zebu cattle and sheep under controlled conditions. *Journal of Agricultural Science, Cambridge* 104, 367–374.

Starling, A.E. (1991) Captive breeding and release. *Ornis Scandanavica* 22, 255–257.

Steffen, K., Kruska, D. and Tiedemann, R. (2001) Postnatal brain size decrease, visual performance, learning, and discrimination ability of juvenile and adult American mink (*Mustela vison*). *Mammalian Biology* 66, 269–280.

Steingrund, P. and Fernö, A. (1997) Feeding behaviour of reared and wild cod and the effect of learning: two strategies of feeding on the two-spotted goby. *Journal of Fish Biology* 51, 334–348.

Steiniger, R. (1950) Beiträge zur Soziologie und sonstigen Biologie der Wanderratte. *Zeitschrift für Tierpsychologie* 7, 356–379.

Stevens, V.J. (1978) Basic operant research in the zoo. In: Markowitz, H. and Stevens, V. (eds) *Behavior of Captive Wild Animals*. Nelson-Hall, Chicago, Illinois, pp. 209–246.

Stevenson, M.F. (1983) The captive environment: its effect on exploratory and related behavioural responses in wild animals. In: Archer, J. and Birke, L.I.A. (eds) *Exploration in Animals and Humans*. Van Nostrand Reinhold, New York, pp. 176–197.

Stoinski, T.S., Daniel, E. and Maple, T.L. (2000) A preliminary study of the behavioral effects of feeding enrichment on African elephants. *Zoo Biology* 19, 485–493.

Stolba, A. (1981) A family system in enriched pens as a novel method of pig housing. In: *Alternatives to Intensive Housing Systems*. Universities Federation for Animal Welfare, Potters Bar, pp. 52–67.

Stolba, A. and Wood-Gush, D.G.M. (1981) The assessment of behaviour needs of pigs under free-range and confined conditions. *Applied Animal Ethology* 7, 388–389.

Stolba, A. and Wood-Gush, D.G.M. (1984) The identification of behavioural key features and their incorporation as to a housing design for pigs. *Annuale de Recherches Veterinaires* 15, 287–298.

Storz, J.F. (1999) Genetic consequences of mammalian social structure. *Journal of Mammalogy* 80, 553–569.

Stuermer, W., Plotz, K., Wetzel, W., Wagner, T., Leybold, A. and Scheich, H. (1997) Reduced brain size and faster auditory discrimination learning in laboratory gerbils compared to wild Mongolian gerbils (*Meriones unguiculatus*). *Society for Neuroscience* 23, 2067 (Abst.).

Stunz, G.W. and Minello, T.J. (2001) Habitat-related predation on juvenile wild-caught and hatchery-reared red drum, *Sciaenops ocellatus*. *Journal of Experimental Marine Biology and Ecology* 260, 13–25.

Stunz, G.W., Levin, P.S. and Minello, T.J. (2001) Selection of estuarine nursery habitats by wild-caught and hatchery-reared juvenile red drum in laboratory mesocosms. *Environmental Biology of Fishes* 61, 305–313.

Suboski, M.D. and Templeton, J.J. (1989) Life skills training for hatchery fish: social learning and survival. *Fisheries Research* 7, 343–352.

Suleman, M.A., Yole, D., Wango, E., Sapolsky, R.M., Carlsson, H.-E. and Hau, J. (1999) Peripheral blood lymphocyte immunocompetence in African green monkeys (*Cercopithecus aethiops*) and the effects of capture and confinement. *In Vivo* 13, 25–28.

Suleman, M.A., Wango, E., Farah, I.O. and Hau, J. (2000) Adrenal cortex and stomach lesions associated with stress in wild male African green monkeys (*Cercopithecus aethiops*) in the post-capture period. *Journal of Medical Primatology* 29, 338–342.

Sullivan, J.P., Grasman, K.A. and Scanlon, P.F. (1992) Effects of handling and pair management on reproduction in Japanese quail (*Coturnix coturnix*). *Theriogenology* 37, 877–883.

Sundell, K., Dellefors, C. and Björnsson, B.T. (1998) Wild and hatchery-reared brown trout, *Salmo trutta*, differ in smolt related characteristics during parr–smolt transformation. *Aquaculture* 167, 53–65.

Swain, D.P. and Riddell, B.E. (1990) Variation in agonistic behavior between newly emerged juveniles from hatchery and wild populations of coho salmon, *Oncorhynchus kisutch*. *Canadian Journal of Fisheries and Aquatic Sciences* 47, 566–571.

Swain, D.P., Riddell, B.E. and Murray, C.B. (1991) Morphological differences between hatchery and wild populations of coho salmon (*Oncorhynchus kisutch*): environmental versus genetic origin. *Canadian Journal of Fisheries and Aquatic Sciences* 48, 1783–1791.

Swaisgood, R.R., White, A.M., Zhou, X., Zhang, H., Zhang, G., Wei, R., Hare, V.J., Tepper, E.M. and Lindburg, D.G. (2001) A quantitative assessment of the efficacy of an environmental enrichment programme for giant pandas. *Animal Behaviour* 61, 447–457.

Symons, P.E.K. (1971) Behavioural adjustment of population density to available food by juvenile Atlantic salmon. *Journal of Animal Ecology* 40, 569–586.

Szabó, T. (2001) Hormonally induced ovulation of northern pike via sustained-release vehicles. *North American Journal of Aquaculture* 63, 137–143.

Takahashi, L.K. and Blanchard, R.J. (1982) Attack and defense in laboratory and wild Norway and black rats. *Behavioural Processes* 7, 49–62.

Tamazouzt, L., Chatain, B. and Fontaine, P. (2000) Tank wall colour and light level affect growth and survival of Eurasian perch larvae (*Perca fluviatilis* L.). *Aquaculture* 182, 85–90.

Tanabe, Y. (2001) The roles of domesticated animals in the cultural history of the humans. *Asian-Australian Journal of Animal Science* 14 (Special Issue), 13–18.

Tannenbaum, J. (1991) Ethics and animal welfare: the inextricable connection. *Journal of the American Veterinary Medical Association* 198, 1360–1376.

Tave, D. (1993) *Genetics for Fish Hatchery Managers*. Van Nostrand Reinhold, New York.

Tchernov, E. and Horwitz, L.K. (1991) Body size diminution under domestication: unconscious selection in primeval domesticates. *Journal of Anthropological Archaeology* 10, 54–75.

Temple, S.A. (1978) Reintroducing birds of prey to the wild. In: Temple, S.A. (ed.) *Endangered Birds: Management Technologies for Preserving Threatened Species*. University of Wisconsin, Madison, Wisconsin, pp. 355–363.

Templeton, A.R. and Read, B. (1984) Factors eliminating inbreeding depression in a captive herd of Speke's gazelle (*Gazella spekei*). *Zoo Biology* 3, 177–199.

Tennessen, T. and Hudson, R.J. (1981) Traits relevant to the domestication of herbivores. *Applied Animal Ethology* 7, 87–102.

Terlouw, C.E.M., Lawrence, A.B. and Illius, A.W. (1991) Influences of feeding level and physical restriction on the development of sterotypies in sows. *Applied Animal Behaviour Science* 42, 981–991.

Tessier, N., Bernatchez, L. and Wright, J.M. (1997) Population structure and impact of supportive breeding inferred from mitochondrial and microsatellite DNA analyses in land-locked Atlantic salmon, *Salmo salar*. *Molecular Ecology* 6, 735–750.

Thomas, B.B. and Oommen, M.M. (1999) Reproductive biology of the South Indian gerbil (*Tatera indica cuvieri*) under laboratory conditions. *Mammalia* 63, 341–347.

Thornhill, N.W. (1993) *The Natural History of Inbreeding and Outbreeding*. University of Chicago Press, Chicago, Illinois.

Thorpe, J.E. (1991) Acceleration and deceleration effects of hatchery rearing on salmonid development, and their consequences for wild stocks. *Aquaculture* 98, 111–118.

Thorpe, J.E., Talbot, C., Miles, M.S. and Keay, D.S. (1990) Control of maturation in cultured Atlantic salmon, *Salmo salar*, in pumped seawater tanks, by restricting food intake. *Aquaculture* 86, 315–326.

Tipping, J.M. (2001) Adult returns of hatchery sea-run cutthroat trout reared in a semi-natural pond for differing periods prior to release. *North American Journal of Aquaculture* 63, 131–133.

Tomka, S.A. (1992) Vicuñas and llamas: parallels in behavioral ecology and implications for the domestication of Andean camelids. *Human Ecology* 20, 407–433.

Topál, J., Miklósi, Á. and Csányi, V. (1997) Dog–human relationship affects problem solving behavior in the dog. *Anthrozoös* 10, 214–224.

Treus, V. and Kravchenko, D. (1968) Methods of rearing and economic utilization of the eland in the Askaniyanova Zoological Park. *Symposium of the Zoological Society of London* 21, 395–411.

Troy, C.S., MacHugh, D.E., Bailey, J.F., Magee, D.A., Loftus, R.T., Cunningham, P., Chamberlain, A.T., Sykes, B.C. and Bradley, D.G. (2001) Genetic evidence for Near-Eastern origins of European cattle. *Nature* 410, 1088–1091.

Trut, L.N. (1996) Sex ratio in silver foxes: effects of domestication and the *star* gene. *Theoretical and Applied Genetics* 92, 109–115.

Trut, L.N. (1999) Early canid domestication: the farm-fox experiment. *American Scientist* 87, 160–169.

Tsakas, S.C. and Zouros, E. (1980) Genetic differences among natural and laboratory-reared populations of the olive-fruit fly, *Dacus oleae* (Diptera: Tephritidae). *Entomologia Experimentalis et Applicata* 28, 268–276.

Tsutsumi, H., Morikawa, N., Niki, R. and Tanigawa, M. (2001) Acclimatization and response of minipigs toward humans. *Laboratory Animals* 35, 236–242.

Ucko, P.J. and Dimbleby, G.W. (1969) *The Domestication and Exploitation of Plants and Animals*. Duckworth, London.

Utter, F. (1998) Genetic problems of hatchery-reared progeny released into the wild, and how to deal with them. *Bulletin of Marine Science* 62, 623–640.

Van der Meeren, G.I. (2000) Predation on hatchery-reared lobsters released in the wild. *Canadian Journal of Fisheries and Aquatic Sciences* 57, 1794–1803.

Van Eenennaam, J.P., Webb, M.A.H., Deng, X., Doroshov, S.I., Mayfield, R.B., Cech, J.J. Jr, Hillemeier, D.C. and Willson, T.E. (2001) Artificial spawning and larval rearing of Klamath River green sturgeon. *Transactions of the American Fisheries Society* 130, 159–165.

Van Heezik, Y. and Seddon, P.J. (1998) Ontogeny of behavior of hand-reared and hen-reared captive houbara bustards. *Zoo Biology* 17, 245–255.

Van Lawick-Goodall, J. (1968) The behaviour of free-living chimpanzees in the Gombe Stream Reserve. *Animal Behaviour Monograph* 1, 161–301.

Van Oorschot, R.A.H., Williams-Blangero, S. and VandeBerg, J.L. (1992) Genetic diversity of laboratory gray short-tailed opossums (*Monodelphis domestica*): effect of newly introduced wild-caught animals. *Laboratory Animal Science* 42, 255–260.

van Praag, H., Kempermann, G. and Gage, F.H. (2000) Neural consequences of environmental enrichment. *Nature Reviews – Neuroscience* 1, 191–198.

Van Reenen, C.G., Meuwissen, T.H.E., Hopster, H., Oldenbroek, K., Kruip, T.A.M. and Blokhuis, H.J. (2001) Transgenesis may affect farm animal welfare: a case for systematic risk assessment. *Journal of Animal Science* 79, 1763–1779.

Van Vuren, D. and Hedrick, P.W. (1989) Genetic conservation in feral populations of livestock. *Conservation Biology* 3, 312–317.

Vargas, A. and Anderson, S.H. (1999) Effects of experience and cage enrichment on predatory skills of black-footed ferrets (*Mustela nigripes*). *Journal of Mammalogy* 80, 263–269.

Veissier, I., Gesmier, V., Le Neindre, P., Gautier, J.Y. and Bertrand, G. (1994) The effects of rearing in individual crates on subsequent social behaviour of veal calves. *Applied Animal Behaviour Science* 41, 199–210.

Verspoor, E. (1988) Reduced genetic variability in first-generation hatchery populations of Atlantic salmon (*Salmo salar*). *Canadian Journal of Fisheries and Aquatic Sciences* 45, 1686–1690.

Vignes, S., Newman, J.D. and Roberts, R.L. (2001) Mealworm feeders as environmental enrichment for common marmosets. *Contemporary Topics in Laboratory Animal Science* 40, 26–29.

Vilà, C., Savolainen, P., Malconado, J.E., Amorim, I.R., Rice, J.E., Honeycutt, R.L., Crandall, K.A., Lundeberg, J. and Wayne, R.K. (1997) Multiple and ancient origins of the domestic dog. *Science* 276, 1687–1689.

Vilà, C., Leonard, J.A., Götherström, A., Marklund, S., Sandberg, K., Lidén, K., Wayne, R.K. and Ellegren, H. (2001) Widespread origins of domestic horse lineages. *Science* 291, 474–477.

Villavaso, E.J. and McGovern, W.L. (1986) Attractiveness and competitiveness of laboratory versus native strain of boll weevils (Coleoptera: Curculionidae). *Journal of Economic Entomology* 79, 1240–1243.

Vincent, R.E. (1960) Some influences of domestication upon three stocks of brook trout (*Salvelinus fontinalis* Mitchill). *Transactions of the American Fisheries Society* 89, 35–52.

Vinke, C.M. (2001) Some comments on the review of Nimon and Broom on the welfare of farmed mink. *Animal Welfare* 10, 315–323.

Waddington, C.H. (1961) Genetic assimilation. *Advances in Genetics* 10, 257–293.

Wakahara, M. (1996) Heterochrony and neotonic salamanders: possible clues for understanding the animal development and evolution. *Zoological Science* 13, 765–776.

Wallace, M.E. (1981) The breeding, inbreeding and management of wild mice. *Symposium of the Zoological Society of London* 47, 183–204.

Waltz, E.C. (1976) Domestication effects on the anti-predator behavior and vulnerability to predation of Norway rats (*Rattus norvegicus*) in encounters with European polecats (*Mustela putorius furo*). MS thesis, State University of New York College of Environmental Science and Forestry, Syracuse, New York.

Wang, L. and White, R.J. (1994) Competition between wild brown and hatchery greenback cutthroat trout of largely wild parentage. *North American Journal of Fisheries Management* 14, 475–487.

Wang, N., Hayward, R.S. and Nolte, D.B. (2000) Effects of social interaction on growth of juvenile hybrid sunfish held at two densities. *North American Journal of Aquaculture* 62, 161–167.

Wang, S., Foote, W.C. and Bunch, T.D. (1990) A polymorphic protein character of red blood cells (EP-1) in domesticated and wild sheep. *Journal of Heredity* 81, 206–208.

Waples, K.A. and Stagoll, C.S. (1997) Ethical issues in the release of animals from captivity. *BioScience* 47, 115–121.

Waples, R.S. (1991) Genetic interactions between hatchery and wild salmonids: lessons from the Pacific Northwest. *Canadian Journal of Fisheries and Aquatic Sciences* 48 (Suppl. 1), 124–133.

Waples, R.S. (1999) Dispelling some myths about hatcheries. *Fisheries* 24, 12–21.

Washburn, D.A. and Hopkins, W.D. (1994) Videotape versus pellet-reward preference in joystick tasks by macaques. *Perceptual Motor Skills* 78, 48–50.

Wayne, R.K. (1986) Cranial morphology of domestic and wild canids: the influence of development on morphological change. *Evolution* 40, 243–261.

Weary, D.M., Phillips, P.A., Pajor, E.A., Fraser, D. and Thompson, B.K. (1998) Crushing of piglets by sows: effects of litter features, pen features and sow behaviour. *Applied Animal Behaviour Science* 61, 103–111.

Webb, M.A.H., Van Eenennaam, J.P., Feist, G.W., Linares-Casenave, J., Fitzpatrick, M.S., Schreck, C.B. and Doroshov, S.I. (2001) Effects of thermal regime on ovarian maturation and plasma sex steroids in farmed white sturgeon, *Acipenser transmontanus*. *Aquaculture* 201, 137–151.

Weber, R. (1997) New farrowing pens without confinement of the sow. In: Bottcher, R.W. and Hoff, S.J. (eds) *Livestock Environment V*, Vol. 1. American Society of Agricultural Engineers, St Joseph, Missouri, pp. 280–286.

Webster, A.J.F. (2001) Farm animal welfare: the five freedoms and the free market. *The Veterinary Journal* 161, 229–237.

Wecker, S. (1963) The role of early experience in habitat selection by the prairie deer mouse, *Peromyscus maniculatus bairdii*. *Ecological Monographs* 33, 307–325.

Wedemeyer, G.A. (1997) Effects of rearing conditions on the health and physiological quality of fish in intensive culture. In: Iwama, G.K., Pickering, A.D., Sumpter, J.P. and Schreck, C.B. (eds) *Fish Stress and Health in Aquaculture*. Cambridge University Press, Cambridge, pp. 35–71.

Weil, L.S., Barry, T.P. and Malison, J.A. (2001) Fast growth in rainbow trout is correlated with a rapid decrease in post-stress cortisol concentrations. *Aquaculture* 193, 373–380.

Weir, B.J. (1974) Notes on the origin of the domestic guinea-pig. *Symposium of the Zoological Society of London* 34, 437–446.

Weiss, J.M. (1970) Somatic effects of predictable and unpredictable shock. *Psychosomatic Medicine* 32, 397–408.

Weiss, J.M. (1971) Effects of coping behavior with and without a feedback signal on stress pathology in rats. *Journal of Comparative and Physiological Psychology* 77, 22–30.

Weiss, S. and Schmutz, S. (1999) Performance of hatchery-reared brown trout and their effects on wild fish in two small Austrian streams. *Transactions of the American Fisheries Society* 128, 302–316.

Wemelsfelder, F. (1997) Life in captivity: its lack of opportunities for variable behaviour. *Applied Animal Behaviour Science* 54, 67–70.

West, B. and Zhou, B.-X. (1988) Did chickens go north? New evidence for domestication. *Journal of Archeological Science* 14, 515–533.

West, L.B. (1960) The nature of growth inhibitory material from crowded *Rana pipiens* tadpoles. *Physiological Zoology* 32, 232–239.

Wheeler, J.C. (1995) Evolution and present situation of the South American Camelidae. *Biological Journal of the Linnean Society* 54, 271–295.

White, R.J., Karr, J.R. and Nehlsen, W. (1995) Better roles for fish stocking in aquatic resource management. *American Fisheries Society Symposium* 15, 527–547.

Wielebnowski, N.C. (1999) Behavioral differences as predictors of breeding status in captive cheetahs. *Zoo Biology* 18, 335–349.

Wilkinson, P.F. (1974) The history of muskox domestication. *The Polar Record* 17, 13–22.

Willard, D.H., Bodnar, W., Harrise, C., Keifer, L., Nichols, J.S., Blanchard, S., Hoffman, C., Moyer, M., Burkhart, W., Weiel, J., Luther, M.A., Wildison, W.O. and Rocque, W.J. (1995) Agouti structure and function: characterization of a potent α-melanocyte stimulating hormone receptor antagonist. *Biochemistry* 34, 12341–12346.

Willard, J.G., Willard, J.C., Wolfram, S.A. and Baker, J.P. (1977) Effect of diet on cecal pH and feeding behavior of horses. *Journal of Animal Science* 45, 87–93.

Williams, E.S., Thorne, E.T., Kwiatkowski, D.R., Anderson, S.T. and Lutz, K. (1991) Reproductive biology and management of captive black-footed ferrets (*Mustela nigripes*). *Zoo Biology* 10, 383–398.

Willis, J.H. (1999) The role of genes of large effect on inbreeding depression in *Mimulus guttatus*. *Evolution* 53, 1678–1691.

Wilson, A.C. and Stanley-Price, M.R. (1994) Reintroduction as a reason for captive breeding. In: Olney, P.J., Mace, G.M. and Feistner, A.T.C. (eds) *Creative Conservation: Interactive Management of Wild and Captive Populations*. Chapman and Hall, London, pp. 243–264.

Wilson, R.J., Drobney, R.D. and Hallett, D.L. (1992) Survival, dispersal, and site fidelity of wild female ring-necked pheasants following translocation. *Journal of Wildlife Management* 56, 79–85.

Wimmers, K., Ponsuksili, S., Hardge, T., Valle-Zarate, A., Mathur, P.K. and Horst, P. (2000) Genetic distinctness of African, Asian and South American local chickens. *Animal Genetics* 31, 159–165.

Winskill, L.C., Young, R.J., Channing, C.E., Hurley, J. and Waran, N.K. (1996) The effect of a foraging device (the modified 'Edinburgh Foodball') on the behaviour of the stabled horse. *Applied Animal Behaviour Science* 48, 25–35.

Wolf, K.N., Wildt, D.E., Vargas, A., Marinari, P.E., Ottinger, M.A. and Howard, J.G. (2000) Reproductive inefficiency in male black-footed ferrets (*Mustela nigripes*). *Zoo Biology* 19, 517–528.

Wong, T.T.Y. and Nakahara, L.M. (1978) Sexual development and mating response of laboratory-reared and native Mediterranean fruit flies. *Annals of the Entomological Society of America* 71, 592–596.

Wong, T.T.Y., Couey, H.M. and Nishimoto, J.I. (1982) Oriental fruit fly: sexual development and mating response of laboratory-reared and wild flies. *Annals of the Entomological Society of America* 75, 191–194.

Wood-Gush, D.G.M. (1959) A history of the domestic chicken from antiquity to the 19th century. *Poultry Science* 38, 321–326.

Wood-Gush, D.G.M. and Duncan, I.J.H. (1976) Some behavioural observations on domestic fowl in the wild. *Applied Animal Ethology* 2, 255–260.

Wood-Gush, D.G.M. and Vestergaard, K. (1989) Exploratory behavior and the welfare of intensively kept animals. *Journal of Agricultural Ethics* 2, 161–169.

Worobec, E.K., Duncan, I.J.H. and Widowski, T.M. (1999) The effects of weaning at 7, 14, and 28 days on piglet behaviour. *Applied Animal Behaviour Science* 62, 173–182.

Wright, P. and Spence, A.M. (1976) Changes in emotionality following section of the tractus occipito-mesencephalicus of the barbary dove (*Streptopelia risoria*). *Behavioural Processes* 1, 29–40.

Wright, S. (1977) *Evolution and the Genetics of Populations*, Vol. 3, *Experimental Results and Evolutionary Deductions*. University of Chicago Press, Chicago, Illinois.

Würbel, H. (2001) Ideal homes? Housing effects on rodent brain and behaviour. *Trends in Neurosciences* 24, 207–211.

Würbel, H. and Stauffacher, M. (1997) Age and weight at weaning affect corticosterone level and development of stereotypy in ICR-mice. *Animal Behaviour* 53, 891–900.

Würbel, H. and Stauffacher, M. (1998) Physical condition at weaning affects exploratory behaviour and stereotypy development in laboratory mice. *Behavioural Processes* 43, 61–69.

Würbel, H., Stauffacher, M. and von Holst, D. (1996) Stereotypies in laboratory mice – quantitative and qualitative description of the ontogeny of 'wire-gnawing' and 'jumping' in Zur:ICR and Zur: ICR nu. *Ethology* 102, 371–385.

Wyban, J.A. and Sweeney, J.N. (1991) *Intensive Shrimp Production Technology: the Oceanic Institute Shrimp Manual*. The Oceanic Institute, Honolulu, Hawaii.

Xu, Z., Primavera, J.H., de la Pena, L.D., Pettit, P., Belak, J. and Alcivar-Warren, A. (2001) Genetic diversity of wild and cultured black tiger shrimp (*Penaeus monodon*) in the Philippines using microsatellites. *Aquaculture* 199, 13–40.

Yin, S. (2002) A new perspective on barking in dogs. *Journal of Comparative Psychology* 116, 189–193.

Yoerg, S.I. (1994) Development of foraging behaviour in the Eurasian dipper, *Cinclus cinclus*, from fledging until dispersal. *Animal Behaviour* 47, 577–588.

Young, P.S. and Cech, J.J. Jr (1993) Effects of exercise conditioning on stress responses and recovery in cultured and wild young-of-the-year striped bass, *Morone saxatilis*. *Canadian Journal of Fisheries and Aquatic Sciences* 50, 2094–2099.

Young, R.J., Carruthers, J. and Lawrence, A.B. (1994) The effect of a foraging device (the 'Edinburgh foodball') on the behaviour of pigs. *Applied Animal Behaviour Science* 39, 237–247.

Youngson, A.F. and Verspoor, E. (1998) Interactions between wild and introduced Atlantic salmon (*Salmo salar*). *Canadian Journal of Fisheries and Aquatic Sciences* 55, 153–160.

Yu, M.-L. and Perlmutter, A. (1970) Growth inhibiting factors in the zebrafish, *Brachydanio rerio*, and the blue gourami, *Trichogaster trichopterus*. *Growth* 34, 153–175.

Zeuner, F.E. (1963) *A History of Domesticated Animals*. Harper and Row, New York.

Zohar, Y. and Mylonas, C.C. (2001) Endocrine manipulations of spawning in cultured fish: from hormones to genes. *Aquaculture* 197, 99–136.

Zohary, D., Tchernov, E. and Horwitz, L.K. (1998) The role of unconscious selection in the domestication of sheep and goats. *Journal of Zoology, London* 245, 129–135.

Subject Index

abandonment (desertion) of offspring *see* Maternal behavior

activity *see* locomotor activity

age effects
    reproductive success 52, 139–140
    response to novelty 105–106
    survival of captive-reared animals reintroduced to wild 190, 193–194, 196, 198–201
    see also early experience effects

age structure of wild vs. captive populations 135, 141

aggressive (agonistic) behavior 16, 22, 23, 32–33, 45–46, 60, 68–69, 70, 80, 81, 85, 88, 89, 94, 101, 102, 103–104, 126–128, 131, 133, 134, 135–138, 139–140, 141, 146, 147, 156, 157, 158, 159, 161, 164, 166, 173, 175–178, 181, 190–191, 210, 211, 214, 215, 225

agility (mobility) 23, 24, 28, 64, 70, 81, 170, 191, 192, 196, 225

agouti (pelage color) gene locus 16–17

alarm signals 14, 166, 194–196, 199–200

artificial selection
    for behavior 12, 17, 25, 35, 43, 45–46, 48, 70, 78–79, 80, 81, 84, 85, 88, 112, 114–119, 120–121, 126, 127–128, 129, 136–137, 140, 154, 159, 162, 167, 168, 169–170, 171–172, 173–174, 180, 216, 225

for egg production 84–85, 124

general 24–26, 30, 35–36, 43–50, 51, 55, 56, 68, 69, 70, 72, 73, 74–75, 82, 84–86, 92–93, 94, 126, 136, 139, 162, 168, 183, 185, 186, 225–226

for growth 46, 48, 78, 79, 80, 81, 84–85, 86, 92, 138, 162, 170–171, 178, 225

for milk yield 46, 80, 92, 225–226

for physiological traits 43–44, 47, 55, 64, 75, 78, 79–80, 81, 83, 84, 92, 127–128, 129, 162, 168, 169, 171, 173, 209, 216, 225–226

for structural traits 26, 27, 35, 43–44, 45, 47, 48, 55, 73, 78, 79, 83, 84–85, 86, 91, 92, 93, 162, 178, 225

see also reproductive success in captivity, artificial selection; taming, selection for

audition 88, 89, 126, 216

behavior *see* aggressive behavior; artificial selection, for behavior; burrowing; climbing; dominance, social; drinking; emotional reactivity; fearfulness; feeding, behavior; flight distance; geotaxis; grooming; learning; locomotor activity;

behaviour *continued*
  maternal behavior; mating
    systems; novelty, response to;
    predation; reintroductions,
    preconditioning for release; sexual
    behavior; shelter, effect on
    behavior; social organization;
    socialization; stress, behavioral
    response to; subordinate animals,
    behavior; swimming; taming;
    vocalizations; welfare, atypical
    behaviors
biological control (invertebrates) 7, 8–9,
    29, 59, 145
body size
  adrenal 147
  brain and skull size 5, 83, 84, 85,
    87–91, 172
  eye 83, 91
  general 5, 25, 34, 68–69, 78, 79, 81,
    83, 84, 85–86, 91, 126–127,
    134, 137, 162, 171, 178, 190,
    191
  other morphological characteristics
    68–69, 83, 91–92, 93
  reproductive organs 79, 92, 168
  see also growth
brain biochemistry 17, 19, 43, 80, 93, 113,
    123, 127–129, 219
brain development and function 83, 89,
    91, 155, 189, 204, 220
brain (and skull) size *see* body size
burrowing behavior 162

capture (by man) 14, 16, 24, 26, 52, 57,
    77, 105–106, 109–110, 123,
    138, 147, 150, 151, 189, 205–210,
    221
capture (by predators) *see* predation
catecholamines *see* brain biochemistry
chemical signals *see* olfaction
climbing 36, 102, 149–150, 162–163,
    179
co-adapted gene complexes 37, 48, 50
cognitive abilities *see* learning
color
  environmental preferences 159–160
  feeding responses 99–100

recognition of humans (clothing)
    126
social interactions 47, 138, 191
see also pelage, coloration
conditioning *see* learning

density *see* population, density
development *see* early experience,
    behavior, sexual maturation;
    growth; hatchery rearing;
    human–animal interactions,
    hand-rearing, socialization;
    isolation effects on behavior;
    learning; reintroductions,
    preconditioning for release; sexual
    development; shelter, effect on
    behavior, effect on maturation;
    socialization; taming, experiential
    factors; welfare, atypical
    behaviors, environmental
    enrichment, handling of juveniles
disease 5, 21–22, 26, 55, 110–111, 112,
    133, 183, 186–187, 201–202, 203,
    204, 210, 211, 216–217
domestic phenotype 10–12, 16, 24, 28,
    129, 147, 148, 161, 180, 181
domestication
  adaptation to captive environment
    1, 10–12, 21–29, 56, 91, 93,
    95–96, 105–106, 113,
    146–147, 161–162, 183,
    223–225; *see also* climate;
    drinking; feeding; humans;
    humidity; infectious diseases
    and parasites; light;
    photoperiod; predation;
    reproductive success in
    captivity; salinity; shelter;
    social environment; substrate;
    temperature
  adaptation to humans 1, 6, 113–129,
    134, 147, 168, 184; *see also*
    flight distance; human–
    animal interactions; taming
  animal welfare, effect on 1, 204, 209,
    211, 221, 226–227
  approaches to the study of
    domestication

comparison of wild and domestic
stocks 13–15
domestication genes, search for
16–17
hybridization of wild × domestic
stocks 18–19
longitudinal approach
(monitoring generation by
generation changes)
15–16
aquacultural enterprises 6
behavior, effect on *see* behavior
definition of 10–12
a developmental phenomenon 10–12,
19, 22, 184
an evolutionary process 10–12, 89,
161–162, 183, 186, 227
genetic mechanisms influencing
the domestication process
see artificial selection;
inbreeding; genetic drift;
natural selection in captivity;
relaxation of natural selection
genetic variation, effect on 72–78,
185–186, 226–227; *see also*
genetic variability
genotype–environment interactions,
effect on 174–175, 178–180
'loss' of behavioral patterns
161–165
morphological traits, effect on 83,
84–92, 93–94
physiological traits, effect on 83, 88,
89, 92, 93–94, 127–129, 147,
162
pre-adaptations for 21–29, 132, 140,
141, 185, 223–225, 226
list of traits – invertebrates 28
list of traits – vertebrates 23
products from domestic animals 1, 4,
6, 7, 8–9, 21–22, 46, 48–49;
see also artificial selection, on
egg production, on growth,
on milk production, on
structural characteristics
rate of change during domestication
14, 15, 16, 44, 183, 187
reasons for 4–6, 7, 8–9, 21–22

reintroductions, effect of
domestication on success of
see reintroductions
and resource allocation 96–97, 133;
see also resource allocation
theory
self-domestication 25–26, 29
societal effects 1, 6
species domesticated, historical
aspects
alpaca 2
bee 27–28
capybara 6, 130
carp 6
cat 21, 24
cattle 2
chicken 2, 21
cicada 7
cricket 7
dog 1–2, 24–26, 45
eland 6
ferret 21
fox 114
fruit fly 7
gerbil 38
giant pouched rat 6
goat 2
guinea pig 2
hamster 38, 113
horse 2, 85
invertebrates 7, 8
livestock 2
llama 2
mink 87, 88
mouse 2, 5, 13–14
musk ox 6
pig 2, 86
prawn 6
rabbit 2
rat 5
salmon 6
sheep 2
silkworm 2
turkey 2, 21
water buffalo 21
taxonomic designations (distinction
between wild and domestic
animals) 3–4

dominance hierarchy 23, 81–82, 131, 132,
        133, 157
dominance, social 14, 18, 68–69, 70,
        81, 103–104, 127, 131–132,
        133, 136–138, 139–140, 157,
        158, 168, 170, 173, 175–178,
        180, 191
    see also aggressive behavior;
            dominance hierarchy;
            subordinate animals,
            behavior
dominant genes 18, 19, 31
dopamine *see* brain biochemistry
drinking 63–64, 85, 101, 121, 130, 143,
        149, 152, 157, 173, 194, 210,
        227

early experience effects
    behavior 82, 110, 120, 121–123, 127,
            133, 134, 135–136, 139,
            146–147, 150–151, 155–156,
            162–164, 171–172, 174–175,
            176, 191, 192, 193–194,
            196, 198–201, 202, 206, 210,
            211, 213–214, 216–217,
            218–220
        see also learning
    sexual maturation 57, 61, 68, 80, 85,
            169, 170
        see also sexual development
    welfare 206, 210, 211, 213–215,
            216–217, 218–220
        see also welfare, environmental
                enrichment
emotional reactivity 22, 43, 69, 82, 88, 89,
        112, 117–119, 177, 180
    see also fearfulness; novelty, response
        to
enrichment (physical and social
        environments) *see* welfare,
        environmental enrichment
exploration 79, 103, 150, 151, 156, 160,
        166, 171, 191, 193–194
    see also locomotor activity

fearfulness 11–12, 45–46, 54, 80,
        114–127, 128–129, 133, 146–147,

            148, 164, 176–177, 180, 194–197,
            199–201, 206, 215–216, 225
    see also emotional reactivity; novelty,
            response to
feed efficiency (conversion) 79–80, 84, 92,
        156
feeding
    behavior 16–17, 23–24, 28, 36,
            60–61, 63–64, 79–80, 81, 82,
            91, 96–106, 110, 111, 114,
            119, 132, 134, 136–137, 145,
            150, 154, 155, 156, 157, 158,
            164, 165, 166, 176–177, 185,
            189, 190–191, 192, 193,
            197–199, 200–201, 206–207,
            211, 213, 225, 227
    competition for food 81–82,
            103–104, 119, 130, 132,
            136–137, 138, 154, 158,
            176–177, 191, 192
    delivery methods (of food) in
            captivity 79–80, 101–103,
            114, 142, 153, 154, 155, 156,
            158–159, 164, 165, 166, 191,
            192, 200–201, 218–219
    deprivation (including self-imposed)
            36, 61, 64, 105–106, 111,
            152, 170, 171, 194, 206–207,
            212, 218
    diets fed to captive animals 24, 28,
            52, 58–60, 60–61, 63–64,
            79–80, 93, 96–100, 110, 111,
            139, 185, 190–191, 192,
            197–199, 205, 211, 219, 227
    general 10, 22, 23, 24, 25, 26, 28, 52,
            63–64, 68, 76, 78, 81, 85,
            93–94, 95–106, 108, 121,
            143, 149, 166, 173, 184–185,
            205, 210, 211, 218, 227
    hand-feeding 100–101, 114, 122,
            164, 165, 184–185, 189,
            200–201, 206–207
    schedules of feeding 101
    see also growth; predation, predatory
            capabilities
feralization *see* reintroductions, feralization
flight distance 113–114, 120, 121–122,
        123, 129
    see also taming

foraging behavior *see* feeding, behavior
founder effect 33, 35, 36, 38, 39, 40, 55,
    74, 77, 187
fresh water, adaptation to 6, 76, 137, 189

genetic drift 30, 35, 38–42, 55, 70–71, 72,
    186
genetic variability
    in captive populations 37, 38, 39–40,
        41–42, 48, 55, 72–78, 82,
        186, 209
    estimation (measurement) of 39
    genetic drift effects 38, 39–42, 72
    inbreeding effects on 30–31, 32, 35,
        39, 42
    population differences 14, 38, 39–42,
        72–78
    see also domestication, effect on
        genetic variation; founder
        effect; heritability;
        heterozygosity, genetic
        variation; hybridization,
        heterozygosity and genetic
        variation
geotaxis (response to gravity) 48
gonad size and function 26–27, 57, 80, 92,
    128, 139, 144, 168, 169, 205, 206,
    208, 225
grooming 17, 18, 65, 114, 123, 147, 154,
    157
growth 16, 26, 32, 46, 60, 68, 69, 80,
    81–82, 84–85, 92, 97–99, 132,
    138, 145, 147, 155–156, 158,
    159–160, 170–171, 172, 178–179,
    180–181, 188, 189, 190–192, 208,
    209, 214, 225, 226
    see also body size

habituation *see* learning
hand-rearing *see* human–animal
    interactions, hand-rearing
hatchery rearing (aquatic species, mostly
    fish) 6, 14, 15, 27, 39–40, 43–44,
    47–48, 69, 73–74, 77, 78, 93–94,
    99, 108, 109–110, 136–138, 154,
    158–160, 170–171, 186, 187–188,

189, 190–191, 192–193, 194–196,
    197, 208–209
health *see* disease; parasites; welfare
heart size and function 64, 91–92, 109,
    151–152
heritability 114, 120, 186, 209
heterosis 18, 31, 34, 36
    see also hybridization; outbreeding
heterozygosity
    genetic variation, relation to 30–31,
        34, 36–37, 39–40, 40–42, 48,
        73–74, 75–78
    phenotypic characters, effect on 16,
        32, 36–37, 81, 111
    see also heterosis; hybridization;
        genetic drift; inbreeding;
        outbreeding
history of domestication; *see species names
    listed under* domestication
hormonal induction of breeding *see*
    reproductive success in captivity,
    hormonal
human–animal interactions
    with captive animals
        aggression toward humans 23,
            80, 126–127, 134, 222
        breeding 10, 48–49, 55, 74–75,
            80, 84, 114–115, 167,
            171, 173, 180
        capture and handling *see* capture
            (by humans); welfare,
            capture, handling and
            confinement, handling
            of juveniles
        commensal/symbiotic
            relationships 6, 24–26,
            29, 65, 74, 166, 167, 184
        discrimination of individual
            humans 126
        fear of humans (taming) 11–12,
            16–17, 23, 28, 45–46, 52,
            80, 84, 113–129,
            146–147, 148, 166, 171,
            173, 177, 180, 196,
            205–206, 215, 216–217
        food provisioning 10, 96–103
        hand-rearing 121, 134, 163–164,
            170, 198, 199–201, 222

human–animal interactions *continued*
    husbandry 10, 65, 68, 69, 94,
        95, 107, 134, 143, 166,
        167, 173, 222
    socialization of companion
        animals 24–26, 173–174,
        222
    see also flight distance; learning;
        socialization; taming
reintroductions to wild
    human influence, lack of 3, 182,
        183, 184–185
    with wild animals 63
humidity 44, 64, 75, 143, 145, 147
hybridization
    behavior, productivity and fitness,
        effect on 18, 36–37, 43, 45,
        56, 60, 64–65, 82, 97,
        107–108, 127, 144, 154, 168,
        176, 185
    in breed formation 25, 45
    heterozygosity and genetic variation,
        effect on 14, 19, 34, 36–37,
        72, 82, 186–187
    see also heterosis; heterozygosity;
        outbreeding

imprinting *see* socialization
inbreeding
    and artificial selection 35–36
    general 30–37, 42, 70–71, 187
    genetic variability, effect on 30–31,
        36–37, 39, 40, 76, 77–78
    inbreeding depression 19, 31–32,
        34–37, 55, 57, 76, 85, 111,
        112, 226
    phenotypic traits, effect on 16, 32–33,
        34, 85, 111, 226
    population size, role of 33–34, 35,
        42, 55, 186, 187
infanticide *see* maternal behavior;
    mortality, infanticide
infertility *see* reintroductions, reproductive
    success; reproductive success in
    captivity
isolation effects on behavior
    physical 127, 130, 134, 153, 173

    social 119, 130, 133–136, 141, 153,
        164, 218–219, 220
    visual 137, 139, 158, 191, 214

K- vs. r-selection (reproduction) 85–86

learning 10–12, 35, 36, 65, 91, 96,
    100–101, 102–103, 107, 108, 109,
    110, 112, 113, 120–122, 126, 127,
    134, 136–137, 143, 147, 150, 153,
    155, 156, 164–165, 166, 173, 174,
    175, 189–190, 192–201, 202,
    209–210, 215, 216–217, 227, 228
light 44, 98, 145–146, 159–160, 166
litter (clutch) size 32, 34, 53–54, 57, 68,
    124, 214, 226
locomotor activity
    environment and management effects
        19, 44, 52, 65, 85, 88, 93,
        103, 105–106, 132, 133, 134,
        142, 149–151, 153, 156, 164,
        166, 171, 178, 180, 192, 193,
        200, 205, 210, 212–213, 215,
        218, 220
    feeding and food deprivation effects
        101, 103, 105–106, 145,
        152
    general 23, 28, 79–80, 81, 170
    physiological factors 64–65, 85, 88,
        93, 110–111, 149, 151–152,
        170, 192, 208, 211
    wheel-running 17, 36, 45, 61, 64–65,
        105–106, 149–150, 152, 178
    see also stress; welfare, atypical
        behaviors

management of captive animals *see* capture
    (by man); early experience effects;
    feeding, delivery methods in
    captivity, diets fed to captive
    animals, hand-feeding, schedules
    of feeding; hormonal induction of
    breeding; learning; locomotor
    activity, environment and
    management effects; photoperiod;

reintroductions, dietary considerations, genetic and demographic considerations, habitat considerations, preconditioning for release; reproductive success in captivity, technology; taming; temperature; welfare, capture, handling and confinement, conservation and captive-breeding programs, environmental enrichment, management techniques, quality of care

maternal behavior 52, 53, 54–55, 60, 62, 65, 67–68, 130–131, 134, 163–164, 215

maternal effects (on development) 18–19, 115, 117, 206

mating behavior *see* sexual behavior

mating systems 23, 42, 66–67, 68, 70, 85, 92, 139–140, 141, 168

metabolism 19, 55, 59–60, 85, 98, 127–128, 144, 197–198, 208
    see also growth; brain biochemistry

mortality
    captive animals 6, 29, 35, 51, 52, 55, 59, 64, 65, 68, 69, 75, 76, 78, 80, 91, 92, 95, 97–98, 137, 142, 145, 157, 159, 168, 173, 218
    diet-related 97, 99
    inbreeding depression, associated with 31–32, 34, 35, 55, 75
    infanticide or abandonment of offspring 52, 53, 54–55, 68, 130–131, 153, 214, 215
    predation, related to 109–110
    see also reintroductions, survival

natural selection
    on captive-reared animals reintroduced to nature 184, 186, 187, 194
    in captivity 22, 25, 30, 34, 35, 39, 40–41, 47, 51–62, 63, 64, 66–67, 68, 69–70, 72, 80, 85–86, 95, 97, 106, 112, 138, 157, 159, 168, 169–170, 173,

183, 186, 187, 194, 209, 225–226, 227
    general 18, 34, 35, 47, 48–49, 68, 85–86, 95, 149, 157, 175, 183, 209

neoteny 167, 168, 172–174, 181

nomenclature *see* domestication, taxonomic designations

novelty, response to 17, 18–19, 23, 61, 63–64, 70, 80–81, 96, 105–106, 109, 110–111, 112, 120, 122, 127, 131, 142, 147, 150–151, 154–156, 160, 164, 165, 166, 175–177, 178, 180, 181, 189–190, 190–191, 196, 201, 202, 205–207, 208–209, 212–213, 214, 216, 227
    see also emotional reactivity; fearfulness

odor *see* olfaction

olfaction 7, 83, 89, 91, 100, 126, 128, 166, 194–196

open-field test 16–17, 147, 178–179, 80–81

outbreeding 30, 31, 34, 36–37, 40, 72
    see also genetic variability; heterosis; hybridization

parasites 7, 52, 111, 201–202, 210, 211, 215

pelage (and skin)
    coloration 16–17, 19–20, 23, 25, 26, 47, 55, 84, 93, 138, 191
    thickness 83, 144

pheromones *see* olfaction

photoperiod 26, 143, 145–146, 149, 170

physical condition
    endurance 64–65, 70, 151–152, 189, 191, 208, 215, 227
    vigor, general 19, 30, 31, 34, 36, 37, 42, 56, 85, 112, 149, 190, 191, 204, 227
    see also agility; hybridization; locomotor activity

pleiotropy 16–17, 79, 85

polygenic inheritance 17, 18, 19

populations
    density effects 59, 66, 97, 108, 130,
            133, 141, 157, 158, 160, 166
    isolation, effects of 34, 41, 186
    size 30, 33–34, 40–42, 74, 78, 186,
            188, 203, 226
pre-adaptations (for domestication) *see*
        domestication, pre-adaptations for
predation
    anti-predatory capabilities 63, 64, 69,
            85, 107–110, 112, 127, 146,
            154, 159, 188–189, 190, 191,
            192–193, 194–197, 199–200,
            215–216
    general 2, 7, 70, 78, 85, 88, 94, 95,
            107–110, 180, 184, 202, 210,
            227
    predatory capabilities 61, 62, 97, 98,
            103, 110, 112, 127, 167, 174,
            189, 192, 193–194, 198
productivity *see* artificial selection, for egg
        production, growth, milk yield;
        domestication, products from
        domestic animals; litter (clutch)
        size; taming, relation to
        production; welfare, production
        efficiency and profitability

reintroductions
    age and size effects 190
    dietary considerations
        change in diet 190–191, 192,
            197–198
        food provisioning by humans,
            post-release 189
        prey-capture techniques 192,
            198–199
    disease and parasite transmission
        201–202
    examples of 183–184
    feralization
        biological traits, effects on
            87–88, 184, 185
        definition 182–184
        examples of 183–184
        pre-adaptation for 185–186,
            189–190

    prevention of 27
    rate of 3, 184–185
    genetic and demographic
        considerations 183, 185–188
    growth effects 192
    habitat considerations
        acclimation to release site
            189–190, 193–194
        availability of habitat 188–189
    preconditioning for release
        anti-predator
            behavior 194–197,
                199–200, 227
        to foraging techniques 193–194,
            198–199, 200, 227
        to habitat 193–194
        to natural diets 197–198
        parent vs. artificial
            rearing 199–201
        prey-capture techniques *see*
            reintroductions, dietary
            considerations
        rearing with wild conspecifics
            201
    predation on released animals
        188–190, 191, 192–193, 197,
        200
    reasons for reintroducing animals in
        nature 5–6, 182, 183, 227
    reproductive success 5–6, 78, 182,
        184, 185, 186, 187, 190–191,
        197, 200, 202, 227–228
    survival (in nature) 5–6, 44, 161–162,
        182, 184, 185, 186–190, 191,
        192, 193–203, 227–228
    wild populations, effects on
        competition offered by
            reintroduced animals
            186–187, 190–191
        interbreeding (hybridization) and
            genetic influences 185,
            186–187
relaxed selection 30, 44, 48, 60, 63–71,
        75, 80, 84, 85, 97, 106, 107, 144,
        152, 165, 167, 168, 169–170, 173,
        175
reproductive success (failure) in captivity,
        factors related to

artificial selection 47–48, 54, 68, 70, 79–80, 81, 85–86, 92, 112, 168
   emotional reactivity and taming 25, 112, 129, 180
   environmental 26, 144, 145–146, 147, 214, 226
   experiential 52, 56–57, 91, 94, 95, 122–123, 129, 133, 137–138, 139–140, 146, 156, 163–164, 167–168, 173, 180, 206, 209, 214, 218, 226
   general 22, 24, 47–48, 78, 79, 92, 95, 171, 174
   hormonal 26–27, 69, 122–123, 168, 206, 209
   inbreeding 31–32, 55, 226
   natural selection in captivity 16, 42, 47, 48–49, 51–55, 56–60, 61–62, 69–70, 85–86, 112, 209
   nutrition 58–60, 97, 167–168, 226
   relaxation of natural selection 66–69, 70, 80, 91
   reproductive behaviors 23, 32, 42, 52–53, 54–55, 66–69, 81, 85, 97, 112, 122–123, 134, 137–138, 139–140, 156, 163–164, 168, 173
   technology 5, 24, 26–27, 66, 81, 138
resource allocation theory 79–80, 82, 85, 87–88, 94, 225–226
restocking *see* reintroductions

salinity 26, 31, 76
salt water, adaptation to 6, 31, 76, 94, 137, 189
selection (genetic), general 30, 40, 63, 70–71, 82, 83, 184
serotonin *see* brain biochemistry
sex structure of wild vs. captive populations 135, 141
sexual behavior 23–24, 32, 37, 40, 42, 44, 45, 59, 64, 65–67, 82, 85, 92, 97, 134, 135, 138, 139–140, 146, 156, 163, 164, 165–166, 168, 170, 180, 212
   see also mating systems

sexual development 5, 44, 59, 68, 80, 85, 143–144, 145, 146, 156, 167–168, 169–170, 173, 181
shelter (in captive environment)
   effect on behavior 11–12, 139, 146–147, 193–194, 196, 214, 215, 221
   effect on maturation 147, 171
   effect on reproduction 68, 139, 146–147
   general resource 121, 130, 146, 149, 178, 184–185, 191, 210, 212, 214, 221
   use as refuge 11–12, 107, 143, 144, 146–147, 188–189, 193–194, 196, 214, 215
social behavior *see* behavior; human–animal interactions; social organization; socialization; taming
social organization *see* dominance hierarchy; territoriality
socialization 19, 26, 121–122, 133–134, 171–172, 173, 183, 196, 200
   see also human–animal interactions, socialization; taming
stereotyped behavior *see* welfare, atypical behavior
stress
   behavioral response to 190, 196, 206–207, 208, 217–221
   change in living environment 151, 190, 206, 208, 214, 227–228
   influence of stress on expression of inbreeding depression 34
   physical and environmental stressors 34, 93, 189, 205, 206, 208, 210, 216–217, 228
   physiological response to 117–119, 123, 128, 151, 156, 164, 190, 205–209, 212, 216–217
   psychological stressors 52, 100, 112, 117–119, 123, 129, 130, 132, 133, 134, 146, 164, 180, 190, 196, 205, 206–209, 210, 211, 212, 214, 216–217, 218–220, 228
   selection for stress responsiveness 115, 177, 209, 225

stress *continued*

    stress reduction 134, 156, 167, 177,
        180, 189, 190, 208, 209–210

    from weaning 110–111, 210, 214,
        218–219

    see also taming; reproductive success in
        captivity; welfare, atypical
        behavior

subordinate animals, behavior 14, 81,
    103–104, 132, 136–138, 141, 157,
    173, 175–178, 210, 215

    see also aggressive behavior;
        dominance, social

substrate effects 26, 108, 153–154, 156,
    158–159, 220–221

substrate (for oviposition) 58, 97

swimming

    endurance 64, 93–94, 189, 191, 208

    in feeding 154

    general 107, 159, 208, 212–213

    in thermoregulation 145

taming, tameness, tameability

    defined 113–114, 119

    experiential factors contributing
        to 11–12, 113–114,
        116–117, 120–122, 123–127,
        146–147, 171–172, 200, 202,
        216–217

    genetic factors and heritability 114,
        116–117, 118, 120

    see also taming, selection for

    modification with drugs and brain
        lesions 128–129

    negative aspects 126–127

    physiological basis *see* brain
        biochemistry

    as a pre-adaptation for domestication
        21–22, 23–24, 24–26, 126

    relation to animal welfare 114,
        216–217

    relation to domestication 10, 11–12,
        23–24, 113–129

    relation to emotional reactivity 112,
        117, 119, 180, 216–217

    relation to neoteny 171–172,
        173–174

    relation to pelage coloration 16–17

relation to productivity 124–126

relation to reproduction 122–123,
    168, 169

selection for 16, 43, 45–46, 80, 81,
    84, 114–115, 118, 120, 126,
    127–128, 168, 169, 171,
    173–174, 180

taxonomic designations (distinction
    between wild and domestic
    animals) 3–4

temperature 26, 44, 76, 143–145, 146,
    147, 189, 224

territoriality 23, 24, 132–133, 137, 138,
    141, 156–157, 158, 166, 175–176

tonic immobility 205–206

training *see* learning

vigor *see* physical condition

vision 12, 79, 83, 89, 91, 104, 126,
    134–135, 137, 139, 147, 158, 171,
    172, 191, 192, 195, 196, 214

visual isolation *see* vision

vocalizations 16, 32, 37, 78–79, 123,
    166–167, 178

water column, response to 98, 139, 154,
    158, 159

water current, response to 158, 189, 191,
    194, 208

welfare

    adaptation to captive environment 1,
        180, 205, 210–211, 223–225,
        227

    artificial selection

        correlated effects related to
            welfare 50, 80, 81, 92,
            225–226

        selection for welfare-relevant
            variables 115, 177, 209,
            225–226

    atypical (stereotyped) behaviors

        dietary factors 100, 219

        environmental deprivation and
            enrichment 134, 150,
            153, 154, 155, 160, 167,
            213, 217–218, 220–221,
            222

examples 217–221
feeding methods and schedule
100, 101–103, 218
genetic predispositions 217, 220
weaning and social isolation
effects 133–134, 218–219
wild-caught rodents, lack of
atypical behaviors in
captivity 219–220
behavioral and physiological needs
100, 140, 211–213, 223
capture, handling and confinement,
effects of 100, 105–106,
113–114, 124–126, 150, 151,
180, 205–210, 211
conservation and captive-breeding
programs 227–228
density-related effects 157, 160; *see
also* populations, density
effects
dietary factors 106, 211, 219,
225–226
disease *see* disease
environmental enrichment 134–135,
142–143, 146–147, 148,
149–150, 153–156, 212–216,
220–221

ethical concerns 204, 221, 223, 226,
227–228
general 68, 180, 204, 221, 228–229
genetic modification (engineering)
226–227
handling of juveniles, positive
effects 121–123, 216–217
management techniques 35,
110–111, 112, 113–114,
134–135, 146–147, 157,
209–210, 211–216, 218–219,
221–223, 225–226, 227–228
parasites *see* parasites
pre-adaptations, importance of 211,
223–225, 227
production efficiency and
profitability 92, 124–126,
155–156, 221–222, 225–227
quality of care 35, 124–126, 221–223
see also domestication, effects on
welfare; reproductive success
in captivity; stress; taming,
relation to welfare

zoos 3, 5, 24, 32, 87, 101, 139–140, 153,
157, 228

Species Index

amphibians and reptiles
    alligator 192
    anole 205
    frog 205–206
    komodo dragon 146
    rattlesnake 109

birds
    bustard 200
    buzzard 189
    canary 78
    chicken 2, 3, 21, 45, 46, 66, 67, 73, 76, 79–80, 84–85, 92, 97, 99, 104, 114, 120, 124–125, 126, 153, 157, 162, 183, 200, 217, 222, 223, 225
    condor 139, 200–201
    crane 100–101, 139, 151, 170, 189, 200–201
    dove 129
    duck 19, 89–90, 91, 129, 139, 168, 171, 185
    eagle 190
    falcon 164
    finch 78–79
    general 22, 69, 90, 91, 146, 186, 198, 228
    goose 90, 170, 199
    heron 109
    jungle fowl 2, 3, 73, 79–80, 84–85
    ostrich 99–100
    owl 109, 198
    parrot 122, 216–217
    partridge 53, 151–152, 191, 192, 196, 197–198, 199–200
    pheasant 109, 189, 199, 200, 201
    pigeon 90
    quail 57, 115, 140, 170, 185, 196, 201, 225
    robin 196
    turkey 2, 14, 21, 67, 81, 90, 162, 163, 168, 170, 220–221
    vulture 190

crustaceans
    crab 108, 194
    crayfish 8
    general 81
    lobster 8, 188, 189
    prawn 6, 8, 77, 81–82, 168
    shrimp 8, 27, 43, 54, 97–98, 99, 111
    water flea 8

fish
    bass 151, 187
    carp 6, 98–99, 111
    cichlid 41–42
    cod 110, 111, 190, 192
    damselfish 205
    drum 109–110

flounder 27, 97–98, 99, 108, 145,
    159, 194
general 6, 26, 27, 81, 111, 146, 158,
    159, 170, 186, 197–188, 189,
    192–193, 197, 206, 209
goby 192
grouper 26
guppy 31, 76
haddock 99
Leporinus 27
lumpfish 99
minnow 197
perch 159–160
pickerel 196
pike 27, 195
pinfish 109
pomfret 97–98
salmon
    Atlantic 14–15, 40, 47–48, 77,
        78, 107, 109, 136–138,
        145, 158, 159, 170–171,
        186, 191
    chinook 187, 194, 196, 208
    chum 193
    coho 93–94, 110, 136, 137, 138,
        187, 189, 191, 196, 208
    general 6, 27, 43, 69, 138,
        156–157, 187, 194, 202,
        206
    honmasu 192
    masu 154
    pink 187
sculpin 108
seabream 77, 186, 225
snapper 208
stickleback 197
sturgeon 27, 144, 158, 206–207
sunfish 99, 132
swordtail 194
tilapia 27
trout
    brook 39–40, 136, 144, 196
    brown 14, 69, 108, 171, 187,
        188, 189, 190–191, 192
    cutthroat 194
    rainbow 107–108, 178, 195,
        206, 208, 209
    steelhead 107–108, 158, 194
turbot 99, 105

whitefish 99
wolffish 98

insects
    Adoxophyes 145
    ant 8
    apple maggot 97
    beetle 7, 8
    big-eyed bug 60–61, 98, 110
    blowfly 75–76
    boll weevil 7, 165–166
    bollworm 7
    cabbage looper 59
    cicada 7
    cochineal 9
    cockroach 7, 8
    Colpoclypeus 145
    corn earworm 166
    cricket 7
    general 4, 7, 8, 47, 75, 97, 142, 186
    grasshopper 7, 8
    fly
        house 7, 33, 36, 40
        melon 44
        parasitic 9
        pomace 9
    fruit fly
        Caribbean (*Anastrepha*) 58–59
        Mediterranean (*Ceratitis*) 58,
            65–66, 97
        olive (*Dacus*) 59–60, 62, 78, 144
        oriental (*Drosophila*) 5, 7, 19, 29,
            32, 33–34, 37, 40, 41, 44,
            48, 55, 59, 61, 64, 70, 75,
            76, 131–132, 139
    honeybee 8, 27–29
    lac insect 9
    locust 143–144
    louse 7
    mosquito 7, 76–77
    moth 8
    pea aphid 60, 110
    screwworm 7
    silkworm 2, 9
    soybean looper 40
    termite 8
    tobacco budworm 59, 60, 110
    tobacco hornworm 96–97

insects *continued*
    walking stick 8
    wasp 9
    waterbug 8

invertebrates, other
    Aeromonas 202
    Bonamia 202
    clam 8
    cuttlefish 8
    earthworm 8
    fluke 8
    Gyrodactylus 202
    nematode 111
    oyster 8, 36–37, 73–74, 202
    palolo worm 8
    planaria 8
    rotifer 97–98
    sea cucumber 9
    sea urchin 9
    snail 8, 76, 185
    spider 8
    squid 8
    Tribolium 5

mammals
    alpaca 2, 87
    antelope 6, 35–36, 111
    auroch 73
    baboon 134
    bear 103
    buffalo (bison) 21, 22
    Calomys 5
    capybara 6, 130–131
    cat 12, 21, 24, 87, 164, 183, 184
    cattle 2, 3, 46, 65, 67, 73, 74, 80, 92, 104, 114, 122, 124–126, 133, 183, 222, 223, 225–226
    cheetah 54, 156
    chimpanzee 101, 131, 133–134, 155, 156, 157, 163
    coyote 87, 172, 193
    deer 22, 31, 110–111, 122, 126, 167–168, 209
    deermouse 17, 52, 67–68, 105, 109, 146, 150–151, 152, 174–175

dingo 184
dog 1, 3, 12, 24–26, 32, 45, 47, 65, 69–70, 74–75, 84, 87, 92, 101–102, 121, 127, 133, 147, 164–165, 166–167, 168, 170, 172, 172–174, 183, 184, 196, 219, 222
dolphin 228
donkey (burro) 87, 183
elephant 101
elk 22
ferret 21, 52–53, 109, 193–194, 198–199
fishing cat 103
fox 14, 16, 45–46, 80, 81, 84, 93, 114–116, 122, 127–128, 139, 147, 168, 169, 171–172, 206, 215–216
general 12, 19, 22, 24, 74, 83, 186, 198
gerbil 38, 39, 52, 57, 87, 91, 146–147, 171
giraffe 100
goat 2, 67, 87, 107, 115–118, 120, 183
gorilla 163–164
guanaco 87, 132, 133
guinea pig 2, 166, 167, 177–178
hamster 38, 113, 193, 198
horse 2, 21, 22, 67, 73, 85, 87, 101–102, 104, 183, 219, 222
jackal 87
kangaroo 157
kinkajou 102
lemming 175
leopard cat 103
llama 2, 87, 132–133
marmoset 93
mink 73, 80, 87, 88, 89, 91–92, 114–118, 120–121, 127–128, 129, 145, 146, 180, 212–213, 216, 219
monkey 37, 102, 133, 134–136, 154–155, 156, 164, 205, 206–207, 209–210, 216
mouse, house 2, 5, 13–14, 16, 17, 19, 32–33, 36, 45, 52, 57, 61, 63–64, 64–65, 67, 75, 82, 96,

109, 133, 144, 149–150, 154, 157, 177, 198, 218
mouse, striped 220
musk ox 6
ocelot 24
orangutan 153–154
Oryzomys 52
possum 206
prairie dog 193–194, 198
pronghorn antelope 22
rabbit 2, 3, 55, 83, 84, 100, 126, 147
rat, giant pouched 6
rat, Norway 3, 5, 14, 16–17, 18–19, 52, 53–54, 56–57, 60, 67, 80–81, 88, 89, 104–105, 109, 122–123, 126, 127–128, 129, 146, 147, 149, 152, 162–163, 169–170, 175–177, 178–180, 183–184, 216, 217, 222
rat, roof 128–129, 218–219, 220, 221
rat, water 5
reindeer (caribou) 3, 202
rhinoceros 139–140, 206

rodents, general 11–12, 24, 64, 67, 89, 146, 149, 155, 171, 217–218
seal 126, 221
sheep 2, 3, 67, 82, 86, 89, 93, 107, 114, 117, 119, 121, 122, 126, 127, 135–136, 183, 225
shrew 51, 52, 57, 140
swine 2, 3, 4, 46, 48–49, 60, 67, 83, 86, 87, 88, 89, 92, 97, 101, 114, 120, 123–125, 126, 144, 153, 155–156, 158, 183, 213–215, 216–217, 225, 226
tamarin 54–55, 189–190, 193, 197, 198
vicuna 132, 133
whale 228
wolf 1, 3, 24–26, 32, 34, 65, 69, 74, 75, 87, 92, 93, 120, 121, 127, 133, 164–165, 166–167, 168, 170, 172–174, 184
zebra 223